Rhododendrons in Horticulture
and Science

Rhododendrons in Horticulture and Science

Papers presented at the International Rhododendron Conference, Edinburgh, 2002

Edited by

George Argent and Marjory McFarlane

Royal Botanic Garden Edinburgh

Published by the Royal Botanic Garden Edinburgh, 20A Inverleith Row, Edinburgh EH3 5LR, UK (www.rbge.org.uk)

Text © Royal Botanic Garden Edinburgh 2003

Plates © photographers as listed in plate credits (p. viii)

ISBN 1 872291 49 X

Copyedited and typeset by Erica Schwarz, Scientific Editorial Services, Midlothian, UK (e-mail: e.schwarz@btinternet.com)

Printed and bound by The Charlesworth Group, Huddersfield, UK (tel: 01484 517077)

Cover illustration: *Rhododendron acrophilum*. This species is known only from Mt. Mantalingahan near the southern tip of the island of Palawan in the Philippines. Originally described in 1953, it was not recollected until an expedition from the Royal Botanic Garden Edinburgh revisited the locality and brought back living materials in 1992. It is a delightful dwarf species in cultivation, flowering freely several times a year. (Watercolour painted at the Royal Botanic Garden Edinburgh by Eve Bennett. © Eve Reid Bennett.)

Contents

Acknowledgements

Many people have helped in the production of this volume. The Library staff of the Royal Botanic Garden Edinburgh have never failed to cheerfully put their professional knowledge at our disposal during our editorial searches, and we are especially grateful for expert help to Jane Hutcheon, George Forrests, Graham Hardy, Ruth Hourston, Lynda Marquis and Leonie Paterson. Similarly, computer support has been readily forthcoming – and at times desperately needed! – and for this we are deeply indebted to Len Scott and Duncan Reddish. John-Paul Shirreffs has used his usual skills and flair to produce the cover for this book, and greatly helped with previous graphic work connected with the conference. The horticultural staff displayed our collections to the very best advantage: Louise Galloway, Alistair Reed and Paul Smith made special efforts to re-landscape the vireya house and produce new informative display labels for the plants. Expert advice and help has also been readily forthcoming from Alan Bennell, Barbara Campbell, David Chamberlain, Brian Coppins, Stephan Helfer, Fiona Inches, David Long, Ida Maspero, Douglas McKean, Mary Mendum and Debbie White. Lastly, thanks to all the lecturers who have found the time to provide text and illustrations for these varied talks.

George Argent and Marjory McFarlane

Plate credits

Cox: Plate 1, David Chamberlain; Plate 2, Debbie White.

Pradhan: Plate 1, Tony Schilling; Plate 2, L. Baldock.

Chamberlain: Plate 1, Peter Cox; Plates 2–9, David Chamberlain.

Helfer: Plates 1 and 2, Stephan Helfer.

Schepker & Werbeck: Plates 1 and 2, Hartwig Schepker.

Thornley: Plates 1–3, Michael Thornley.

Robinson: Plates 1 and 2, Michael Robinson.

Rankin: Plates 1 and 2, David Rankin.

Dixon et al.: Plates 1–3, Geoffrey Dixon.

Gardiner: Plate 1, Geoffrey Dixon.

Fairweather: Plates 1–4, Christopher Fairweather.

Argent: Plates 1, 3 and 4, George Argent; Plate 2, Debbie White.

De Zhu & Paterson: Plates 1–9, David Paterson.

Smith: Plates 1 and 2, Graham Smith.

Main: Plates 1–4, John Main.

Rotherham: Plates 1–4, Ian Rotherham.

Hootman: Plates 1–6, Steve Hootman.

Eeckhaut et al.: Plate 1, Tom Eeckhaut.

Grayer & Miller: Plate 1, RHS Photo Library.

Warwick: Plates 1–20, Maureen Warwick.

Conference photograph: Debbie White.

Introduction

Professor Stephen Blackmore

Regius Keeper & Director, Royal Botanic Garden Edinburgh

Following 20 years of development of the *Rhododendron* collections in Edinburgh and at its Specialist Gardens, and considerable progress here and around the world in research on the genus, it was thought appropriate to initiate the second Edinburgh International Conference. The genus has clearly established itself on both the scientific and horticultural international circuits since the Royal Horticultural Society meeting in London in 1949 and the first designated International Conference on the genus in New York in 1978. The first one in Edinburgh followed this in 1982, followed by three in Australia – Wollongong, Birnie and Melbourne – and one in New Zealand. There have also been the annual meetings of the American Rhododendron Society, all of which could be called International and which included the Oban meeting in Scotland of 1996, the first ARS meeting outside the North American continent. The genus is still one of the most popular in temperate horticulture, and its large size, great diversity and geographical range have made it a classic for many kinds of fundamental and applied research and biodiversity and conservation studies, many strands of which are demonstrated in the papers included within this volume. Edinburgh is justly proud of its association with the genus, which goes back over 100 years. The last 20 years have maintained this interest with explorations associated with the *Flora of Bhutan*, the re-establishing of links with China, which are stronger now than they have ever been, and many expeditions to SE Asia which have brought into cultivation almost half the known vireya species from the Malesian archipelago.

I am grateful to our co-sponsors of this conference, the Royal Horticultural Society and the Scottish Chapter of the American Rhododendron Society, who both made magnificent contributions to the overall success of the meeting. Many people have worked hard over a long period to make the conference a success and it has been a truly collaborative effort. Of special note because of their largely voluntary contributions are firstly Marjory McFarlane, who used her previous conference-organising experience to very good effect and worked tirelessly to co-ordinate the many facets that need attention to keep the delegates happy and the programme running smoothly, as it undoubtedly did; and secondly Eve Bennett, who organised the concurrent 'Rhododendrons in Art' exhibition which so elegantly achieved its aim of putting together the finest collection of rhododendron-associated art under one roof.

It is of course the purpose of such a conference to bring together people of diverse backgrounds and individual expertise but united in their passion for the genus. This promotes discussion of the ideas presented, and we hope that the delegates went away stimulated to do more and even better work. It also promotes friendship and collaboration: a measure of the success of any conference such as this is as much what grows out of it as what actually occurs at the time.

'Riddle of the Tsangpo Gorges': Retracing Frank Kingdon-Ward's 1924 expedition

Kenneth N. E. Cox

Glendoick Gardens, Glendoick (West Lodge), Perth PH2 7NS, UK

The little-known Tsangpo Gorge in southeast Tibet is the world's deepest gorge and, in the early part of the 19th century, when 'discovering' big waterfalls was all the rage, rumours abounded about a giant waterfall hidden within its unexplored depths. Frank Kingdon-Ward set about finding it. He wasn't the only one. This 19th-century obsession with the Tsangpo River saw some incredible feats of exploration. With local cannibalistic tribes and treacherous terrain, the history of exploration in this area is riddled with wonderful stories.

Frank Kingdon-Ward was one of the great explorer–plant-hunters. Many of the plants he brought back are still in cultivation today, for example the now-famous blue poppy, *Meconopsis betonicifolia*. This paper describes following in his footsteps to research a new edition of Kingdon-Ward's book *Riddle of the Tsangpo Gorges* with two Americans, Ken Storm and Ian Baker, and finding many of his plants still growing in the wild, exactly where Kingdon-Ward described them. Not only was Kingdon-Ward's journey retraced but the team penetrated further into the unknown in search of the mysterious waterfall.

Background

Frank Kingdon-Ward, one of the most important of the Asiatic plant-hunters, led his most significant expedition, to the Tsangpo Gorges region of southeast Tibet, in 1924–25. Myself and two American explorers, Kenneth Storm Jr. and Ian Baker, retraced Ward's journey and ventured further into this little-known land. I edited a fully illustrated new edition, published in 2001, of Ward's classic book *Riddle of the Tsangpo Gorges*, describing his 1924–25 expedition.

The little-known Tsangpo Gorge in southeast Tibet is considered one of the world's deepest. In the early part of the 19th cen-

tury, when 'discovering' big waterfalls was all the rage, rumours abounded about a giant waterfall hidden within its unexplored depths. Several explorers set out in a bid to find it. Some made it back to India, while others were killed by the local tribes who fiercely guarded the Assam Himalayan valleys leading into Tibet.

This 19th-century obsession with the Tsangpo River saw some incredible feats of exploration. There is, for example, the extraordinary story of Pandit Kintup, sent by the British into unknown territory to throw some logs into the water to find out if this was the same river as the great Bramaputra. He spent four years on the quest and

Plate 1 Collecting rhododendrons in the field.

managed to throw the logs into the river as planned. Unfortunately, in the meantime the British official who had sent Kintup had retired to Europe and died. With local cannibalistic tribes and treacherous terrain, the history of exploration in this area is riddled with wonderful stories.

In 1924, plant-hunter Frank Kingdon-Ward set out to botanise the area and to confirm or deny that the great waterfall existed. By this time Kingdon-Ward had already proved himself in Yunnan and Burma as one of the great explorer–plant-hunters, but the 1924 expedition was the most ambitious and daring of all. Many significant plants were collected for the first time. Most famously, he introduced the blue poppy, *Meconopsis betonicifolia* Franch. Rhododendrons discovered and introduced by Kingdon-Ward from the 1924–25 expedition included: *R. exasperatum* Tagg, *R. montroseanum* Davidian, *R. scopulorum* Hutch., *R. megacalyx* Balf.f., *R. fragariflorum* Kingdon-Ward, *R. cinnabarinum* Hook.f. ssp. *xanthocodon* (Hutch.) Cullen (Concatenans Group), and *R. parmulatum* Cowan. Other famous plants included *Berberis temolaica* Ahrendt, *B. calliantha* Mullig., *Primula florindae* Kingdon-Ward and *P. cawdoriana* Kingdon-Ward.

One of the most memorable of Ward's plant-collecting stories concerns 'Orange Bill', Ward's nickname for the distinctive form of *R. cinnabarinum* he found in Pemako, southeast Tibet in 1924. Ward describes the discovery:

'I caught sight of something a little distance away which astonished no less than it delighted me; it was a rhododendron with bright orange flowers, that rarest of all colours. But it was no easy matter to reach the spot. It grew some way up the slope, in the midst of this tanglewood, which for long held me at bay. But in the end

I reached the prize, to find a rhododendron of the 'Cinnabarinum' type dangling orange tubes before my fascinated eyes: Orange Bill, the prince of orange rhododendrons (*R. cinnabarinum* ssp. *xanthocodon* Concatenans gp.) K.W. 5874. The young foliage is verdigris coloured, and the bushes gleam shrilly against the snowy cliffs. But I had a really desperate time getting seed of it. It grew, as I say, well up the steep slope and often out of reach on the cliffs above. I went after it on October 22nd during a heavy snowstorm and got a few capsules. On October 26th I tried again; by this time the bushes were well snowed up, but I got some more seed, and it is a relief to think that the seeds are germinating, considering the awful strain on my temper while struggling in that accursed cold muddle.' (Cox *et al.*, 2001)

Retracing the expedition

In 1995, David Burlinson, from the travel company Exodus, and myself managed to get permission to take an expedition into the Tsangpo Gorges region to retrace part of Ward's journey and to look for plants which he had recorded but not introduced. The 1995 trip was very successful and led to two more in subsequent years. We found the area largely unchanged and unspoilt and found many of Ward's plants growing in exactly the locations he described them.

In 1996 we met up with an American expedition exploring the inner Tsangpo Gorges, led by Ian Baker and Kenneth Storm Jr. We soon established that we formed part of a mutual and very exclusive fanclub for Frank Kingdon-Ward and his long out-of-print classic (and our guidebook of the day) *Riddle of the Tsangpo Gorges* (Kingdon-Ward, 1926). Since my first visit to southeast Tibet, I had always dreamed of producing a new edition of Ward's book, complete with colour pictures of much of what is described. Over a few beers in a run-down, over-crowded hotel in Ningtri, southeast Tibet, this idea moved a step closer to possibility.

Jean Rasmussen, Frank Kingdon-Ward's widow, now living in Eastbourne and almost 80 years old, was very supportive of the project and once we managed to get the copyright returned to her, it was simply a matter of finding a publisher. I had been most impressed by the production values of Roy Lancaster's magnificent *Travels in China* (Lancaster, 1989) and approached

Plate 2 *Rhododendron cinnabarinum.*

the same publishers, Antique Collector's Club. Eventually the deal was done and the book began to take shape. Ward's original text forms the bulk of the book, with over 200 accompanying photographs, from both the 1924–25 and the more contemporary expeditions. Ian Baker, an expert on Tibetan Sacred Art, contributed a chapter on the religious significance of the Tsangpo Gorges as a place of pilgrimage and mythology, and Ken Storm wrote a detailed and meticulously researched history of the 19th- and early 20th-century attempts to explore the gorge. I concentrated on a short biography of Kingdon-Ward and on the often strained relationship with Ward's companion on the expedition, Lord Cawdor.

The last part of the new edition of *Riddle of the Tsangpo Gorges* describes the exploration the three contemporary authors have undertaken, in particular the journeys into areas which Ward did not reach. The province of Pome, which forms the eastern boundary to the Tsangpo Gorges, was long a goal of both Ward and later Ludlow & Sherriff, but none was granted permission. In 1996 and 1997, I led two expeditions into Pome on the Showa La and Dashing La passes, finding several new and exciting plants such as *Omphalogramma tibeticum* H.R.Fletcher and the elusive red lily, *Lilium paradoxum* Stearn. Through my three expeditions to this area, we managed to collect several rhododendron species which Ward discovered but had not managed to introduce. These included *R. bulu* Hutch., a member of the Lapponica SS, *R. dignabile* Cowan, essentially a glabrous version of *R. beesianum* Diels, *R. cephalanthum* Franch., Nmaiense Group and the true *R. laudandum* Cowan and var. *temoense* Kingdon-Ward ex Cowan & Davidian.

Ken Storm and Ian Baker on their five expeditions managed to penetrate further into the gorge than Ward. Perhaps most exciting, in 1997 and 1998 they explored part of the river where the steepness of the terrain forced Ward to miss an entire section of the gorge. Here, Baker and Storm recorded a waterfall, completely unknown in the west. A National Geographic film crew, who documented the descent to, and measurement of, the waterfall accompanied the 1998 expedition. Unfortunately the press release and documentary made rather grandiose claims to have 'discovered' a 'new' waterfall. This angered the Chinese who claimed they had already documented it. Needless to say, both the Chinese and the Americans were entirely dependent on local hunters who had always known about the falls.

Kingdon-Ward's legacy

In a long life of plant-collecting, Kingdon-Ward made over 20 expeditions to Yunnan, Burma, Tibet, India and Sri Lanka. He covered thousands of lonely miles, suffered a great deal from illness, disease (including malignant tertiary malaria) and accidents, and fought to overcome or ignore his fear of heights, rope-bridges and snakes. Unlike most of his fellow collectors, who left much of the work to their local staff, Ward did most of the collecting and pressing work himself. He left a legacy of over 20 books, most describing his expeditions. Compared with the rather turgid prose of most of Ward's fellow plant-hunters (the exception being Reginald Farrer who gilded every possible lily) Ward's writing is lively, often poetic, and full of incident as well as detailed background information. He learned several languages, was well versed in, and fascinated by, geology and had a great curiosity for the customs of the people he lived amongst.

It is hard to be precise about how many species of plants Kingdon-Ward discovered and introduced. Some of the 100 or so rhododendron species he was credited with have been subsequently downgraded taxonomically ('lumped' or 'sunk' where they are considered to be merely variations on a species described earlier). But many of his *Rhododendron* introductions are still widely grown, none more so than *R. wardii* W.W.Sm. Some Kingdon-Ward introductions such as the blue poppy, *Meconopsis betonicifolia*, and *Lilium mackliniae* Sealy are first-class garden plants, as Ward guessed they would be, while others are mainly prized by collectors and connoisseurs. Many beautiful plants, especially amongst the *Primula* species, proved to be difficult to cultivate and survived only a short time in western gardens (again Ward usually knew when this would be the case). And thankfully he even left one or two plants for those who have followed him to discover and introduce. In his book *Plant Hunting in China* (Cox, 1945), my grandfather Euan Cox, who knew Ward well, gives an insight into what really motivated him:

> 'As Ward once wrote, it is his profession to collect seeds and herbarium material, but the real work of exploration is his hobby. He is and always has been much more than a great plant collector. He tells me that he is prouder of the Royal Medal of the Royal Geographical Society and the Livingstone Medal of the Scottish Royal Geographical Society than of his three gold medals for horticulture.'

Frank Kingdon-Ward discusses his motivations as a lifelong plant-hunter in his book *Pilgrimage for Plants* (Kingdon-Ward, 1960):

> 'I should find it difficult now to give any one simple reason ... the pleasure and the pride gradually increased ... I can only suppose that some element of romance must have also lain behind it. Plant hunting has always seemed to be a romantic occupation – and romance must be a part of life, or people would not cling to it so irrationally, even when the pursuit of it gives so much unease. ... No sooner was one journey finished ... "thank heavens *that's* over!" and assured myself, "never again", than all the troubles and difficulties fell away, were forgotten ... and I began to long for the hills again.'

Concluding comment

The splendour of its scenery and the richness of its unique flora and fauna have astonished all who have been to the Tsangpo Gorges region of Tibet. It is essential that pressure be brought to protect the area for future generations. The Chinese media have reported a plan for a vast hydroelectric dam on the Tsangpo River, which would have catastrophic effects on the environment. The Monba and Lopa tribesmen who live in the gorge are currently being forced to leave by the Chinese, in order to 'protect the area'. We feel this is ill advised, as these tribes have lived there for generations and are the true guardians of the sacred land of Pemako and the only true guides to its secrets.

References

Cox, E. H. M. (1945). *Plant Hunting in China*. London: Collins, 230pp.

Cox, K., Storm, K., Jr. & Baker, I. (2001). *Riddle of the Tsangpo Gorges*. Original text by Frank Kingdon-Ward, edited by Kenneth Cox. Woodbridge, Suffolk: Antique Collector's Club, 319pp.

Kingdon-Ward, F. (1926). *Riddle of the Tsangpo Gorges*. London: Edward Arnold, 324pp.

Kingdon-Ward, F. (1960). *Pilgrimage for Plants*. London: G.G. Harrop, 191pp.

Lancaster, R. (1989). *Roy Lancaster Travels in China*. Woodbridge, Suffolk: Antique Collector's Club.

Biological control of invasive alien weeds using fungi, with particular reference to *Rhododendron ponticum* in the British Isles

Harry C. Evans

CABI Bioscience, Silwood Park, Ascot, Berks SL5 7TA, UK

A number of non-indigenous plant species introduced into the British Isles as ornamentals during the 19th century have, after a long period of naturalisation, become invasive in both urban and natural habitats. Depending on the interpretation, such species have collectively been termed alien weeds, plant invaders or, less provocatively, non-native plants (neophytes). *Rhododendron ponticum* is now firmly placed within this group, which also includes *Buddleja davidii* (butterfly bush), *Heracleum mantegazzianum* (giant hogweed), *Impatiens glandulifera* (Himalayan balsam) and *Fallopia* (*Reynoutria*) *japonica* (Japanese knotweed). *Rhododendron ponticum* is still on an invasive front and poses a threat to SSI sites, as well as to the endemic flora and fauna, especially in small island systems such as the Hebrides and Lundy. Conventional control methods are extremely expensive and, at best, contain rather than eliminate the problem. Biological control offers a safer, and potentially cheaper, sustainable management option. Two strategies are reviewed: classical, involving the introduction and release of exotic, coevolved or biotrophic fungi, and inundative, whereby indigenous, typically necrotrophic fungi are mass-produced and formulated as mycoherbicides, and applied in much the same way as chemical pesticides.

From a risk–benefit analysis, based on a mycological survey in the native range of *R. ponticum* in the Algarve region of Portugal, it is concluded that the classical biological control approach presents too many risks, given that over 500 rhododendron species and their hybrids are currently in cultivation in the British Isles. This is despite well-documented successes against similar woody invasive species in other parts of the world when this strategy has been employed: for example, control of *Acacia saligna* with an Australian gall-forming rust in South Africa. However, the inundative or mycoherbicide approach is considered to be a viable alternative, based on the use of indigenous, basidiomycete wood-rotting fungi as stump treatments. This has been exploited successfully for invasive, alien tree species in both the Netherlands and South Africa following the marketing of the mycoherbicides BioChon® and Stumpout®, respectively. Such fungi, including *Chondrostereum purpureum*, are currently being evaluated against *R. ponticum* in the UK.

Biological control: A viable management strategy?

In order to address this question, general principles need to be introduced, as well as working examples from other countries. This is because – in the UK at least – there is no history of legislation for the implementation of biological control as a tool for the management of alien weeds and exotic pests *sensu lato* (Tatchell, 1996). The jury capable of giving a judgement on such initiatives is not just out in the UK, it has yet to be convened.

The philosophy

Traditionally, classical or inoculative biological control has been the strategy favoured for the management of invasive, alien weeds by pioneering countries, such as Australia, New Zealand, South Africa and the USA. In fact, it pre-dates the use of chemical herbicides but has slid into relative obscurity since their advent. However, the sheer scale of many alien weed infestations, and their invasion of what are now recognised as environmentally sensitive habitats or of land of low commercial value, often precludes the employment of conventional control methods based on economic and/or safety reasons. Potentially, the classical approach offers a cheaper, safer and more sustainable alternative. In certain cases, biocontrol can provide 'the silver bullet solution', in others it needs to be integrated into a weed management programme; depending on the target weed, the invaded ecosystem and, as always, on the economics. Above all, it should be transparent and based on good science: if it won't work or the risks are too high, because of the lack of sufficiently aggressive or specific natural enemies, then this should be spelled-out early in any assessment of management options in order

to avoid the inevitable disappointments and bad press which some ill-conceived or badly presented programmes have generated. Indeed, even supposedly reputable, popular scientific journals have not been averse to jumping on the 'finger-pointing' band wagon whenever a biocontrol project has proven or been (mis)interpreted to have unforeseen and unwanted ecological impacts (Evans, 2000; New Scientist, 2000).

The concepts, principles and protocols of classical biological control have been developed and refined over the past century by entomologists using arthropod natural enemies, and the resultant successes and failures of over 600 introductions have been well documented (Julien & Griffiths, 1998; McFadyen, 1998). However, it is only within relatively recent times that fungi have been exploited for the biological control of weeds (Evans *et al.*, 2001; Evans, 2002). The philosophy which underpins this approach to the management of alien weeds, and exotic pests in general, is based on the premise that such non-native organisms or immigrants arrive in new geographic areas without any, or with only a partial complement, of their coevolved natural enemies. Thus, potentially, they are biologically fitter than their indigenous competitors. In the absence of specialised, as well as of any local opportunistic, natural enemies, and given optimum conditions for growth and fecundity, population explosions are almost inevitable. Thus, in response to such threats, Australia has now established a permitted rather than a prohibited list for all plant introductions, based on a weed risk assessment scoring system (Pheloung, 2001). All imports have to be accompanied by the appropriate documentation and data, in order for the assessment to be made: the ball is now in the court of the importer rather than the quarantine authorities. However, would this system have identified past alien invasions of poorly known plant species, some of which are rare or even endangered in their native ranges (Barreto & Evans, 1988, 1997)? Indeed, some of the most unlikely and useful plant species have become highly invasive weeds, especially in delicately balanced, endemically rich ecosystems, such as *Eucalyptus* and *Pinus* spp. in the fynbos of South Africa (Henderson, 1998), and guava (*Psidium guajava* L.) and the quinine tree (*Cinchona succirubra* Pav.) in the Galàpagos Islands (Cronk & Fuller, 1995; Tye, 2001).

The classical approach aims to restore the balance by introducing natural enemies from the native range or centre of origin of the target pest following well-established protocols (FAO, 1996). The emphasis here is on coevolution, in that the natural enemy is not only specific to its host, or to related species, but also that it is in a genetically stable and hence sustainable association. In contrast, however, chemical pesticides have been described as evolutionarily unstable or evanescent (Holt & Hochberg, 1997). As previously mentioned, a single agent may suffice, and spectacular successes have been achieved by entomologists, notably employing weevil biocontrol agents (McFadyen, 1998). For other more problematic, typically woody, weeds, a suite or guild of natural enemies, including both arthropods and fungi, may be required (Julien & Griffiths, 1998).

In certain cases, and this applies to several invasive, alien weeds in the British Isles, classical biological control is not a viable management option. Usually for reasons of safety and following a risk assessment (Evans, 2000), an alternative biocontrol approach, involving the release or application of large numbers of an indigenous, generalist natural enemy, can be considered. This is inundative biological control which, essentially, is feasible only for readily mass-produced organisms, such as necrotrophic fungi, whereby a formulated product, in this case a mycoherbicide, is developed usually as part of a commercial venture.

The practice

Thus far, the mycota exploited for the classical biological control of alien weeds comprises little more than 25 species. The majority of these are rust fungi because these highly specialised biotrophs possess the traits required of a classical agent: high specificity, virulence, and efficient long-distance dispersal mechanisms (Evans, 2002). However, the results have been impressive, with low inputs yielding high returns. An example is a Mediterranean rust introduced into Australia in the mid-1970s against its coevolved host, *Chondrilla juncea* L. (skeleton weed), a major invasive alien species in the wheat belt of New South Wales. The introduced rust had an almost immediate impact, with weed populations rapidly being reduced to densities approaching those in its native range. Benefits have been estimated at AU$18 million a year (Burdon *et al.*, 1981), with accrued savings up to the year 2000 being put at AU$260 million (Marsden *et al.*, 1980), in terms of increased crop yields and reduced herbicide use. A similar success story resulted from another pioneering project in which a white smut fungus was released in Hawaii for control of the Mexican mistflower plant, *Ageratina riparia* (Regel) K. & R., a species with a native range restricted to the mountain rivers of Veracruz State where it is of rare occurrence (Barreto & Evans, 1988).

Within the first year there was a significant reduction in weed populations and large areas of invaded upland pasture have since been rehabitated and conservation land reclaimed (Trujillo, 1985). More recently, this success story has been repeated in South Africa and New Zealand, at relatively little additional cost (Evans, 2002).

Amongst the invasive, alien woody plants in South Africa, in addition to *Eucalyptus* and *Pinus* spp., is a complex of Australian *Acacia* spp., whose management is being addressed through various biocontrol initiatives depending on the results of risk assessments and, significantly, on conflicts of interest issues. Thus, for Port Jackson willow, *Acacia saligna* (Labill.) Wendl., a classical biocontrol approach, using rust fungi from Australia, has been pursued so successfully that tree populations have been reduced by 90–95% of the original densities and the weed is now considered to be under complete control (Morris, 1997, 1999). Interestingly, the gall-forming rust initially appeared to be having no impact on its host but, after an eight-year lag period, the rust built up to epiphytotic proportions, with up to 1500 shoot galls per tree, effectively stopping growth and leading to tree decline and death. However, other members of the genus – the wattle species *A. mearnsii* De Wild., *A. melanoxylon* R.Br. and *A. pycnantha* Benth. – as well as being invasive, have high commercial value (paper, tannins), presenting problems for classical biological control. Nevertheless, by introducing a gall wasp from Australia which attacks only the seed pods, subsequent invasions of the fragile ecosystems have been halted without affecting the commercial plantation interests. In addition, the existing invasions are being eliminated through the use of a locally produced mycoherbicide based on an indigenous, wood-rotting fungus. Large labour forces are cutting out the trees but, without this inundative biocontrol agent, the cut stumps would simply re-sprout and probably add to rather than reduce the problem (Zimmermann & Neser, 1999). Thus, based on the experiences of other countries, as well as from promising on-going programmes (Evans, 2002), there is sufficient evidence to think that classical biological control is a viable management strategy for invasive, alien weeds in the British Isles.

Alien weeds in the British Isles: prospects for biological control with emphasis on *Rhododendron ponticum*

> 'O thou weed,
> Who are so lovely fair and smell'st so sweet,
> That the sense aches at thee,
> Would'st thou had'st never been born!'
> (Shakespeare, *Othello*)

Thus are the conflicts of interest which entangle the weed species in this section, several of which are included in the Wildlife and Countryside Act 1981, and which also have been assessed as biocontrol targets by CABI Bioscience (formerly International Institute of Biological Control, IIBC). Unlike most of the weeds discussed above, they do not infest vast, contiguous tracts of land, but each in their own right constitutes a dangerous addition to the flora of the British Isles.

Rhododendron ponticum L.

The absence of a vernacular 'weedy' name reflects the conflicts of interest associated with this ornamental shrub introduced into the British Isles more than 200 years ago from the Iberian Peninsula (Bean, 1976). Its native range extends in disparate popula-

11

tions from this region to the mountains of northwest Turkey (Cross, 1975; Milne & Abbott, 2000). For many years, *R. ponticum* (see Note 1) was more or less restricted to the gardens of large houses and estates, where it was also valued as cover for game. In more recent times, with the reduction of the attendant workforces, this plant has been allowed to re-seed and spread unchecked (Thomson *et al.*, 1993). Previously, in the well-maintained estates, seedlings would have been removed manually and the populations would have stabilised and certainly would have been non-invasive. The warmer winters and wetter climates of the western British Isles appear to favour the fecundity, spread and establishment of the plant. Thus, within the past 40–50 years, *R. ponticum* has moved from being a prized garden plant to being an invasive Category 4 species (Cronk & Fuller, 1995) (see Note 1). This category denotes the invasion of SSI sites and nature reserves. However, in some regions and locations *R. ponticum* could be classified in the highest Category (5) since it is threatening endemic species with extinction, as for example in western Ireland (Cross, 1981; Kelly, 1981; Foley, 1990), western Scotland (Cronk & Fuller, 1995), and Lundy Island, where it poses a direct threat to the endemic Lundy cabbage, *Coincya wrightii* (O.Schulz) Stace, and its two coevolved chrysomelid beetles (Compton & Key, 2000). In addition, *R. ponticum* is now having a significant economic impact on commercial forestry operations (Edwards *et al.*, 2000).

Pollen analyses reveal that *R. ponticum* occurred in the western British Isles during the interglacial periods but disappeared after the last ice age (Cross, 1975). Thanks to man's intervention, the 'native' has returned without its natural enemies: a potent mix!

Current management practices include both mechanical and chemical methods, which are labour intensive and difficult to sustain over large areas of infestation, and the costs involved can be daunting (Gritten, 1992). Gritten (1995) put the total cost for management of *R. ponticum* in Snowdonia National Park at £45 million, whilst admitting that these efforts are not even keeping pace with the spread of the weed. Research continues on improving the efficiency of existing treatments (Edwards *et al.*, 2000).

The classical approach

As pinpointed by the molecular work of Milne & Abbott (2000), the British introductions of *R. ponticum* originated from Portugal (Serra do Monchique, Algarve). This area was visited recently and a brief mycological survey was undertaken in the small, isolated populations still remaining. The results of the survey (present author, unpublished), and additional records (Farr *et al.*, 1996), indicate that there is a highly specialised and potentially exploitable mycota associated with this plant in its native range. However, even the most basic pest risk assessment will show that this approach is untenable on host-range testing alone (Evans, 2000), given that over 500 rhododendron species are under cultivation in the British Isles (Mabberley, 1997), quite apart from the putative hybrids (Milne *et al.*, 1999).

The inundative approach

The proposed strategy will be based on developing an indigenous pathogen for use as a site-selective stump treatment, adopting the approach, risk assessment and technology which has already been implemented successfully in the Netherlands and South Africa for woody invasive, alien weeds. The Dutch product BioChon® (Koppert Biological Systems) is based on a mycelial formulation of the wood-rotting basidiomycete *Chondrostereum purpureum* (Pers.: Fr.) Pouzar. It is employed as a stump treatment for control of North American *Prunus* spp., as well as *Populus* spp., invading both natural and commercial forests in the Netherlands (de Jong, 2000). Ironically, this pathogen is the cause of silver-leaf disease of numerous woody plants worldwide, and, until relatively recently, was still a notifiable disease of stone fruits in Britain (Buczacki & Harris, 1981). However, the decision by the Dutch Plant Protection Authorities to approve the product was made not on data relating to host range but on epidemiology, following simulated and actual modelling of inoculum dispersal (de Jong *et al.*, 1990). A similar product (EcoClear®) is also being developed in Canada, targeted at woody invasives in public rights-of-way and amenity areas (Wall, 1997), whilst Stumpout®, based on a locally common, wood-rotting fungus (*Cylindrobasidium laeve* (Pers.: Fr.) Chamuris), is being used in South Africa to eradicate invasive Australian wattle (*Acacia*) species (Morris *et al.*, 1999).

Ad hoc surveys are being undertaken in the UK by CABI Bioscience to build-up an inventory of potential candidate fungi. *Chondrostereum purpureum* has been reported previously on *R. ponticum* in the British Isles (Strouts & Winter, 2000), and, pending further funding, this will be the primary candidate selected for pathogenicity screening, and evaluation as a mycoherbicide.

What would be the advantages of a mycoherbicide over conventional treatments? It is difficult to provide a definitive economic answer at the moment, and a pragmatist would need additional hard evidence before considering that investment in this approach is worthwhile. However, patently, the Dutch and South African biocontrol initiatives suggest that systemic herbicides were not providing acceptable control, in preventing re-sprouting of the cut stumps through kill of the root systems. In Canada, the motives for turning to biological control are governed also by legislation: over the next two to three years, chemical pesticides will be phased out for use in public areas, as well as in forestry. On the basis of current circumstantial evidence, the biocontrol approach should offer an environmentally safer, and potentially cheaper, alternative for the management of *R. ponticum*: more economic because only a single application is envisaged, perhaps incorporated with the pruning treatment. A possible additional advantage may be that the relatively slow-acting process of death and decay of the cut vegetation will be more ecologically in tune with the re-colonisation process. There is evidence that *R. ponticum* is allelopathic, releasing toxic metabolites to suppress competitors (E. Nilsen, personal communication), and that the gradual changes afforded by the mycoherbicide approach would also result in the breakdown of these metabolites as the native micro- and macroflora builds up.

Buddleja davidii Franch. (butterfly bush)

Of Tibetan-Chinese origin, this plant has become a firm favourite, both as an ornamental and as a butterfly attractant, since its introduction into the British Isles in the 1890s, leading Mabey (1998) to describe it as the saviour of the butterfly populations in urban areas. This large shrub readily colonises waste ground, but it is in the railway systems that the plant has had the most impact, causing physical damage to the infrastructure by penetrating the fabric of bridges and buildings. Control is difficult, since mechanical removal increases structural damage and herbicide use is problematic, particularly near railway property bordering water courses.

As with *R. ponticum*, the considerable conflicts of interest preclude the use of the classical biocontrol approach, although an ecological survey in its native range may provide valuable insights into its invasiveness, which could benefit its future management. Funding has been released by Railtrack (now Network Rail) for CABI Bioscience to undertake a feasibility study for the development of a mycoherbicide targeted at *B. davidii* and other woody invasives damaging the railway infrastructure. Potential candidates among wood-rotting basidiomycetes have been isolated from dying plants and expectations are high for the future of this product, and even the product name ('Budder-off'!) has been selected in anticipation.

Heracleum mantegazzianum Somm. & Lev. (giant hogweed)

This lays claim as the tallest herbaceous plant growing in the British Isles (Mabey, 1998), and was introduced during the 19th century as a showy ornamental for Victorian estates (Tiley *et al.*, 1996). Recently, it formed the frontispiece of a Kew Gardens 'glossy' publication drawing attention to the problem of invasive, alien weeds (Anon., 2001). The main environmental impact is due to its suppressive nature, which causes displacement or even eradication of native herbaceous species with a subsequent loss of biodiversity and a reduction in conservation value, as well as increasing erosion (Wadsworth *et al.*,

2000). However, its overriding claim to fame is the phytophototoxic sap, which causes severe blistering that can leave permanent scars (photodermatitis). This characteristic led to its rise to public prominence in the 1970s (Mabey, 1998), and even to the penning of a surprisingly erudite pop song, 'The Return of the Giant Hogweed' [Genesis, 1972], which amazingly managed to squeeze the Latin binomial into the chorus line! Its recent increase in invasiveness, both in the British Isles and in continental Europe, is due to a change in land use patterns, partly as a result of the European Union (EU) Common Agricultural Policy, which has encouraged farmers to leave large areas of land fallow, especially along rivers and streams, and partly following increased urban development.

A consortium of European agencies, including CABI Bioscience, is being funded by the EU to develop an action plan for management of *H. mantegazzianum*. Several surveys have been conducted recently in the northwest region of the Caucasus, both in Russia and in Georgia, with CABI Bioscience being charged with evaluating biological control potential and a team of Danish, German and Russian scientists investigating ecological aspects. Thus far, a number of candidate, seemingly coevolved fungal pathogens, including species of *Phloeospora*, *Ramularia* and *Septoria*, have been collected and are currently being screened in high-containment quarantine in the UK. These were all isolated from seedlings and first-year plants, and initial field evidence indicates that this is a highly vulnerable stage of the life-cycle and that this should be targeted for any biological control initiative, rather than the older, second- or third-year plants. At present, mycoherbicides are considered to be a much less attractive proposition compared with the classical approach.

Impatiens glandulifera Royle (Himalayan balsam)

At present, this striking annual herb is not covered by any UK legislation, but, like giant hogweed, it is an increasingly important and pernicious invader of riparian habitats, forming monocultures and displacing native vegetation (Collingham *et al.*, 2000; Wadsworth *et al.*, 2000). Absence of control at the flowering period in 2001, because of foot-and-mouth restrictions, could have serious knock-on implications in the future, as also could the human vectoring of the plant, recently reported by Rotherham (2001). Preliminary literature surveys for natural enemies have been undertaken by CABI Bioscience (Fowler & Holden, 1994), but there is little published data, and clearly information is wanting on this plant in its native range in the forests of the Indo-Pakistan Himalayas (Gupta, 1989). Certainly, in the British Isles the insect guild of natural enemies is impoverished (Beerling & Perrins, 1993). However, there are encouraging developments on a related species, *Impatiens parviflora* DC., of Central Asiatic origin, which is now an invasive alien weed in the forests of eastern Europe. A biotrophic rust fungus, *Puccinia komarovii* Tranz., has spread with its host and has been shown to be impacting severely on the weed population dynamics; it has been described as an important regulating agent in Slovakia (Bacigalova *et al.*, 1998). Species of the genus *Puccinia* have been used successfully as classical biocontrol agents of weeds (Evans, 2002), and within this complex of Himalayan *Impatiens* species there is almost certainly a complex of highly host-specific rust species. The classical approach,

therefore, offers potential, once conflicts of interest have been resolved and a preliminary risk assessment undertaken.

Fallopia (*Reynoutria*) *japonica* (Houtt.) Ronse Decr. (Japanese knotweed)

As related by Mabey (1998) and Bailey & Conolly (2000), since its introduction in the 1800s this herbaceous, rhizomatous perennial has turned from a prized ornamental into what is now regarded as the most pernicious weed in the British Isles. It has now been recorded in most of the 10-km grid squares. *Fallopia* (*Reynoutria*) *japonica* is a clonal plant, described as the 'biggest female in the world' (Bailey & Conolly, 2000), and spreads through dispersal of rhizome fragments in water, or, more usually, following anthropogenic transport of soil (Beerling *et al.*, 1994). It is illegal to cause knotweed to grow in the wild, and disposal of infected soil in deep landfill sites is so expensive that the development value of invaded land can be reduced dramatically (estimated at £500K per hectare). If it is not removed efficiently, as developers have found, the consequences can be catastrophic since knotweed can push through tarmac and even concrete. Weed invasions of this species are most severe in South Wales, and also include natural riparian ecosystems, and the Welsh Development Agency (WDA) recently funded CABI Bioscience to carry out an exploratory survey in Japan. The plant is relatively uncommon in Japan but has a wide distribution, from sea level to over 2000 m. In these elevated, volcanic habitats on rocky laval screes it displays the ability to colonise which allows it to grow on building sites in the British Isles, and in urban situations in general. *Fallopia* (*Reynoutria*) *japonica* was found to be attacked by a range of insects and fungal pathogens, and one of these, a rust fungus belonging to the genus *Puccinia*, is currently being held in high-risk, quarantine containment at CABI Bioscience, pending the release of further funding to allow in-depth screening of it as a classical biocontrol agent to be undertaken. An evaluation of local (UK) pathogens is also being considered for potential development of a mycoherbicide in urban situations, where rapid and permanent control, through killing of the rhizome system, is essential. It should be stressed here that a coevolved agent released for classical biological control will not eliminate its host from an invaded ecosystem: a balance will be achieved as the pest population and its competitive edge are reduced in favour of the indigenous flora and fauna. Thus, in the case of Japanese knotweed, for example, these two biocontrol approaches can be used in tandem. The coevolved rust agent spreads naturally within weed populations and gradually impacts negatively on the pest population dynamics, while the opportunistic, virulent local pathogen is formulated and applied as a herbicide to eradicate the weed from high-priority development sites. As shown earlier, the lag phase for classical agents to impact may be prolonged (5–10 years for some rusts), but population decline can be dramatic in the long term: 95–99% reduction for both the skeleton weed and *Acacia* rusts in Australia and South Africa, respectively (Evans, 2002).

Discussion

For all the weeds included here, the concept of using biological control for their management is of relatively recent origin and therefore public awareness, as well as official legislation, need to be addressed before this strategy can be taken for granted in the British Isles: the learning curve may be steep. The lessons from a previous, pioneering biocontrol venture involving CABI Bioscience are still being learned. The target weed was bracken (*Pteridium aquilinum* (L.) Kuhn) in the British Isles and the search for natural enemies focused on the Southern Hemisphere. A suite of potential arthropod agents was identified and a bracken-specific, phytophagous, noctuid moth, *Conservula cinisigna*, was eventually selected and imported into the UK from South Africa. However, because of political, legal, environmental and socio-economic problems, detailed by Lawton (1988), the agent was never released and the moth died in quarantine along with the project. Conflicts of interest and the economics of control were highlighted, especially the fact that the calculated rate of invasion was significantly lower than that predicted at the start of the project (Lawton, 1988).

Have we moved on since then and does bracken merit re-evaluation as a target for biological control? The giant hogweed and Japanese knotweed projects give cause for optimism in that the classical biocontrol approach for the management of invasive, alien weeds in the British Isles is at last receiving the attention, but not necessarily the funding, that it deserves from the experiences in other regions. Judging from recent research, as well as from direct enquiries to CABI Bioscience, bracken does appear to be viewed as an increasing problem invasive weed of upland or hill farming (Robinson, 1995), and lowland conservation land (Mitchell *et al.*, 1997), in the British Isles. From earlier work in Southern Africa it is evident that classical biological control is a feasible management strategy. This is because although bracken is a British native, it is depauperate in specific or coevolved natural enemies. The only explanation is that whilst bracken survived in Britain during the glacial periods, probably in isolated pockets, its associated arthropods and pathogens did not, but persisted in the more southerly range of its distribution. Whether or not the seemingly intractable conflicts of interest can ever be resolved is debatable.

References

ANONYMOUS (2001). The making of monsters. *Kew* Spring 2001: 24–26.

BACIGALOVA, K., ELIAS, P. & SROBAROVA, A. (1998). *Puccinia komarovii*, a rust fungus on *Impatiens parviflora* in Slovakia. *Biologia, Bratislava* 53: 7–13.

BAILEY, J. P. & CONOLLY, A. P. (2000). Prize winners to pariahs – a history of Japanese knotweed s.l. (Polygonaceae) in the British Isles. *Watsonia* 23: 93–110.

BARRETO, R. W. & EVANS, H. C. (1988). Taxonomy of a fungus introduced in Hawaii for biological control of *Ageratina riparia* (Eupatorieae; Compositae), with observations on related weed pathogens. *Trans. Br. Mycol. Soc.* 91: 81–97.

BARRETO, R. W. & EVANS, H. C. (1997). Role of fungal biocontrol of weeds in ecosystem sustainability. In: PALM, M. E. & CHAPELA, I. H. (eds) *Mycology in Sustainable Development*, pp. 183–210. Boone, NC: Parkway Publishers Inc.

BEAN, W. J. (1976). *Trees and Shrubs Hardy in the British Isles*, 8th edition. London: J. Murray.

BEERLING, D. J. & PERRINS, J. M. (1993). Biological flora of the British Isles: *Impatiens glandulifera* Royle. *J. Ecol.* 81: 367–382.

BEERLING, D. J., BAILEY, J. P. & CONOLLY, A. P. (1994). Biological flora of the British Isles: *Fallopia japonica* (Houtt.) Ronse Decraene. *J. Ecol.* 82: 959–979.

BUCZACKI, S. & HARRIS, K. (1981). *Collins Guide to the Pests, Diseases and Disorders of Garden Plants.* London: W. Collins.

BURDON, J. J., GROVES, R. H. & CULLEN, J. M. (1981). The impact of biological control on the distribution and abundance of *Chondrilla juncea* in south-eastern Australia. *J. Appl. Ecol.* 8: 957–966.

COLLINGHAM, Y. C., WADSWORTH, R. A., HUNTLEY, B. & HULME, P. E. (2000). Predicting the spatial distribution of non-indigenous riparian weeds: issues of spatial scale and extent. *J. Appl. Ecol.* 37: 13–27.

COMPTON, S. G. & KEY, R. S. (2000). *Coincya wrightii* (O.E. Schulz) Stace (*Rhynchosinapis wrightii* (O.E. Schulz) Dandy ex A.R. Clapham). *J. Ecol.* 88: 535–547.

CRONK, Q. C. B. & FULLER, J. L. (1995). *Invasive Plants: The Threat to Natural Ecosystems Worldwide.* London: Chapman & Hall.

CROSS, J. R. (1975). Biological flora of the British Isles: *Rhododendron ponticum* L. *J. Ecol.* 63: 345–364.

CROSS, J. R. (1981). The establishment of *Rhododendron ponticum* in the Killarney oakwoods, S.W. Ireland. *J. Ecol.* 69: 807–824.

EDWARDS, C., CLAY, D. V. & DIXON, F. L. (2000). Stem treatment to control *Rhododendron ponticum* under woodland canopies. *Aspects Appl. Biol.* 58: 39–46.

EVANS, H. C. (2000). Evaluating plant pathogens for biological control of weeds: an alternative view of pest risk assessment. *Austral. Pl. Path.* 29: 1–14.

EVANS, H. C. (2002). Biological control of weeds. In: KEMPKEN, F. (ed.) *The Mycota XI : Agricultural Applications*, pp. 135–152. Berlin: Springer-Verlag.

EVANS, H. C., GREAVES, M. P. & WATSON, A. K. (2001). Fungal biocontrol agents of weeds. In: BUTT, T. M., JACKSON, C. & MAGAN, N. (eds) *Fungi as Biocontrol Agents*, pp. 169–192. Wallingford, Oxon.: CABI Publishing.

FAO (1996). *International Standards for Phytosanitary Measures: Import Regulations.* Rome: Secretariat of the International Plant Protection Convention.

FARR, D. F., ESTEBAN, H. B. & PALM, M. E. (1996). *Fungi on Rhododendron: A World Reference.* Boone, NC: Parkway Publishers Inc.

FOLEY, C. (1990). *Rhododendron ponticum* at Killarney National Park. In: *Biology and Control of Invasive Plants*, pp. 62–63. Ruthin, Wales: Richards, Morehead & Laing.

FOWLER, S. V. & HOLDEN, A. N. G. (1994). Classical biological control for exotic invasive weeds in riparian and aquatic habitats. In: DE WAAL, L. C., CHILDS, L. E., WADE, P. M. & BROCK, J. H. (eds) *Ecology and Management of Invasive Riverside Plants*, pp. 173–182. London: John Wiley & Sons.

GRITTEN, R. H. (1992). The control of *Rhododendron ponticum* in the Snowdonia National Park. *Aspects Appl. Biol.* 29: 279–286.

GRITTEN, R. H. (1995). *Rhododendron ponticum* and some other invasive plants in the Snowdonia National Park. In: PYSEK, P., REJMANEK, M. & WADE, M. (eds) *Plant Invasions – General Aspects and Special Problems*, pp. 213–219. Amsterdam: SPB Academic Publishing.

GUPTA, R. K. (1989). *The Living Himalayas 2: Aspects of Plant Exploration and Phytogeography*. New Delhi: Today & Tomorrows Printers.

HENDERSON, L. (1998). Invasive alien woody plants of the southern and south-western Cape region, South Africa. *Bothalia* 28: 91–112.

HOLT, R. D. & HOCHBERG, M. E. (1997). When is biological control evolutionarily stable (or is it)? *Ecology* 78: 1673–1683.

DE JONG, M. D. (2000). The BioChon story: deployment of *Chondrostereum purpureum* to suppress stump sprouting in hardwoods. *Mycologist* 14: 58–62.

DE JONG, M. D., SCHEEPENS, P. C. & ZADOKS, J. C. (1990). Risk analysis for biological control: a Dutch case study in biocontrol of *Prunus serotina* by the fungus *Chondrostereum purpureum*. *Plant Dis.* 74: 189–194.

JULIEN, M. H. & GRIFFITHS, M. W. (1998). *Biological Control of Weeds: World Catalogue of Agents and their Target Weeds*, 4th edition. Wallingford, Oxon.: CABI Publishing.

KELLY, D. L. (1981). The native forest vegetation of Killarney, south-west Ireland: an ecological account. *J. Ecol.* 69: 437–472.

LAWTON, J. H. (1988). Biological control of bracken in Britain: constraints and opportunities. *Phil. Trans. R. Soc. Lond. B* 318: 335–355.

MABBERLEY, D. J. (1997). *The Plant Book*, 2nd edition. Cambridge: Cambridge University Press.

MABEY, R. (1998). *Flora Britannica*. London: Chatto & Windus.

MARSDEN, J. S., MARTIN, G. E., PARHAM, D. J., RISDILL-SMITH, J. J. & JOHNSON, B. E. (1980). *Returns on Australian Agriculture Research*. Canberra: CSIRO Publishing Division.

MCFADYEN, R. E. C. (1998). Biological control of weeds. *Rev. Entomol.* 43: 369–393.

MILNE, R. I. & ABBOTT, R. J. (2000). Origin and evolution of invasive naturalized material of *Rhododendron ponticum* L. in the British Isles. *Mol. Ecol.* 9: 541–556.

MILNE, R. I., ABBOTT, R. J., WOLFF, K. & CHAMBERLAIN, D. F. (1999). Hybridization among sympatric species of *Rhododendron* (Ericaceae) in Turkey: morphological and molecular evidence. *Amer. J. Bot.* 86: 1776–1785.

MITCHELL, R. J., MARRS, R. H., LE DUC, M. G. & AULD, M. H. D. (1997). A study of succession on lowland heaths in Dorset, southern England: changes in vegetation and soil chemical properties. *J. Appl. Ecol.* 34: 1426–1444.

MORRIS, M. J. (1997). Impact of the gall-forming rust fungus *Uromycladium tepperianum* on the invasive tree *Acacia saligna* in South Africa. *Biological Control* 10: 75–82.

MORRIS, M. J. (1999). The contribution of the gall-forming rust fungus *Uromycladium tepperianum* (Sacc.) McAlp. to the biological control of *Acacia saligna* (Labill.) Wendl. (Fabaceae) in South Africa. In: OLCKERS, T. & HILL, M. P. (eds) *Biological Control of Weeds in South Africa (1990–1998)*, pp. 125–128. Pretoria: Entomological Society of Southern Africa.

MORRIS, M. J., WOOD, A. R. & DEN BREEYEN, A. (1999). Plant pathogens and biological control of weeds in South Africa: a review of projects and progress during the last decade.

In: OLCKERS, T. & HILL, M. P. (eds) *Biological Control of Weeds in South Africa (1990–1998)*, pp. 129–137. Pretoria: Entomological Society of Southern Africa.

NEW SCIENTIST (2000). When good bugs turn bad. 15 January 2000: 30–33.

PHELOUNG, P. C. (2001). Weed risk assessment for plant introductions to Australia. In: GROVES, R. H., PANETTA, F. D. & VIRTUE, J. G. (eds) *Weed Risk Assessment*, pp. 83–92. Canberra: CSIRO Publishing Division.

ROBINSON, R. C. (1995). Bracken: incumbent on the open hill. In: SMITH, R. T. & TAYLOR, J. A. (eds) *Bracken: An Environmental Issue*, pp. 205–213. Aberystwyth: International Bracken Group.

ROTHERHAM, I. D. (2001). Himalayan Balsam – the human touch. In: BRADLEY, P. (ed.) *Exotic Invasive Species – Should We Be Concerned?*, pp. 41–50. Proceedings of the 11th Conference of the Institute of Ecology and Environmental Management, Birmingham, April 2000. Winchester: IEEM.

STROUTS, R. G. & WINTER, T. G. (2000). *Diagnosing Ill-Health in Trees*, 2nd edition. London: HMSO.

TATCHELL, G. M. (1996). Supporting the introduction of exotic biological control agents for U.K. horticulture. *BCPC Symposium Proceedings* No. 67: 101–111. Farnham, Surrey: BCPC.

THOMSON, A. G., RADFORD, G. L., NORRIS, D. A. & GOOD, J. E. G. (1993). Factors affecting the distribution and spread of *Rhododendron* in North Wales. *J. Environ. Manag.* 39: 199–212.

TILEY, G. E. D., DODD, S. & WADE, P. M. (1996). Biological flora of the British Isles: *Heracleum mantegazzianum* (Sommier & Levier). *J. Ecol.* 84: 297–319.

TRUJILLO, E. E. (1985). Biological control of Hamakua – pamakani with *Cercosporella* sp. in Hawaii. In: DELFOSSE, E. S. (ed.) *Proceedings of the Sixth International Symposium on Biological Control of Weeds*, pp. 661–671. Ottawa, Canada.

TYE, A. (2001). Invasive plant problems and requirements for weed risk assessment in the Galapagos Islands. In: GROVES, R. H., PANETTA, F. D. & VIRTUE, J. G. (eds) *Weed Risk Assessment*, pp. 153–175. Canberra: CSIRO Publishing Division.

WADSWORTH, R. A., COLLINGHAM, Y. C., WILLIS, S. G., HUNTLEY, B. & HULME, P. E. (2000). Simulating the spread and management of alien riparian weeds: are they out of control? *J. Appl. Ecol.* 37: 28–38.

WALL, R. E. (1997). Fructification of *Chondrostereum purpureum* on hardwoods inoculated for biological control. *Canad. J. Pl. Path.* 19: 181–184.

ZIMMERMANN, H. G. & NESER, S. (1999). Trends and prospects for biological control of weeds in South Africa. In: OLCKERS, T. & HILL, M. P. (eds) *Biological Control of Weeds in South Africa (1990–1998)*, pp. 165–173. Pretoria: Entomological Society of Southern Africa.

Note

1 Milne & Abbott (2000) have shown that a significant proportion of the so-called naturalised *R. ponticum* in Britain exhibits additional genetic variation resulting from hybridisation with *R. catawbiense* in particular, but also with *R. maximum* and other unknown species. Introgression with *R. catawbiense* could be expected to confer increased cold tolerance and is apparently more marked in Scottish populations.

Unique anatomical traits in leaves of *Rhododendron* section *Vireya*: a discussion of functional significance

Erik T. Nilsen

Professor of Biology, Virginia Tech, Blacksburg, VA 24061, USA

Plants have suites of traits that make them adapted to their natural habitats. These adaptive traits regulate the performance of species in natural and garden environments. Identifying adaptive traits and their distribution among species of *Rhododendron* is important to understand the evolution of adaptive traits in plants in general, the production of plants in nurseries, the husbandry of species in gardens, and the conservation of *Rhododendron* species in natural areas. *Rhododendron* is an excellent system for identifying adaptive traits because this one genus has an exceptional diversity of species growing from the arctic to the tropics and from sea level to the alpine zone. Probing the diversity of *Rhododendron* species for adaptive traits has been a focus of research for the Plant Ecological Physiology laboratory at Virginia Tech for more than 15 years.

While investigating adaptive traits of *Rhododendron* species leaf anatomy in relation to the latitude and elevation of the native ranges, unusually large cells were observed below the epidermis in species of section *Vireya*. These 'giant cells' (idioblasts) were found in 100% of the 26 species of *Vireya* and 0% of the 27 non-*Vireya* species studied. Additional studies of 52 species growing at the Royal Botanic Garden Edinburgh verified these initial findings. Therefore, idioblasts may be an excellent taxonomic marker for *Vireya*. The abundance, distribution, and size of idioblasts varied among *Vireya* species, indicating that this cell type may have different adaptive significance in different species. Electron microscopy images indicated that the idioblasts are live cells with a large central vacuole and in some species the central vacuole contains mucilage. The presence of mucilage suggests idioblasts may influence leaf water relations. Moreover, idioblasts of *R. zoelleri* form a structure that suggests glandular secretion. Further research on cell constituents, leaf water relations, and herbivore defence is needed before the functional significance of these cells can be deduced.

Introduction

The amazing capability of plants to grow in all corners of the world is a testimony to their unbounded ability for adaptation to the varied array of environmental conditions on this planet. Each species, or ecotype, has a suite of adaptive traits that make them suited to specific environmental attributes (Nilsen & Orcutt, 1996). Therefore, we are able to understand plant native ranges and growth requirements by studying their suite of adaptive traits. Adaptive traits can be identified in the behavioural, morphological, anatomical, or physiological processes of plants. Study of these adaptive traits and their significance to plant performance in varied ecosystems is the realm of the 'Ecological Physiologist'.

The term 'environment' in this work refers to both the physical/chemical aspects of space around a plant and the biotic influences by microbes, fungi, plant, and animals that impact plants (Orcutt & Nilsen, 2000). Therefore, the significance of plant adaptive traits must be considered in the light of both biotic and abiotic aspects of the environment.

Research on the significance of adaptive traits is clearly important for understanding the basic biology of plants. However, application of this information to aspects of human society is equally important. For example, conservation of endangered species requires understanding the requirements for growth and performance, which are based on adaptive traits. Limiting the impact of invasive species is accomplished by understanding adaptive traits of those species, particularly in relation to biological control agents. Many pharmaceutical products are derived from compounds produced by plants as a defence against herbivores. Selection, propagation, and production of plant material in nurseries or gardens are dependent upon understanding growth requirements, which are regulated by adaptive traits.

Rhododendron is an excellent system for studying adaptive traits. Within this single phylogenetic group there is a wide diversity of species (c.1000 species) containing a wealth of genetic information (Chamberlain *et al.*, 1996). The broad diversity of growth forms (trees, shrubs, cushion plants, and epiphytes) and large number of habitats used by these species (arctic, alpine, temperate, subtropical, and tropical) provides an abundance of ecological and physiological traits

for analysis. Moreover, interactions between *Rhododendron* species and other plants and animals through competition, allelopathy, pollination, and herbivory provide a robust suite of biotic interactions for analysis.

The Plant Ecological Physiology Laboratory personnel at Virginia Tech have been studying adaptive traits of *Rhododendron* species over the past 15 years. Our research has grown from studies of one species in one habitat to studies of many species across continents. Some example adaptive traits we have studied are:

1 The importance of leaf movements to cold stress tolerance in *Rhododendron maximum* L. (Nilsen, 1985, 1991, 1992).

2 The value of variation in leaf survivorship patterns for light stress tolerance by *R. maximum* (Nilsen, 1986; Nilsen *et al.*, 1988).

3 The effect of water stress and light intensity on the photosynthesis and growth of several *Rhododendron* species (Nilsen & Bao, 1987; Nilsen *et al.*, 1988).

4 The importance of stem hydraulic properties for tolerance of freeze–thaw and drought in several species of temperate *Rhododendron* (Lipp & Nilsen, 1997; Cordero & Nilsen, 2002).

5 The correlation between leaf anatomical and morphological traits with latitude and longitude in a wide diversity of *Rhododendron* species growing in a common garden.

The overall goal of this laboratory is to develop *Rhododendron* as an important model system for understanding the diversity and adaptive significance of plant traits.

In this paper, focus will be on a current study covering comparative leaf anatomy in *Rhododendron* species with particular emphasis on species of *Rhododendron* subgenus *Rhododendron* section *Vireya*. This section is comprised of species located in the tropical and subtropical highlands of Malesia and surrounding areas. There are currently 310 recognised species in section *Vireya* (Chamberlain *et al.*, 1996), although many changes in taxonomy are likely because of the limited research on this group. Section *Vireya* is identified by floral, seed, and leaf scale traits (Sleumer, 1966). However, these identifying traits overlap considerably with those of species from other sections of *Rhododendron* (Sleumer, 1966). Moreover, taxonomic traits are extremely variable among species of *Vireya*. In fact, there is only one trait (two-tailed seeds) that can be considered synapomorphic for *Vireya* (Sleumer, 1980). The goal of this project was to identify distinguishing leaf morphological and anatomical traits of *Rhododendron* species from a wide range of latitudes (with a focus on *Vireya*) that could be used for further study on adaptive significance of those traits. We hypothesised that:

1 Leaf trait variation for *Vireya Rhododendron* species (mostly tropical) would be as great or greater than the variation in leaf traits for non-*Vireya Rhododendron* species (widely distributed). This hypothesis is based on the supposition that the diversity of habitats in the Malesian tropics would result in as much diversity as that found in the rest of the habitats occupied by *Rhododendron* species.

2 Leaf traits of *Rhododendron* species would correlate with the central latitude and central altitude of the species' native range. This hypothesis is based on the supposition that plant leaf morphology and anatomy is adapted to regional climatic conditions.

Species selected and methods

Species selected

Species selection for this study was based on several criteria: (i) individuals of all species must be located at one common garden; (ii) the native distributions of the selected species should include those from a wide variety of latitudes; (iii) the selected species should include members of all seven subsections of *Vireya*; and (iv) the species selected should include non-*Vireya* species from a wide latitudinal range. The 53 species (Table 1) were selected (before seeing them) from the species list of the Rhododendron Species Foundation/Botanical Garden in Federal Way, Washington, USA (RSF). The native range for each species was obtained from the Biological Recording Units (BRUs defined in Hollis & Brummitt, 1992) previously reported for *Rhododendron* species (Chamberlain *et al.*, 1996). Justin Moat (Royal Botanic Gardens, Kew) provided updates to the *Rhododendron* BRU list, and specific latitude and longitude values for the BRUs. Species range descriptions (particularly elevation information) were attained from various published sources (Sleumer, 1966; Davidian, 1982, 1989, 1992; Cox, 1985, 1990; Feng, 1992; Cox & Cox, 1997; Feng & Yang, 1999).

Tissue collection and analysis

Two recently mature leaves were collected from five individuals of each species where possible and shipped to the Plant Ecological Physiology Laboratory at Virginia Tech, Blacksburg, Virginia, USA for anatomical and morphological analyses. Upon arrival each leaf was measured (length, width, area). Therefore, mean morphology data are based on a sample size of 10 per species. One of the five leaves was selected randomly (random number generator) for anatomical analyses. A section of the mid leaf from the margin to the midvein was removed and preserved in FAA (10% formalin, 50% ethanol,

5% acetic acid, 35% water) for anatomical studies. Preserved sections ($n = 5$ per leaf) were dehydrated in an alcohol gradient, infiltrated with paraffin oil, embedded in paraffin, and sectioned on a rotary microtome at 10 micron width. The sections were double stained with fast green and safranin-O in order to differentiate vascular tissues from ground tissues (Ruzin, 1999). Following the staining protocol, all sections were permanently mounted, resulting in five slides per collected leaf and 25 slides per species. Photomicrographs were made of each slide and anatomical measurements of these photomicrographs were made using image analysis software (SCION image, Scion Corporation, Frederick, Maryland).

An additional leaf from each *Vireya* species was prepared for transmission electron microscopy (TEM). Fresh leaf tissues were fixed in a gluteraldehyde fixative (5% gluteraldehyde, 4.4% formaldehyde, 2.75% picric acid, in 0.05% Na cacadylate). Following fixation, samples were washed in three changes of Na cacadylate for 15 minutes each, and washed in 15, 30, 50, and 70% ethanol, post fixed for 2 hours in osmium tetroxide (1% osmium in 0.1 M Na cacadylate), then washed for 3 hours in 0.1 M Na cacadylate. Tissues were then block stained with 2% uranyl acetate in 70% ethanol for 1 hour. The tissues were equilibrated in 95% ethanol, 100% ethanol, propylene oxide, and then spurs medium before incubation for 48 hours in spurs. Samples were sectioned on an ultramicrotome, placed on grids, and observed with an electron microscope (JEOL TEM 100cx II, Tokyo, Japan).

Samples from some of the selected *Vireya* species also were prepared for scanning electron microscopy (SEM) to study the adaxial (top) epidermal cell architecture. Fresh tissues

were prepared as for the TEM samples until post fixing in osmium tetroxide. The samples were equilibrated in 70, 95, and 100% ethanol, and critical point dried at 10–120°C with 10 changes of carbon dioxide. All samples for SEM were mounted in a paradermal orientation on SEM stubs with silver paint. Just before observing (Phillips SEM 505, Hillsborough, Oregon) the samples were sputter coated with gold. Wax coatings on the adaxial leaf surface precluded any clear observations of cell architecture. Here we present only one image taken of the cut edge of one *R. javanicum* (Blume) Benn. leaf.

Additional leaf measurements

In order to expand the significance of this study to understanding the ubiquity of idioblasts (exceptionally large cells in the epidermis or mesophyll) in *Vireya* species, leaves from an additional selection of *Vireya* species (52 including a 17-species overlap with the RSF collection) growing in the Royal Botanic Garden Edinburgh (RBGE) were evaluated. Free-hand, paradermal sections were made through leaves collected from each greenhouse-grown species. Here we present the species list, idioblast average diameter and the relative abundance of idioblast cells in the leaves of each species (Table 4).

Results and discussion

Leaf morphology

The species selected for this study had wide variation in leaf dimensions. The average dimensions (± 1 standard deviation) for leaves of these species was: leaf length → 1.1 cm ± 0.5 (*R. gracilentum* F.Muell.) to 41.5 cm ± 20.1 (*R. sinogrande* Balf.f. & W.W.Sm.); leaf width → 0.2 cm ± 0.1

Table 1 Species utilised in the study of leaf anatomy of *Rhododendron* with a focus on section *Vireya*, and their taxonomic designation (based on Chamberlain *et al.*, 1996). All species were growing at the Rhododendron Species Foundation/Botanical Garden in Federal Way, Washington, USA. Species are listed alphabetically within subsections.

Species	Authority	Subgenus	Section	Subsection
arboreum ssp. *cinnamomeum*	(Lindl.) Tagg (1930)	*Hymenanthes*	*Ponticum*	*Arborea*
adenopodum	Franch. (1895)	*Hymenanthes*	*Ponticum*	*Argyrophylla*
barbatum	Wall. ex G.Don (1834)	*Hymenanthes*	*Ponticum*	*Barbata*
sinogrande	Balf.f. & W.W.Sm. (1916)	*Hymenanthes*	*Ponticum*	*Grandia*
tsarensis	Cowan (1937)	*Hymenanthes*	*Ponticum*	*Lanata*
pachysanthum	Hayata (1913)	*Hymenanthes*	*Ponticum*	*Maculifera*
aureum	Georgi (1794)	*Hymenanthes*	*Ponticum*	*Pontica*
brachycarpum	D.Don ex G.Don (1834)	*Hymenanthes*	*Ponticum*	*Pontica*
catawbiense	Michx. (1803)	*Hymenanthes*	*Ponticum*	*Pontica*
caucasicum	Pall. (1788)	*Hymenanthes*	*Ponticum*	*Pontica*
degronianum	Carriere (1869)	*Hymenanthes*	*Ponticum*	*Pontica*
degronianum ssp. *yakushimanum*	(Nakai) H.Hara (1986)	*Hymenanthes*	*Ponticum*	*Pontica*
hyperethrum	Hayata (1913)	*Hymenanthes*	*Ponticum*	*Pontica*
macrophyllum	D.Don ex G.Don (1834)	*Hymenanthes*	*Ponticum*	*Pontica*
makinoi	Tagg (1927)	*Hymenanthes*	*Ponticum*	*Pontica*
maximum	L. (1753)	*Hymenanthes*	*Ponticum*	*Pontica*
ponticum	L. (1762)	*Hymenanthes*	*Ponticum*	*Pontica*
smirnowii	Trautv. (1885)	*Hymenanthes*	*Ponticum*	*Pontica*
ungernii	Trautv. (1885)	*Hymenanthes*	*Ponticum*	*Pontica*
williamsianum	Rehder & E.H.Wilson (1913)	*Hymenanthes*	*Ponticum*	*Williamsiana*
carolinianum	Rehder (1912) = *R. minus*	*Rhododendron*	*Rhododendron*	*Caroliniana*
minus var. *chapmanii*	(A.Gray) Duncan & Pullen (1962)	*Rhododendron*	*Rhododendron*	*Caroliniana*
minus var. *minus*	Michx. (1792)	*Rhododendron*	*Rhododendron*	*Caroliniana*
lapponicum	(L.) Wahlenb. (1812)	*Rhododendron*	*Rhododendron*	*Lapponica*
groenlandicum	(Oeder) Kron & Judd (1990)	*Rhododendron*	*Rhododendron*	*Ledum*

ferrugineum	L. (1753)	*Rhododendron*	*Rhododendron*	*Rhododendron*
hanceanum	Hemsl. (1889)	*Rhododendron*	*Rhododendron*	*Tephropepla*
correoides	J.J.Sm. (1915)	*Rhododendron*	*Vireya*	*AlboVireya*
malayanum	Jack (1822)	*Rhododendron*	*Vireya*	*MalayoVireya*
bryophilum	Sleumer (1960)	*Rhododendron*	*Vireya*	*PhaeoVireya*
konori	Becc. (1878)	*Rhododendron*	*Vireya*	*PhaeoVireya*
rarum	Schltr. (1918)	*Rhododendron*	*Vireya*	*PhaeoVireya*
superbum	Sleumer (1960)	*Rhododendron*	*Vireya*	*PhaeoVireya*
kawakamii	Hayata (1911)	*Rhododendron*	*Vireya*	*PseudoVireya*
sororium	Sleumer (1958)	*Rhododendron*	*Vireya*	*PseudoVireya*
herzogii	Warb. (1892)	*Rhododendron*	*Vireya*	*SiphonoVireya*
loranthiflorum	Sleumer (1935)	*Rhododendron*	*Vireya*	*SolenoVireya*
maius	(J.J.Sm.) Sleumer (1960)	*Rhododendron*	*Vireya*	*SolenoVireya*
aurigeranum	Sleumer (1960)	*Rhododendron*	*Vireya*	*Vireya*
blackii	Sleumer (1973)	*Rhododendron*	*Vireya*	*Vireya*
burtii	P.Woods (1978)	*Rhododendron*	*Vireya*	*Vireya*
celebicum	(Blume) DC. (1839)	*Rhododendron*	*Vireya*	*Vireya*
crassifolium	Stapf (1894)	*Rhododendron*	*Vireya*	*Vireya*
gracilentum	F.Muell. (1889)	*Rhododendron*	*Vireya*	*Vireya*
lochiae	F.Muell. (1887)	*Rhododendron*	*Vireya*	*Vireya*
longiflorum	Lindl. (1848)	*Rhododendron*	*Vireya*	*Vireya*
javanicum	(Blume) Benn. (1838)	*Rhododendron*	*Vireya*	*Vireya*
macgregoriae	F.Muell. (1891)	*Rhododendron*	*Vireya*	*Vireya*
polyanthemum	Sleumer (1963)	*Rhododendron*	*Vireya*	*Vireya*
praetervisum	Sleumer (1973)	*Rhododendron*	*Vireya*	*Vireya*
rugosum	Low ex Hook.f. (1852)	*Rhododendron*	*Vireya*	*Vireya*
stenophyllum	Hook.f. ex Stapf (1878)	*Rhododendron*	*Vireya*	*Vireya*
zoelleri	Warb. (1892)	*Rhododendron*	*Vireya*	*Vireya*

(*R. stenophyllum* Hook.f. ex Stapf.) to 15.9 cm ± 6.3 (*R. sinogrande*); leaf surface area → 0.3 cm^2 ± 0.1 (*R. gracilentum*) to 1030 cm^2 ± 950.9 (*R. sinogrande*); leaf thickness → 0.14 mm ± 0.01 (*R. groenlandicum* (Oeder) Kron & Judd) to 0.57 mm ± 0.05 (*R. degronianum* Carrière ssp. *yakushimanum* (Nakai) H.Hara). The association between the product of length times width and leaf area was very strong among all species

(Figure 1A), indicating that measurements of leaf dimensions can be used as a proxy for leaf area. Moreover, the variation of leaf surface area among the selected *Vireya* species was similar to that for the selected non-*Vireya* species (except for *R. sinogrande*). However, when considered as a group the selected *Vireya* species had a significantly lower leaf area and leaf length than the selected non-*Vireya* species group (Table 2). Moreover, leaf thickness of the selected

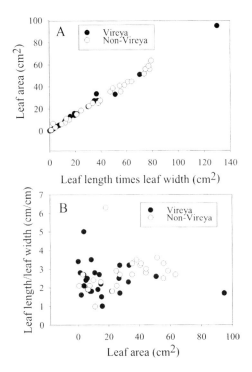

Figure 1 Measurements of leaf dimensions for *Rhododendron* species growing at the Rhododendron Species Foundation/Botanical Garden, Federal Way, Washington, USA. (A) The relationship between leaf area and the product of leaf length and width. One outlier (*R. sinogrande*; leaf area = 1030 cm²) is not shown. (B) The relationship between leaf length divided by leaf width and leaf area. One outlier (*R. stenophyllum*; leaf length/width = 25.6 cm/cm) is not shown.

Vireya species as a whole was greater than that for the selected non-*Vireya* species (Table 2). Also, the selected group of *Vireya* species invested a higher proportion of leaf thickness to mesophyll (i.e. lower palisade/mesophyll ratio). In general, the data supported Hypothesis 1 because the variation in leaf morphology and anatomy dimensions was similar for *Vireya* and non-*Vireya* groups. However, based on these selected species, leaves of *Vireya* species (treated as a group) were slightly smaller and thicker (more investment in mesophyll) than those of the non-*Vireya* species group.

The ratio of leaf length to leaf width is an indicator of leaf shape for *Rhododendron* leaves. Linear leaves are represented by high values of this ratio and smaller values indicate a more ovate leaf shape. The ratio of leaf length to width averaged 2.9 for all species measured (Figure 1B) and there was no significant difference in the length/width ratio between the two groups (Table 2). This average ratio value reflects the common lanceolate leaf form of *Rhododendron* species. A few species had very high ratios (*R. makinoi* Tagg = 6.3; *R. stenophyllum* = 25.6). Linear leaf form is indicated by leaf length to width ratios greater than 4.0. A few species had low values for this ratio (*R. blackii* Sleumer = 1.0; *R. williamsianum* Rehder & E.H.Wilson = 0.98). Ovate leaf form is denoted by ratios equal to or less than 1.0. Both the selected *Vireya* and non-*Vireya* species groups had primarily a lanceolate leaf form, and deviations (both towards linear and ovate) were found in both groups with similar frequency (Figure 1B).

Before the study began it was predicted that leaf size would decrease as the native range of the species became more northerly or higher in elevation (Hypothesis 2). Our

Table 2 Mean, standard deviation, and significant differences (Student's *t*-test) of leaf morphological and anatomical dimensions for 26 selected *Rhododendron* species in section *Vireya* and 27 selected *Rhododendron* species not in section *Vireya*. All leaves were collected from plants growing at the Rhododendron Species Foundation/Botanical Garden, Federal Way, Washington, USA. SD = One standard deviation of the mean. * = Significant difference at $P < 0.05$ level. Pal/Mes = The ratio of the thickness of the palisade layer to that of the mesophyll layer.

Trait	*Vireya*		Non-*Vireya*		*t*-test P-value
	Mean	SD	Mean	SD	
Length (cm)	6.5	3.48	11.45	9.79	0.010*
Width (cm)	2.9	1.84	4.1	3.27	0.052
Area (cm²)	17.4	20.1	54.02	132.4	0.084
Length/width	3.32	4.60	2.82	0.97	0.293
Thickness (mm)	0.43	0.101	0.32	0.09	0.003*
Pal/Mes (mm/ mm)	0.91	0.187	1.09	0.244	0.002*

data with *Rhododendron* had much variation in leaf area among species that have native ranges with central latitudes below 400°N and central elevations below 2500 m. However, in both relationships leaf area tended to be constrained to lower values for species with native ranges at the highest latitudes and highest elevations. It is also important to recognise that leaf areas for species with native ranges at low latitudes (100°S to 100°N), mostly *Vireya* species, were as variable as those from higher latitudes (Table 2). The analysis of these data is limited by the use of BRUs and published species elevation descriptions to represent species' native ranges. Using the central latitude for each species range as an index obscures differences in range extremes. Narrowly endemic species may have the same central latitude and elevation as widely distributed species. Moreover, separating the elevation from latitude information eliminates some ecological synergy because the effects of latitude and elevation on major climatic factors are

similar. Thus, a species with a low latitude and high elevation range could experience a similar regional macroclimate as a species from a high latitude and low elevation. As a rule of thumb in ecological literature, an increase of 200 m elevation has a similar effect on regional temperature conditions as an increase of 10° latitude. Therefore, we constructed a combined index (Central latitude + (Central elevation/200)) to account for the additive effects of latitude and elevation on regional temperature conditions of the native range (Figure 2C).

The relationship between leaf area and low values of the combined index (−10 to 10) indicated no association between leaf area and climate. However, leaf area tended to increase in association with an increase in combined index values between 20 and 50 (mostly temperate habitats or high elevation subtropical). At values greater than 50 (mostly arctic habitats), leaf area was small and similar among species. This suggests that leaf area tends to increase with decreasing regional temperature conditions until extremely low regional temperature conditions predominate and leaves are constrained to small sizes. This is not in agreement with the ecological paradigm that leaf size of ever-

green plants decreases for species inhabiting cooler climates. However, we must bear in mind the caveat that these data represent only 53 species, and more species should be added to this study to strengthen the results.

Leaf anatomy

Photomicrographs of non-*Vireya* leaf anatomy showed an ordinary anatomical structure for these taxa. We did note variation in the numbers of epidermal layers (1–3), the number of palisade layers (1–5), the presence and types of scales and trichomes, the ratio of palisade to mesophyll thickness (0.6–1.5), and the abundance and placement of fibres in the vascular traces for non-*Vireya* species. The strongest relationship between native range and anatomical metrics was an increase in the number of epidermal cell layers with an increase in central elevation of the native range (Figure 3). Multiple epidermal layers may serve as a protective mechanism against high intensity of UV radiation at higher elevation (Caldwell, 1979).

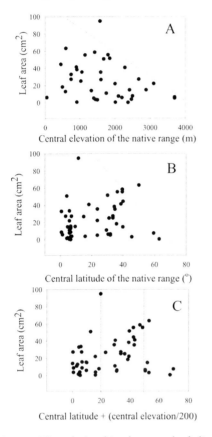

Figure 2 The relationships between leaf dimensions of 53 *Rhododendron* species growing at the Rhododendron Species Foundation/Botanical Garden, Federal Way, Washington, USA are compared with aspects of the species' native ranges. (A) Average mature leaf area plotted against central elevation of each species' native range. (B) Average mature leaf area plotted against the central latitude of each species' native range. (C) Average mature leaf area plotted against an index composed of central latitude and central elevation characteristics of each species' native range. Dashed lines represent important constraining characteristics of the relationships.

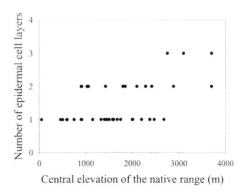

Figure 3 The number of epidermal cell layers on the adaxial surface of 53 species growing at the Rhododendron Species Foundation/Botanical Garden, Federal Way, Washington, USA plotted against the central elevation of each species' native range.

In contrast, *Vireya* species leaves had an additional cell type that is not usually found in normal leaf anatomy. All species of *Vireya* that we measured had 'idioblasts' (Lersten & Curtis, 1995) in or just below the epidermis (Figure 4A). Idioblasts were present in or below the adaxial epidermis in all *Vireya* species. In some species, idioblasts were also found above the abaxial (bottom surface) epidermis, usually clustered near leaf surface scales (Figure 4B). Average idioblast cell volume was calculated for each species using the geometric formula for an oblate spheroid ($4/3 \, \pi a^2 b$), where a = major semiaxis and b = minor semiaxis (Selby, 1972). The average volume of the idioblast cells ranged from 0.50 ± 0.14 to $5.96 \pm 0.30 \times 10^{-5}$ mm^3 with a mean of $2.31 \pm 1.32 \times 10^{-5}$ mm^3 (Table 3).

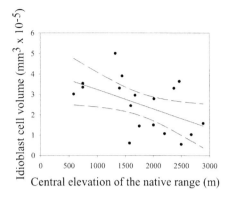

Figure 5 Average idioblast cell volume plotted against mean elevation of the native ranges for 53 species of *Rhododendron* growing at the Rhododendron Species Foundation/Botanical Garden, Federal Way, Washington, USA. The regression line ($r^2 = 0.25$) and 95% confidence limits are shown.

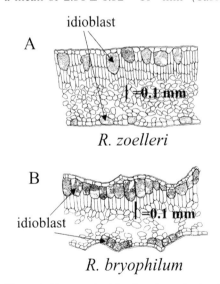

Figure 4 Camera lucida drawings (artist Lara Call) for leaf cross-sections of two species of *Rhododendron* section *Vireya* growing at the Rhododendron Species Foundation/Botanical Garden, Federal Way, Washington, USA. (A) *R. zoelleri*, showing scattered idioblasts (cells with stippled interiors) of adaxial and abaxial leaf surface. (B) *R. bryophilum* Sleumer, showing a continuous layer of idioblasts on the adaxial surface and clumped idioblasts on the abaxial surface.

The abundance of adaxial idioblasts varied from a continuous layer in some species to scattered cells in other species (Figure 4). Species with large and abundant idioblasts were: *R. bryophilum* Sleumer, *R. gracilentum*, *R. rarum* Schltr., *R. sororium* Sleumer, and *R. stenophyllum*. Species with small and occasional idioblasts were: *R. rugosum* Low ex Hook.f., *R. polyanthemum* Sleumer, *R. konori* Becc., and *R. lochiae* F.Muell. Based on these species, relative idioblast size and abundance does not associate with subsections or with geographical range. The volume and abundance of idioblasts in *Vireya* leaves was regressed against all leaf morphological and anatomical characters, as well as the central elevation and the central latitude of the native range. Idioblast abundance or volume was not significantly associated with any other morphological or anatomical traits that were measured. Moreover, idioblast abundance or volume was not associated with central latitude of the native range. However, idioblast volume

29

Table 3 Anatomical characteristics of idioblasts in leaves from *Rhododendron* section *Vireya* plants growing at the Rhododendron Species Foundation/Botanical Garden, Federal Way, Washington, USA. Species are listed alphabetically. Numbers in parentheses are standard deviations. $n = 5$ (one leaf from each of five individuals).

Species	Idioblast cell diameter (mm)	Idioblast cell volume (mm³ × 10⁻⁵)	Idioblast abundance
aurigeranum	0.036 (0.006)	0.50 (0.14)	Common
blackii	0.050 (0.06)	1.56 (0.14)	Occasional
bryophilum	0.065 (0.014)	3.90 (0.02)	Abundant
burttii	0.061 (0.013)	2.15 (1.30)	Abundant
celebicum	0.050 (0.006)	1.07 (0.38)	Common
correoides	0.041 (0.004)	0.67 (0.42)	Abundant
crassifolium	0.067 (0.011)	2.95 (0.89)	Common
gracilentum	0.075 (0.002)	3.29 (0.24)	Abundant
herzogii	0.067 (0.007)	2.52 (2.44)	Common
javanicum	0.052 (0.008)	2.44 (1.33)	Common
kawakamii	0.061 (0.023)	3.30 (3.01)	Common
konori	0.043 (0.012)	0.61 (0.26)	Occasional
lochiae	0.048 (0.009)	0.74 (0.66)	Occasional
longiflorum	0.064 (0.022)	3.34 (2.27)	Abundant
loranthiflorum	0.062 (0.008)	3.01 (0.21)	Abundant
macgregoriae	0.055 (0.013)	1.45 (0.99)	Abundant
maius	0.049 (0.015)	1.58 (0.26)	Common
malayanum	0.042 (0.006)	5.96 (0.30)	Common
polyanthemum	0.056 (0.003)	1.50 (0.10)	Occasional
praetervisum	0.068 (0.017)	3.82 (1.96)	Common
rarum	0.064 (0.010)	3.63 (0.94)	Abundant
rugosum	0.037 (0.004)	0.54 (0.14)	Occasional
sororium	0.057 (0.016)	2.17 (1.30)	Abundant
stenophyllum	0.062 (0.006)	2.78 (0.85)	Abundant
superbum	0.047 (0.009)	1.03 (0.44)	Common
zoelleri	0.062 (0.015)	3.53 (2.00)	Common

tended to decrease in leaves from plants with a high elevation native range (Figure 5). This decrease in idioblast volume was not directly associated with a general decrease in cell size (personal observation). Therefore, based on this result, idioblast significance to leaf structure and function may increase at lower elevation.

All species of *Vireya* (except two) that were hand sectioned at RBGE had obvious idioblasts present (Table 4). I was not able to make good enough hand sections for *R.*

micromalayanum Sleumer or *R. saxifra-goides* J.J.Sm. in order to verify the presence of idioblasts. These species had particularly small leaves that made hand sectioning difficult. Therefore, microtome cross-sections will be required to verify the presence of idioblasts in these species. It is also interesting to note that *R. saxifragoides* is the only species observed with stomata on both leaf surfaces.

Idioblast cell diameter was recorded for 17 species growing at both the RGBE and the RSF. Measurements made by hand section at the RBGE resulted in slightly larger values for idioblast cell diameter than microtome cross-sections of leaves from species growing at the RSF (Figure 6). However, the relative sizes of idioblasts among species at the RSF and the RBGE were similar, as shown by the positive and significant regression. The regression does not pass through the origin,

which means there is not a one to one relationship between idioblast cell diameter at the two common gardens. This variation in cell size between the two common gardens could reflect differences in the environment at the two sites or differences in the genotypes of the individuals grown at the two sites. However, it is realised that idioblast cell size was assayed with different techniques between the two common gardens and this could have caused the differences in magnitude and slope of the relationship. Further studies on plants in their native habitat are required to determine the magnitude of environmental effect on idioblast cell size.

Further characterisation of the idioblast cells for species from the RSF was made using electron microscopy. Scanning electron microscopy clearly indicated that idioblasts had a large central vacuole and that the vacuole contained material other than simply an aqueous solution. Transmission electron microscopy images demonstrated that in one species (*R. sororium*) the central vacuole of idioblasts contained mucilage. In another species (*R. zoelleri* Warb.) the nature of the cells surrounding idioblasts and their cell wall junctions with idioblast cells suggested glandular function. Thus, the ultrastructure of idioblast cells varies widely among species, which suggests that the function of these cells may also be different among species.

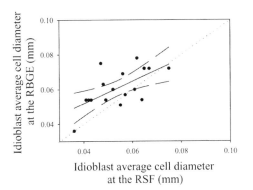

Figure 6 Average idioblast cell diameters for *Rhododendron* species growing at both the Rhododendron Species Foundation/Botanical Garden, Federal Way, Washington, USA and the Royal Botanic Garden Edinburgh, Scotland. The dotted line represents a one to one relationship between the two measurements of average idioblast cell diameter for each species. The regression line (r^2 = 0.40) and 95% confidence limits are shown.

Functional significance of idioblasts

The extent of variation in idioblast cell volume, abundance, and ultrastructure suggests that there are several ecological or physiological significances of these cells. Idioblasts of any one species may have any one or more of the following functional significances.

Table 4 *Rhododendron* section *Vireya* taxa examined for idioblast presence in leaf tissue by paradermal, hand section at the Royal Botanic Garden Edinburgh, Scotland. NA = Information is not available.

Species	Authority	RBGE Acc. no.	Idioblast diameter (mm)	Idioblast abundance
acuminatum	Hook.f.	19891870	0.054	Common
aequabile	J.J.Sm.	19750002	0.039	Common
album	Blume	19882544	0.078	Abundant
alticolum	Sleumer	19875217	0.054	Occasional
apoanum	Stein	19992280	0.042	Occasional
aurigeranum	Sleumer	19832007	0.036	Common
beyerinckianum	Koord.	19882537	0.060	Abundant
brassii	Sleumer	19990806	0.048	Abundant
*brookeanum**	Low ex Lindl.	19801291	0.084	Abundant
bryophilum	Sleumer	19672227	0.072	Abundant
correoides	J.J.Sm.	19941439	0.054	Common
cruttwellii	Sleumer	19750103	0.048	Common
curviflorum	J.J.Sm.	19930948	0.054	Common
fallacinum	Sleumer	19772458	0.042	Occasional
gardenia	Schltr.	19761327	0.042	Occasional
goodenoughii	Sleumer	19772400	0.060	Abundant
gracilentum	F.Muell.	19722842	0.072	Common
herzogii	Warb.	761338	0.072	Common
inconspicuum	J.J.Sm.	19930810	0.048	Common
intranervatum	Sleumer	19622867	0.090	Occasional
jasminiflorum	Hook.	19943001	0.060	Common
javanicum	(Blume) Benn.	19913084	0.075	Abundant
kawakamii	Hayata	19710098	0.060	Abundant
× *keditii*	(Sleumer) Argent, A.L.Lamb & Phillips	19801184	0.054	Occasional
konori	Becc.	19762033	0.054	Occasional
leptanthum	F.Muell.	19630476	0.036	Abundant
leucogigas	Sleumer	19682431	0.072	Abundant
lochiae	F.Muell.	19812937	0.063	Abundant
loranthiflorum	Sleumer	19870129	0.060	Common
lowii	Hook.f.	19820912	0.063	Common
macgregoriae	F.Muell.	19881140	0.051	Abundant
maius	(J.J.Sm.) Sleumer	19861647	0.054	Common
malayanum	Jack	19842313	0.054	Common

meliphagidum	J.J.Sm.	19880517	0.042	Abundant
micromalayanum	Sleumer	19821507	NA	NA
notiale	Craven	19961303	0.042	Occasional
orbiculatum	Ridl.	20000888	0.084	Abundant
polyanthemum	Sleumer	19801305	0.069	Occasional
quadrasianum	Vidal	19972518	0.042	Occasional
rarum	Schltr.	19710243	0.054	Abundant
retusum	(Blume) Benn.	19882543	0.039	Abundant
rhodoleucum	Sleumer	NA	0.066	Common
ruttenii	J.J.Sm.	19880516	0.060	Common
saxifragoides	J.J.Sm.	19930920	NA	NA
sessilifolium	J.J.Sm.	19881434	0.066	Common
× *sheilae*	(Sleumer) Argent, A.L.Lamb & Phillips	801154	0.054	Occasional
solitarium	Sleumer	19681395	0.042	Occasional
sororium	Sleumer	19951613	0.078	Abundant
stenophyllum	Hook.f. ex Stapf.	19952737	0.057	Abundant
superbum	Sleumer	19885037	0.075	Common
taxifolium	Merr.	19922886	0.048	Common
vitis-idaea	Sleumer	19722383	0.084	Abundant
williamsii	Merr. ex Copel.f.	19902927	0.051	Common

*var. *kinabaluense* (Argent, A.L.Lamb & Phillips) Argent.

Idioblasts may behave as do buliform cells in corn leaves by expanding and shrinking in response to water availability. Since the cells are mostly on the adaxial surface, expansion or shrinkage should cause leaf curling as do buliform cells in corn leaves. Leaf curling is well known in *Rhododendron* (Nilsen, 1992) and occurs in leaves from plants in both the *Hymenanthes* and *Rhododendron* subgenera (Nilsen, 1991). However, many species that express extreme leaf curling in *Rhododendron* do not have idioblasts (e.g. *R. catawbiense* Michx. and *R. maximum*). Moreover, none of the *Vireya* species in this study demonstrated any amount of curling in response to freezing or desiccation (data not presented). Therefore, the data do not support an idioblast function similar to buliform cells.

Idioblasts could serve as a water storage system (a succulence component of leaves). This concept is supported by the large central vacuole and the presence of mucilage in this vacuole in some species. Also, the most succulent *Vireya* leaves (thickest leaves with the smallest intercellular air spaces) had the smallest idioblasts (e.g. *R. polyanthemum* and *R. konori*). If idioblasts served as a water storage cell then one would expect few and small idioblasts in leaves that were already succulent due to other anatomical characteristics.

Also, idioblast cells could function as part of a gland that exudes materials on the leaf surface. This potential functionality was supported by the arrangement of cells in the adaxial epidermis of *R. zoelleri*. The apex of each idioblast on the adaxial leaf surface of *R. zoelleri* extended to the leaf surface and was surrounded by a ring of epidermal cells. In addition, TEM images suggested that material oozes out to the epidermis from the region between the upper edges of the idioblasts and the adjacent epidermal cells (personal observation). Moreover, epidermal cells surrounding the apex of the idioblast were connected through highly modified thin regions of cell wall (possibly plasmodesmata fields). This specific epidermal cell arrangement and TEM ultrastructure was found only for leaves of *R. zoelleri* among all the species of *Vireya* sampled. However, this potential idioblast function may be widely spread among *Vireya* species, because hand sections performed on leaves of several species at the RBGE showed similar leaf anatomy to that of *R. zoelleri*.

Safranin-O was retained in the central vacuole of idioblasts in all but one *Vireya* species studied (Nilsen & Scheckler, 2003). Safranin-O preferentially stains nucleoli, chromosomes, lignified cell walls, mucilage, chloroplasts, cutin, chitin, tannin, and suberin (Ruzin, 1999). This histochemical response suggests that idioblasts may contain materials that serve as a chemical or mechanical herbivore defence mechanism.

Conclusion

Hypothesis 1 of our study was supported because the variation in leaf morphology and anatomy was the same or greater for the group of tropical (mostly *Vireya*) species compared with the group of species from all other latitudes. However, the group of *Vireya* species did have smaller and thicker

leaves than the group of non-*Vireya* species. These results were based on a sample size of only 53. It is hypothesised that if the sample size were increased, then the same relationships would hold except for an increase in the number of large leaf sizes for the non-*Vireya* group regions.

This study started with the hypothesis (Hypothesis 2) that leaf morphology and anatomy would be associated with the central elevation and central latitude of the species' native range. It was found that leaf area was constrained to smaller values at high central latitude of the native range (above 400), and the number of epidermal cell layers increased with increasing central elevation of the native range. However, most morphological and anatomical traits had complex relationships with the species' native range information. Further research on plants growing in the field and greater refinement of native range information are required to clarify these complex relationships.

The most important discovery of this study was the presence of idioblasts in leaves of *Vireya* species. Our study of approximately 60 species of section *Vireya* growing at the RSF and the RBGE showed that the presence of idioblasts is a synapomorphic trait for *Vireya*. The variable idioblast structure among species may serve as a diagnostic taxonomic feature through further study. Moreover, there must be several functional significances of idioblasts among species because of the variation in anatomy and ultrastructure documented in this study. Further research on these species in their natural habitats will lead to exciting discoveries about the functional significance of this adaptive trait.

Rhododendron contains a very large wealth of genetic and adaptive variation that is a treasure house for understanding adaptive traits in plants. Understanding those adaptive traits has significance for conservation, forestry, horticulture, and the enthusiast. Scientists, commercial growers, conservationists, and gardeners who work with *Rhododendron* should band together and form a wide coalition to promote the understanding of adaptive traits in the genus *Rhododendron*.

Acknowledgements

I thank Darren DeStefano, Keli Goodman, Brian Hacker, Tina Keesee, Andrea Venetz, and Amber Waller for help with sample preparation and photomicrography. Many thanks go to Rick Peterson (Co-Executive Director of the RSF) for many hours of help collecting leaf samples, making leaf measurements, and helping with logistics at the research garden. Also, many thanks to George Argent for all his advice and help with the outstanding living collection of *Vireya* species at the RBGE. I appreciate the help received with electron microscopy from Kathy Lowe and Virginia Viers at the Morphology Laboratory of the Virginia-Maryland College of Veterinary Medicine. Reviews of early drafts of this manuscript by Duncan Porter and Jonathan Horton are much appreciated. This research was supported by funds from the National Science Foundation and the Virginia Tech College of Arts and Sciences Millennium program.

References

CALDWELL, M. M. (1979). Plant life and ultra-violet radiation: some perspectives in the history of the earth's UV climate. *BioScience* 29: 520–525.

CHAMBERLAIN, D., HYAM, R., ARGENT, G., FAIRWEATHER, G. & WALTER, K. S. (1996). *The Genus Rhododendron. Its classification & synonymy*. Royal Botanic Garden Edinburgh.

CORDERO, R. A. & NILSEN, E. T. (2002). Effects of summer drought and winter freezing on stem hydraulic conductivity of *Rhododendron* species from contrasting climates. *Tree Physiol.* 22: 919–928.

COX, P. A. (1985). *The smaller Rhododendrons*. Portland, OR: Timber Press.

COX, P. A. (1990). *The larger Rhododendron species*. Portland, OR: Timber Press.

COX, P. A. & COX, K. N. E. (1997). *The Encyclopaedia of Rhododendron Species*. Glencarse, Perth: Glendoick Publishing.

DAVIDIAN, H. H. (1982). *The Rhododendron Species. Volume I, Lepidotes*. Portland, OR: Timber Press.

DAVIDIAN, H. H. (1989). *The Rhododendron Species. Volume II, Elepidotes, Part 1 Arboreum – Lacatum*. Portland, OR: Timber Press.

DAVIDIAN, H. H. (1992). *The Rhododendron Species. Volume III, Elepidotes continued, Neriiflorum – Thomsonii, Azaleastrum and Camtscaticum*. Portland, OR: Timber Press.

FENG, G. (1992). *Rhododendrons of China Vol. II*. New York: Science Press.

FENG, G. & YANG, Z. (1999). *Rhododendrons of China Vol. III*. New York: Science Press.

HOLLIS, S. & BRUMMITT, R. K. (1992). *World geographic scheme for recording plant distributions. Plant Taxonomic Database Standards No. 2*. Pittsburgh, PA: Hunt Institute for Botanical Documentation.

LERSTEN, N. T. & CURTIS, J. D. (1995). Two foliar idioblasts of taxonomic significance in *Cercidium* and *Parkensonia* (Leguminosae: Caesalpinioideae). *Amer. J. Bot.* 82: 565–570.

LIPP, C. C. & NILSEN, E. T. (1997). The impact of sub-canopy light environment on the hydraulic vulnerability of *Rhododendron maximum* to freeze–thaw and drought. *Plant Cell Environ.* 20: 1264–1272.

NILSEN, E. T. (1985). Seasonal and diurnal leaf movements of *Rhododendron maximum* L. in contrasting irradiance environments. *Oecologia* 65: 296–302.

NILSEN, E. T. (1986). Quantitative phenology and leaf survivorship of *Rhododendron*

ERIK NILSEN

maximum L. in contrasting irradiance
environments of the Appalachian mountains.
Amer. J. Bot. 73: 822–831.

NILSEN, E. T. (1991). The relationship between
freezing tolerance and thermotropic leaf
movement in five *Rhododendron* species.
Oecologia 87: 63–71.

NILSEN, E. T. (1992). Thermonastic leaf
movements: A synthesis of research with
Rhododendron. Bot. J. Linn. Soc. 110:
205–233.

NILSEN, E. T. & BAO, Y. (1987). The influence
of age, season, and microclimate on the
photochemistry of *Rhododendron maximum*.
I: Chlorophylls. *Photosynthetica* 1: 535–542.

NILSEN, E. T. & ORCUTT, D. M. (1996). *The
Physiology of Plants Under Stress: Abiotic
Factors.* New York: John Wiley & Sons Inc.

NILSEN, E. T. & SCHECKLER, S. E. (2003). A
unique "giant" cell type in leaves of *Vireya*s.
J. Amer. Rhododendron Soc. (January 2003
issue).

NILSEN, E. T., STETLER, D. A. & GASSMAN, C. A.
(1988). The influence of age and microclimate

on the photochemistry of *Rhododendron
maximum* leaves II. Chloroplast structure and
photosynthetic light response. *Amer. J. Bot.*
75: 1526–1534.

ORCUTT, D. M. & NILSEN, E. T. (2000). *The
Physiology of Plants Under Stress: Soil and
Biotic Factors.* New York: John Wiley & Sons
Inc.

RUZIN, S. E. (1999). *Plant Microtechnique and
Microscopy.* New York: Oxford University
Press.

SELBY, S. M. (1972). *Standard Mathematical
Tables, Twentieth Edition.* Cleveland, OH:
CRC Press.

SLEUMER, H. O. (1966). *An account of
Rhododendron in Malesia.* Groningen, The
Netherlands: P. Noordoff Limited.

SLEUMER, H. O. (1980). Past and present
taxonomic systems of *Rhododendron*
based on macromorphological characters.
In: LUTEYN, J. L. & O'BRIEN, M. (eds)
*Contributions Toward a Classification of
Rhododendron*, pp. 19–26. New York: The
New York Botanical Garden.

Wild rhododendrons of Bhutan

Rebecca Pradhan

Royal Society for Protection of Nature, Thimphu, Bhutan

Rhododendron distribution in Bhutan is influenced by altitude, aspect and the east–west longitudinal axis. Species diversity increases from west to east. Diversity in form ranges from dwarf rhododendrons in the alpine meadows to huge trees in subtropical and temperate forests. Of the 46 species found, four are endemic to Bhutan. The uses of rhododendrons in Bhutan are still closely linked to age-old traditional practices. Their ethnobotanical value is exhibited in the day-to-day uses of these plants in the form of local incense, medicine, knife handles and wrapping material.

Introduction

The Kingdom of Bhutan is located in the eastern Himalayas at 26°40′ to 28°15′N latitude and 88°45′ to 92°10′E longitude. It has an area of 46,500 km² and a population of 0.7 million. Some 72.5% of the country's total area is still under forest cover, with 8% forming agricultural fields. The remaining areas constitute snow and glaciers, rock outcrops, landslips, lakes and marshy areas.

Rhododendrons in Bhutan

Descriptions of the rhododendron species can be found in Cox (1979), Pradhan & Lachungpa (1990), Long & Rae (1991) and Pradhan (1999). In Bhutan many species of rhododendrons are found to exist in assemblages that vary according to differences in altitude and aspect and the longitudinal east–west axis. There are 46 species of rhododendrons described in the *Flora of Bhutan* (Long & Rae, 1991). Out of these 46 species, four are endemic to Bhutan. They include: (i) *R. kesangiae* D.G.Long & Rushforth (Plate 1), (ii) *R. pogonophyllum* Cowan & Davidian, (iii) *R. bhutanense* D.G.Long & Bowes Lyon, and (iv) *R. flinckii* Davidian.

Rhododendrons vary in shape, size and form, ranging from low creeping shrubs only 10 cm tall to trees 20 m high and up to 70 cm girth. Some form the understorey of mixed conifer and broad-leaved forests, some form dominant vegetation patches in alpine meadows above the tree line, while yet others are epiphytic. In April to May and June, the forests of Bhutan glow with the blooms of rhododendrons. The colour combinations and the birds they attract make the whole forest appear to sing with joy.

Rhododendrons hybridise readily in the wild, where several species occur together in groups. Common species which hybridise in the wild are *R. thomsonii* Hook.f. with *R. wallichii* Hook.f., *R. cinnabarinum* Hook.f. with *R. wallichii*, and *R. thomsonii* with *R. arboreum* Sm. The hybrids apparently live for only two to three years whereas the parent plants live much longer and predominate.

Plate 1 *Rhododendron kesangiae.*

Rhododendron zones

Subtropical forest, 150–1000 m altitude

The only species found in this zone is *R. arboreum*. It has a wide range of distribution, from 90 to 3300 m. The magnificent flowers of this species signify the arrival of spring in Bhutan.

Warm broad-leaved forest, 1000–2000 m altitude

In this forest environment rhododendrons are mostly mixed with other broad-leaved species. The species to be found are: *R. arboreum*, *R. dalhousiae* Hook.f. var. *dalhousiae*, *R. dalhousiae* var. *rhabdotum* (Balf. & Cooper) Cullen, *R. edgeworthii* Hook.f., *R. grande* Wight, *R. griffithianum* Wight, *R. lindleyi* Moore, *R. maddenii* Hook.f., *R. papillatum* Balf. & Cooper, and *R. virgatum* Hook.f.

Cool broad-leaved forest and conifer forest, 1800–4200 m altitude (temperate zone)

This is the most favourable habitat for most rhododendrons. The assemblages vary from homogeneous forests with one species, to heterogeneous areas with several different rhododendrons growing together. *Rhododendron kesangiae* and *R. hodgsonii* Hook.f. grow in moist habitats with misty weather conditions and are confined to narrow altitudinal bands. These species indicate wet places while *R. arboreum* is more widespread and indicates drier regions. The white-flowered *R. kesangiae* var. *album* Namgyel & D.G.Long is common towards the east while *R. argipeplum* Balf.f. & Cooper replaces *R. barbatum* G.Don in the east.

Rhododendron species found in this forest type are: *R. arboreum*, *R. argipeplum*, *R. baileyi* Balf.f., *R. barbatum*, *R. camelliiflorum* Hook.f., *R. campylocarpum* Hook.f., *R. ciliatum* Hook.f., *R. cinnabarinum*, *R. dalhousiae* var. *dalhousiae*, *R. dalhousiae* var. *rhabdotum*, *R. edgeworthii*, *R. falconeri* Hook.f., *R. fulgens* Hook.f., *R. glaucophyllum* Rehder, *R. grande*, *R. griffithianum*, *R. hodgsonii*, *R. kendrickii* Nutt., *R. kesangiae*, *R. keysii* Nutt. (Plate 2), *R. lanatum* Hook.f., *R. lindleyi*, *R. maddenii*, *R. neriiflorum* Franch., *R. niveum* Hook.f., *R. papillatum*, *R. pendulum* Hook.f., *R. succothii* Davidian, *R. triflorum* Hook.f., *R. tsariense* Cowan, *R. vaccinioides* Hook.f., *R. virgatum*, *R. wallichii*, and *R. wightii* Hook.f.

Alpine zone, 3000–5200 m altitude

This vegetation type is represented by alpine meadows where dwarf and shrubby rhodo-

Plate 2 *Rhododendron keysii*.

dendrons are found. In these open meadows and marshy areas, rhododendrons often actively colonise. In some places the rhododendrons grow so thick as to make it difficult to pass through. These rhododendron thickets provide shelter to game such as musk deer, tragopans and pheasants. The species found here include: *R. aeruginosum* Hook.f., *R. anthopogon* D.Don, *R. bhutanense*, *R. campanulatum* D.Don, *R. cinnabarinum*, *R. flinckii*, *R. fragariiflorum* Kingdon-Ward, *R. lanatum*, *R. lepidotum* G.Don, *R. nivale* Hook.f., *R. pogonophyllum*, *R. pumilum* Hook.f., *R. setosum* D.Don, and *R. thomsonii*.

Figure 1 shows the species distribution across the different zones.

Uses

Several species of rhododendrons are of ethnobotanical value. The leaves of *R. anthopogon*, *R. nivale*, *R. fragariiflorum*, *R. setosum* and *R. lepidotum* are mixed with juniper species to make local incense, widely used in Buddhist monasteries and religious ceremonies.

Rhododendron arboreum and *R. campanulatum* are used in traditional medicine to treat diarrhoea, dysentery, rheumatism and sciatica. The leaves of *R. kesangiae* and *R. hodgsonii* are used to pack yak butter and cheese. Rhododendron wood is used for carving 'Dapa', traditional wooden bowls and containers with lids. In rural areas, farmers make agricultural implements and knife handles out of rhododendron wood because of its smooth, fine-grained timber. The vegetative parts of *R. thomsonii* can be used as a natural insecticide. Almost all the species of rhododendron are used as fuel wood.

Threats and conservation efforts

At present rhododendrons in Bhutan do not face severe threats. However, sometimes infections of rhododendron thrift disease

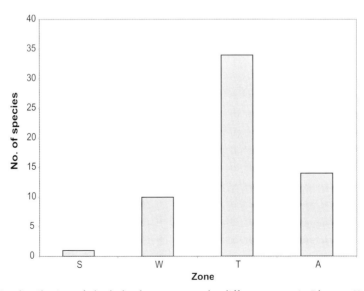

Figure 1 Species distribution of rhododendrons among the different zones in Bhutan. Key to zones: S, subtropical; W, warm broad-leaved; T, temperate; A, alpine.

are seen. At higher altitudes, yak herders set fire to rhododendron patches to make fuel wood. In certain areas natural regeneration is poor because of human interference leading to ecological disturbance.

Although threats to rhododendrons are not a serious issue at present, it would be unwise to remain complacent, and one may speculate as to their future. Fortunately, most of the rhododendron-rich areas have already been included within the National Parks.

In addition, about 2 ha of rhododendron forest in the Thrumshingla National Park, harbouring 22 different species, have been identified as an *in situ* rhododendron garden. It was inaugurated on 2 May 2002 as one of the commemorative activities for the International Year of Mountains 2002.

Acknowledgements

My sincere appreciation to my sponsors The Sibbald Trust and to Eona Aitken, Mary Bates, David Long, Henry Noltie, Mark Watson, and other friends for their encouragement and assistance.

References

Cox, P. A. (1979). *The Larger Rhododendron Species*. Portland, OR: Timber Press.

Long, D. G. & Rae, S. J. (1991). Rhododendrons. In: Grierson, A. J. C. & Long, D. G., *Flora of Bhutan* 2(1): 357–404. Royal Botanic Garden Edinburgh.

Pradhan, R. (1999). *Wild Rhododendrons of Bhutan*. Kathmandu: Quality Printers.

Pradhan, U. C. & Lachungpa, S. T. (1990). *Sikkim-Himalayan Rhododendron*. New Delhi: Thomson Press.

Rhododendrons in the wild: a taxonomist's view

David F. Chamberlain

Royal Botanic Garden Edinburgh, 20A Inverleith Row, Edinburgh EH3 5LR, UK

From rhododendrons in the wild to rhododendrons on the show bench there is quite a gulf. Yet the species names that are applied to individual plants are common to both. The application of these names may depend on detailed laboratory research or on comparison of plants with different provenances in the herbarium and in cultivation. However, the provision of a scientific name ultimately defines a species, subspecies or variety as it occurs in the wild. This paper seeks to show that studies in the field are an integral part of the process that leads to this provision of a name.

Introduction

Rhododendrons are known to hybridise freely, both in the wild and in cultivation; it is also evident that hybridisation has played an important role in the evolution of rhododendrons. While a hybrid may be morphologically intermediate between its two parents, it is not always possible in the herbarium to distinguish such intermediates from those that are variants of a single taxon. Field studies can highlight the frequency of such intermediates and can indicate the physical relationships between the plants in a way that is impossible in the herbarium. Furthermore, a stabilised hybrid population that may now occur outside the range of the parents may be an incipient new taxon.

When a species contains a number of recognisably different elements, field studies may be used to show whether these elements occur in the same populations, or whether they occupy essentially different geographical areas or have different ecological preferences. These studies can have a bearing on the names that are then applied.

When marked morphological discontinuities occur within a group of closely related individuals as they occur in the wild, that group may be separated into two or more definable subgroups, each of which may be given a distinct species name. It is thus important to emphasise that species names describe natural variation within and between natural populations, and that this is not always in line with the variation to be seen in cultivated material.

What's in a name?

The need for a properly classified and named plant surfaces at flower show competitions: consternation can be caused when a plant is assigned the wrong name or is misclassified. So, it is germane to ask what a name actually implies. The question becomes trivial when considering cultivars because there is more or less a right or wrong answer. Either the plant matches exactly a standard reference, or it could be matched by its DNA 'fingerprint'. However, this does not apply to a species, in which there may be considerable variation.

Plate 1 The author in a field of *Rhododendron hippophaeoides* (Zhongdian Plateau, NW Yunnan).

Plate 2 Mountain slope with *Rhododendron temenium* (Mekong–Salween Divide, Meili Shan, NW Yunnan, 3700 m).

Plate 3 *Rhododendron cinnabarinum* ssp. *tamaense* (Dulong–Salween Divide, near Gong Shan, NW Yunnan, c.3500 m).

Plate 4 *Rhododendron temenium* (Meili Shan, NW Yunnan, 3700 m).

Plate 5 *Rhododendron stewartianum*, pink form (Dulong–Salween Divide, near Gong Shan, NW Yunnan, c.3500 m).

Plate 6 Dulong Pass, above Gong Shan, NW Yunnan, c.4000 m.

Unfortunately for the exhibitor, the species name has not been designed to distinguish between plants on the show bench.

So, how do we define our subgroups or species, particularly bearing in mind just how frequent inter-specific hybrids are in the genus *Rhododendron*? There are relatively few species that are so distinct from all the remaining species that they do not hybridise, and cannot be confused with any of them. Two such are *R. schlippenbachii* Maxim. and *R. albrechtii* Maxim. These two species therefore present no problems whether in the wild or on the show bench. However, all too often, related species do hybridise in the wild or have been hybridised in cultivation. Furthermore, even with the most commonly grown species, only a fraction of the variation that occurs in the wild is represented in the world's collections of cultivated rhododendrons.

Hybridisation

Hybridisation is a problem in the study of rhododendron taxonomy as it has played a major role in the evolution of the genus, even in the wild. It is therefore probable that hybrid populations in the wild have become stabilised and established in areas where the parents no longer exist. In such instances it may be appropriate to assign a species name to such populations. But how do we distinguish between populations in which hybrids occur alongside their parents and those in which they do not? It is certain that this cannot be done from plants growing in gardens. It is also likely that the locality information attached to herbarium specimens is not adequate to distinguish between the two situations. There are laboratory techniques that may enable hybrids to be detected in cultivated plants but, ultimately, field studies of

Plate 7 *Rhododendron temenium × R. aganniphum* (Meili Shan, NW Yunnan, 3700 m).

wild populations are often the only practical way to identify what should or should not be considered to be hybrids, and are therefore worthy or not of formal species recognition.

The most clear-cut case that I have come across in the field is *R. × agastum* Balf.f. & W.W.Sm., from the Cangshan in Western Yunnan, China. Reference to the type specimen (though not to the original publication of the name) indicates that George Forrest, who collected it, considered it to be a hybrid of *R. arboreum* Sm. ssp. *delavayi* (Franch.) D.F.Chamb. and a species in subsect. *Fortunea*. In 1981 I visited the type locality of *R. × agastum* and it was obvious that it was a rather rare chance hybrid between two dissimilar parents, ssp. *delavayi* and *R. decorum* Franch. Clearly, it did not have an existence independent of its parents. It follows that the name can be applied only to a hybrid between these two parents. Material in cultivation under the name *R. agastum* is sometimes not this hybrid and is therefore wrongly named.

Rhododendron × erythrocalyx Balf.f. & Forrest is another example of a clear-cut hybrid. This was originally described as a distinct species from the Beima Shan, in NW Yunnan, SW China, along with several other closely allied entities from the same mountain, each of which have been described as separate species. Herbarium studies indicated that these entities might be hybrids between *R. wardii* W.W.Sm. and *R. selense* Franch. This was confirmed when field studies showed that *R. × erythrocalyx* was indeed intermediate between the two putative parents and occurred with both on the Beima Shan.

A more confused example involves the hybrids (or species derived from hybrids) of *R. citriniflorum* Balf.f. & Forrest. Hybrid

populations undoubtedly exist with *R. citriniflorum* var. *citriniflorum*, *R. temenium* Balf.f. & Forrest and probably *R. sanguineum* Franch. as parents on the Meili Shan, on the border between Yunnan and Xizang (Tibet). These give rise to the peach-coloured variants that have in the past been called *R. sanguineum* Franch. var. *didymoides* Tagg & Forrest and *R. cloiophorum* Balf.f. & Forrest. However, the typical *R. citriniflorum* Balf.f. & Forrest, with yellow flowers and a thick, dark indumentum, is now very rare. When I revised subsect. *Neriiflora*, of which these are members, I maintained these colour variants as distinct varieties of *R. sanguineum*. However, it is possible that they are not all that closely related to *R. sanguineum*. In this case field observations indicate likely parents; herbarium studies do not give certain answers.

On the pass into the Dulong Valley, close to the Yunnan/Myanmar frontier, there are extensive and variable populations of the orange- to red-flowered *R. citriniflorum* Balf.f. & Forrest var. *horaeum* (Balf.f. & Forrest) D.F.Chamb. It seems probable that the red-flowered forms are hybrids between var. *horaeum* and *R. sanguineum*, but it is also possible that the orange-flowered var. *horaeum* is itself a hybrid. This Dulong population occurs in an area where *R. sanguineum* is, for the most part, absent. Furthermore, the yellow-flowered var. *citriniflorum* does not appear to occur at all. The implications are that var. *horaeum* does merit recognition in its own right, possibly at the rank of subspecies.

The hybrids of *R. citriniflorum* are morphologically intermediate between the putative parents, but this is not so clear-cut with hybrids of *R. phaeochrysum* Balf.f. & W.W.Sm. Here too a plethora of names exist

Plate 8 Lower Dulong–Salween Divide, near Gong Shan, NW Yunnan, c.2000 m.

for taxa that are obviously closely related to *R. phaeochrysum*. However, some of these differ only subtly in the form of the indumentum on the lower surface of the leaves. Herbarium studies serve only to confuse the picture because it is not possible to elucidate how these different forms relate to one another in natural populations. In the garden these differences can be seen but there is no way to ascertain their significance.

In Western and Northern Sichuan, *R. phaeochrysum* inter-grades with *R. przewalskii* Maxim. A population from the Zheduo Pass in W Sichuan extending over a few hundred square metres contained forms that could be identified as both these species. However, there were forms that did not fit either precisely but could be referred to *R. phaeochrysum* Balf.f. & W.W.Sm. var. *agglutinatum* (Balf.f. & Forrest) D.F.Chamb. and var. *levistratum* (Balf.f. & Forrest) D.F.Chamb. A morphological analysis could lead to two possible scenarios. Either the two species had hybridised in a region where the distributions of the two overlapped, or this was a population that was morphologically variable and intermediate between the two species that were more clear-cut, with *R. przewalskii* to the north and *R. phaeochrysum* to the south. I am not at all sure that this could have been resolved using morphology alone because the differences between the individual plants were small and relatively difficult to define. In this case, laboratory studies involving leaf waxes that are reported elsewhere in this volume (see Rankin, 2003) give results that are easier to interpret. These unequivocally indicate that this population contains two distinct entities and a series of intermediates between them. This suggests that the two species should be maintained as distinct, and queries the

validity of the two varieties mentioned above. *Rhododendron phaeochrysum* also apparently hybridises with *R. aganniphum* Balf.f. & Kingdon-Ward in NW Yunnan and possibly also in Xizang. It remains to be seen whether leaf wax profiles will distinguish between these two sets of hybrids as the intermediates from the two areas are difficult to separate from one another morphologically.

One probable hybrid of *R. forrestii* Diels that was seen in the Dulong area in NW Yunnan differs from its parent in that the flowers were in trusses of three to four; those of *R. forrestii* are generally solitary. This hybrid plant, which I suspect could be referred to *R. chamaethomsonii* (Tagg & Forrest) Cowan & Davidian on account of the flower number, was growing mixed with its putative parent, and physically close to flowering *R. stewartianum* Diels. I have long suspected that *R. chamaethomsonii* var. *chamaethomsonii* is a hybrid of *R. forrestii*. However, it is now impossible to elucidate what the other parent was because there is no information about the co-habiting species. I cannot therefore be sure whether the Dulong plant does equate with *R. chamaethomsonii*.

Populations with apparently intermediate morphology between *R. arizelum* Balf.f. & Forrest and *R. rex* Lévl. ssp. *fictolacteum* (Balf.f.) D.F.Chamb. present an unresolved taxonomic problem that future fieldwork might resolve. The two taxa are distinct over most of their ranges; the former occurs in NW Yunnan and SE Xizang, the latter in W Yunnan and SW Sichuan. The former has ventricose-campanulate, yellow flowers and a dense, characteristic leaf indumentum constructed of deeply fimbriate cup-shaped hairs. The latter has funnel-campanulate white flowers and has a leaf indumentum constructed of only slightly fimbriate hairs.

In the extreme NW of Yunnan, including the Dulong area, there are populations containing a range of intermediates between these two extremes, but in which neither of the putative parents is represented. While these populations probably represent introgression between the two extremes, it is not clear whether they have the potential to evolve into a distinct new species, but it should be noted that they are apparently isolated from both the parents.

Subspecies and varieties

Given that some species are quite variable there may be case for subdivision into subspecies and varieties. Some authors accept that the subspecies is the appropriate rank if two or more morphological entities within a species are isolated from one another by geographical barriers; when they occur together within the same populations these entities are designated as varieties. Furthermore, when the boundaries between species are imprecise, the distributions of different variants can be crucial. It is for instance essential to know whether these variants occur intermixed in the same populations, or whether forms from different isolated areas are for the most part recognisably different.

Variants should be described only when, by so doing, the variation patterns are further elucidated. Populations on the same mountain can be isolated by altitude. This is apparently the case with *R. selense* Franch. ssp. *selense* and ssp. *dasycladum* (Balf.f. & W.W.Sm.) D.F.Chamb. on the Li-ti-ping in Western Yunnan. The former is apparently restricted to higher altitudes than the latter. As a consequence, the two have essentially different flowering times and probably do not often cross-pollinate. On the other hand, *R. selense* Franch. ssp. *jucundum* (Balf.f. & W.W.Sm.) D.F.Chamb., which differs from the other two subspecies in its longer calyx lobes, and in some other minor features, is geographically isolated, occurring further south than the other two subspecies, particularly around Dali, also in Yunnan. The ultimate test is that these three subspecies are isolated from one another, either by geographical distribution or by ecological factors. The morphological differences between the three are small, but are maintained in cultivation and are therefore not just phenotypic responses to the differences in the environment. It is therefore appropriate that they should be maintained as subspecies.

Rhododendron temenium Balf.f. & Forrest exists in three flower-colour variants. Populations on the Meili Shan in NW Yunnan are (hybrids excluded) apparently uniform in having carmine flowers. These are referable to var. *temenium*. However, the few flowering specimens that I have seen above the Dulong Valley are pink-flowered, and should be called var. *dealbatum* (Cowan) D.F.Chamb. It is not clear whether var. *dealbatum* has arisen as a hybrid of var. *temenium*. If that is the case the Dulong population is surprisingly uniform. The indications are that these two forms occur in mixed populations, further north, across the Xizang border. I have therefore maintained the two as varieties rather than subspecies.

Rhododendron hippophaeoides Balf.f. & W.W.Sm. is at present divided into two varieties: var. *occidentale* M.N.Philipson & Philipson is distinguished from var. *hippophaeoides* by its longer style. However, there is an entity that has not yet been described that has large flowers and a relatively dwarf habit. Pure populations of this

Plate 9 Peter Cox holding *Rhododendron glischrum* (Dulong–Salween Divide, near Gong Shan, NW Yunnan).

horticulturally more worthy form occur over several square miles of the Zhongdian Plateau in NW Yunnan. Elsewhere in Yunnan and SW Sichuan Provinces, the flowers are smaller and the plants are more open, growing to a height of 1.5 m or more. I suspect that this form has escaped recognition because the differences that define the Zhongdian variant from var. *hippophaeoides* are more difficult to describe, even though there is apparently very little overlap. I suspect that the Zhongdian variant is an incipient subspecies.

Concluding remarks

Human nature being what it is, biased collections will be made in the field. These can err on the side of emphasising the differences between closely allied entities. Alternatively, there may be sampling that is biased towards all the possible variants, amongst which some will inevitably be hybrids. In the garden this is exaggerated by selection of horticulturally worthy plants and by selection of plants that grow successfully under artificial conditions. For these reasons, herbarium material and plants in cultivation tell us little about naturally occurring populations. As has already been pointed out, scientific Latin names are designed for plants as they occur in the wild.

Herbarium collections come into their own when considering taxa over their complete range. It is generally logistically impossible for one person to sample the possible populations of a species adequately. Laboratory techniques may be able to elucidate variation patterns more precisely than is possible from morphological analysis, but are generally too time-consuming and expensive to pursue unless the questions posed have been narrowed down in other ways. The purpose of this paper has been to demonstrate how field studies can assist the study of variation patterns. However, I hope that it has also demonstrated their limitations. When the variation patterns are as complex as they are in the genus *Rhododendron* a synthesis of different approaches may be required. Only after these variation patterns in wild populations have been clarified can the scientific names be assigned with any degree of certainty.

Reference

RANKIN, D. W. H. (2003). Leaf wax chemistry and *Rhododendron* taxonomy: light in a dark corner. In: ARGENT, G. & MCFARLANE, M. (eds) *Rhododendrons in Horticulture and Science*, pp. 130–139. Royal Botanic Garden Edinburgh.

Ecologically based cultivation techniques for *Vireya* rhododendron species

David Mitchell

Royal Botanic Garden Edinburgh, 20A Inverleith Row, Edinburgh EH3 5LR, UK

During the past 40 years the Royal Botanic Garden Edinburgh has established the largest collection of *Vireya* rhododendron species in the world. Throughout this time considerable changes have taken place in the cultivation techniques used; in recent years these have been driven by field studies carried out in SE Asia. Three expeditions were carried out between 1991 and 1995, to Brunei, Irian Jaya and Mt. Kinabalu (Sabah). Apart from the collection of herbarium and living material, the main aim was the study of microclimates in relation to particular plant species. Throughout the focus was on an examination of environmental parameters. In addition soil structure and nutrient analysis was undertaken. The choice of location for each expedition took into account the need to study a broad spectrum of *Vireya* species and climatic conditions from diverse topographical regions, i.e. tropical rainforest, a mountain range and an isolated mountain peak. On each expedition study plots were established allowing data to be gathered through as wide a band of altitudinal zonation as possible, from lowland tropical forest to the high alpine zone. The data sets were combined to enable comparisons to be made against the environmental data collected from the growing conditions provided within the Research Collections in Edinburgh.

Whilst the most notable difference between conditions in the field and Edinburgh was the winter light difference, the data highlighted our concerns about the temperature of winter irrigation water, the summer glasshouse venting temperature as well as the overall night minimum cultivation temperature. The data also indicated changes which could provide improved conditions for the propagation and raising of young plants. The soil and nutrient data provided a valuable insight into how we could develop our compost and liquid feeding programmes to make them completely peat free, something which has now been achieved. Overall the data collected and the trials undertaken in Edinburgh mean that recommendations for the successful cultivation of *Vireya* rhododendron species can now be made with confidence.

Introduction

The genus *Rhododendron* consists of approximately 850 species (Sleumer, 1966), divided into several subgenera and sections. The largest of these sections, with over 300 species, is *Vireya*, representing approximately one third of the total known *Rhododendron* species. Although these plants are not hardy in cultivation in Northern Europe, vireyas are grown outdoors in the western United

States, Australia and New Zealand, where in recent years an interesting range of hybrids has emerged, improving their popularity.

As part of subgenus *Rhododendron*, vireyas have scales on their leaves; however, their flowers are much more brightly coloured, ranging from bright orange and yellow, into vivid reds, soft pinks and pure white. Many species are also sweetly scented, whilst others have distinct or attractive foliage. Overall they are desirable horticultural plants which deserve to be better known.

They have a wide natural distribution centred mainly in SE Asia, particularly New Guinea, Borneo and Indonesia. However, several outlying species are found in the Himalayas, Peninsular Malaysia, Vietnam and Australia. They occur naturally as either epiphytes or terrestrial plants, with some species occurring in both situations. This varied ecology, combined with a large altitudinal range (sea level to 4000 m), can make many of the species challenging to cultivate successfully.

Historical background

Despite this, during the past 40 years the Royal Botanic Garden Edinburgh has established the largest *Vireya* rhododendron species collection in the world, as part of a wider *Rhododendron* research programme. The earliest introduction was in the late 1800s – *R. javanicum* (Blume) Benn. (Argent & Hamlet, 1985); however, it was not until the 1960s that the build-up began in earnest with the introductions by Mr P. Woods and Mr B.L. Burtt which included *R. leptanthum* F.Muell. and *R. macgregoriae* F.Muell. from New Guinea, followed by *R. crassifolium* Stapf and *R. burtii* P.Woods from Borneo.

Since that time the collection has grown rapidly under the direction of Dr G. Argent,

with many changes taking place in the cultivation techniques used within the collection. These have primarily resulted from the direct participation of various horticultural staff in expeditions studying plants *in situ*. The early results of this work and the initial development of an improved compost were documented by Sinclair (1984).

Microclimate and soil studies

Despite advances, problems still surrounded the long term establishment and propagation of *Vireya* species, and it was in the hope of addressing these issues that information was gathered during three major expeditions over a four-year period: the Royal Geographical Society Brunei Rainforest Project (Mitchell, 1991), the RBGE Trekforce, Irian Jaya Expedition (Mitchell, 1992), and the RBGE Gunung Kinabalu Expedition (Mitchell, 1995). Their specific objective, apart from the collection of herbarium and living material, was to carry out microclimate studies in relation to particular *Vireya* species. Carefully selected and defined plots were established, and during each expedition the following environmental factors were studied in detail: air temperature, relative humidity, soil temperature and light levels. A total of over 2300 environmental readings were taken, plus associated soil and nutrient studies for each plot.

During the four-year expedition programme *Vireya* study plots were established throughout a broad altitudinal range, from the lowland tropical forest zone up to the high alpine zone. At the end of this period the data were amalgamated, providing an insight into the mean climatic parameters for the majority of *Vireya* species. The findings are represented in Figures 1–4. In addition to

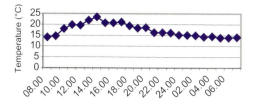

Figure 1 Mean daily air temperature from all expeditions. Minimum recorded air temperature was 10.4°C; maximum recorded air temperature was 34.7°C.

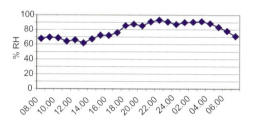

Figure 2 Mean relative humidity from all expeditions. Overall mean relative humidity was 84%.

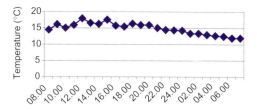

Figure 3 Mean daily soil temperature from all expeditions. Overall mean soil temperature was 16.5°C.

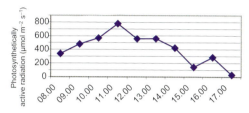

Figure 4 Mean daily light intensity (photosynthetically active radiation) from all expeditions. The regular pattern of daily light intensity seen in this graph, combined with the virtually even day- and night-lengths throughout the year, provide the plants with stable daily photosynthetic radiation levels for growth throughout the entire year. The sudden rise after the dip around 15.00 hours was due to frequent afternoon rain followed by evening sunshine.

the climate information, the soil and nutrient data gathered from the 15 plots proved valuable in providing an insight into how the compost and liquid feeding programmes could be developed to meet the needs of the more difficult species.

Results

Vireyas were found in locations where the soils were nutrient poor, with a low nitrogen and phosphorus content; however, we did note that the potash levels were high in all cases. The pH varied between pH 4 and 6.5, with all the species observed growing well throughout that range. It was also noted that vireyas had a preference for soil pockets with a high organic content, which in many cases was lying over a layer of mixed gravel or sand. Natural regeneration was observed only in areas where the pH was around 6.

In addition, the nitrogen levels here were always higher than in any other parts of the plot. The only exceptions to the above observations were for the species which occurred within the ultrabasic zone, where much further work is required before practical conclusions can be drawn.

Comparisons with environmental conditions at Edinburgh

The figures obtained in the field and represented in the graphs were compared with the environmental conditions being provided at the time of study within the Research Collections in Edinburgh (Argent *et al.*, 1993). The

Edinburgh data were obtained over a two-year period using a Data Hogg monitoring system. Comparison of the air temperature and humidity data proved illuminating, confirming several of our suspicions regarding temperature tolerance and the extremes which vireyas will withstand for short periods. However, the most notable difference between conditions in the field and at Edinburgh was the winter light deficit, which was more extreme than expected. This confirmed the need for a redesign of our existing supplementary lighting systems, and so the Philips Horticultural Lighting SGR system, fitted with SON-T Plus 400 W high-pressure sodium lamps providing the optimum spectrum of light for photosynthesis, was installed. This has made a notable difference to both flower bud initiation and leaf colour, particularly with the young seedlings and year-old plants.

The data also confirmed our concerns about the temperature of the irrigation water, especially during the winter months. Throughout the study, the correlation between soil temperature and rain water temperature was significant, as would have been expected.

In the field the mean average rain water temperature was 18.0°C, resulting in a mean soil temperature of 16.5°C. Upon investigation the temperature of the Edinburgh mains irrigation water was discovered to fluctuate between 5°C and 10°C, depending on the season of the year. Studies showed that this resulted in a mean potting mix temperature of 12°C, with a recorded minimum of 4.8°C during January. Clearly this 4°C mean deficit is detrimental to the cultivation of the collection, especially where young plants are concerned: however, in that case it is easily rectified through the use of clean water butts

which allow the water to reach the ambient temperature overnight. For larger containerised collections, however, this is impractical and the use of in-line warm water heaters should perhaps be considered in such cases for the more difficult species.

As a result of these findings gradual changes were made to the growing environment, particularly to the summer glasshouse venting temperature, as well as to the overall night minimum temperature. We have now reached the point where we are providing the optimum growing environment for *Vireya* species, in a cost effective, environmentally responsible way, with minimum energy wastage.

The current growing regime is as follows: night minimum 7°C, day temperature 10°C with a ventilation temperature of 15°C. Whilst this is balanced for established plants, our field research indicated that adjustments were required to provide improved conditions for the propagation and raising of young plants from seed and cuttings.

Germination

Since it was initially installed much has been done to improve the construction of the seed case originally designed by Rouse (1985) whilst maintaining the integrity of the original design. Currently using standard 150 W fish tank heaters, we hold the humidifying water tank at 21°C and suspend the pots above it on a metal grill with a fan ventilation system placed above water level and below the grill, drawing air in from the outside. This allows a humid atmosphere to be maintained in the case, whilst preventing the formation of large droplets of condensation. In addition the good air movement helps prevent pathological problems occurring with the young seedlings. Furthermore, the case is illuminated, receiving 24-hour supplementary lighting from a series of two warm white and two day-light fluorescent tubes suspended above the transparent lid of the case.

Small-scale trials using the case have shown that excellent germination rates can be obtained for species vireyas, through the open sowing of the seed on riddled Melcourt Propagation Bark sterilised with boiling water. They also indicated a marked improvement in the growth of seedlings which were pricked out in batches into trays rather than singly into pots. Interestingly, we have also observed that the growth was often superior in the seedlings which were left rather than pricked out. While this practice is somewhat impractical, on occasions these seedlings have been most useful in securing the establishment of a problem batch of seedlings.

Vegetative propagation

Vegetative propagation is generally carried out during the period September to October, using semi-ripe nodal cuttings with two nodes where possible. For thicker-stemmed species, e.g. *R. hellwigii* Warb., wounding is beneficial, providing a larger surface area for root initiation. Additionally the cuttings are dipped in Synergol diluted at 3:1 before insertion. Trials have indicated Melcourt Propagation Bark to be the most successful rooting medium. Cuttings are placed in either a mist bench or a closed case with a basal temperature of 21°C. As with the seedlings, supplementary lighting is provided to compensate for the short day-length and poor light quality experienced in northern latitudes. A marked improvement in rooting and establishment of cuttings is obtained as a direct result of the lighting, justifying the additional cost. Following rooting, which

should occur after 8–10 weeks, the cuttings are removed from the rooting environment and hardened off for at least two weeks in an open case with basal heat and supplementary lighting as before. At this stage they are potted into the standard *Vireya* mix; however, great care must be taken as they are very vulnerable due to the exceptionally fine, delicate roots which are easily dislodged from the cutting. After potting they are returned to the open case where they will stay until they establish and show signs of new growth.

The evolution of the compost

From early beginnings some 30 years ago through contributions from different staff members, experimentation and co-operation from industry, our *Vireya* cultivation medium has evolved from a loam based compost, through a peat based formula and finally to the totally peat free medium used today.

The final development of the peat free medium came after the compost trials, primarily as a result of environmental concerns regarding the long term cultivation of the collection and the use of peat within botanic gardens.

The extensive compost trials carried out in conjunction with Melcourt Industries, in which nine different composts – ranging from peat based, peat free, to coir and loam based – were tested, confirmed that purely bark based mediums were viable. Although we were now confident that bark based composts could provide suitable stable growing conditions for a wide range of *Vireya* species, this still had to be tested on the main Research Collection which contained more species than had previously been trialled. Slowly, over a period of two years the new bark based, peat free medium was introduced to the collection, with the observations made during the field work and trials being used to refine the medium in response to what was being observed in the living collection.

Finally, following careful experimentation on the main living Research Collection we can now confidently recommend our standard *Vireya* species mix for a wide range of species, both terrestrial and epiphytic as well as from all altitudinal zones. This consists of two bulky organics and two additives, combined as follows per 100 litre mix:
- 60 litres Melcourt Propagation Bark
- 40 litres Melcourt Potting Bark
- 240 g Magnesium Limestone
- 60 g Fritted Trace Elements.

We find this provides good growth and rooting and, more importantly, that it has good longevity, reducing the need for frequent repotting. It is, however, important that the compost is not regarded as a stand alone cultivation solution, rather it is part of a cultivation regime closely linked to a liquid feeding programme Our liquid feeding programme was developed through further experimentation and trials. These were designed to identify the performance of various liquid feeds, with the results of the trial again being imple-

mented on the larger collection slowly over a period of time. Our main aim was to find a liquid feed which would provide vigorous but non-etiolated growth, combined with good leaf colour and flower bud formation. The results of the trial are shown in Table 1.

As a direct result of this experimentation we abandoned the previous liquid feeding regime in which Maxicrop Natural was applied every three weeks out of four, for the growing season from April to September, with a feed of Maxicrop Natural with Sequestered Iron in the fourth week. In its place we now use Vitafeed 1:1:1, which is used weekly throughout the same period, with a supplementary Polyverdol Foliar Feed (a trace element supplement) being used monthly throughout the growing season. Overall we have seen a marked improvement with this programme: it is providing excellent results, with good quality flowers, steady growth and good foliage colour in the vast majority of species. However, we have noted slight chlorosis in some older plants in the past growing season but this is perhaps to be expected in mature plants (10 years plus) with restricted root growth.

Table 1 Liquid feed trials.

Product	16-week growth R. gracilentum	16-week growth R. macgregoriae	Overall grading (1 = best)
Vitafeed	76.5 mm	236.5 mm	1
Solufeed	80.5 mm	213.5 mm	2
Algoflash	70 mm	184 mm	3
Phostrogen	77.5 mm	242.5 mm	4
Peters Professional	51 mm	210 mm	5
Maxicrop Natural	34.5 mm	46.5 mm	6
Chem Pak 8	53.5 mm	203.5 mm	7
Maxicrop Tomato	56 mm	227.5 mm	8
Control (no feed)	63 mm	31 mm	9

Collection establishment and status

Clearly the establishment of a species collection is about much more than successful cultivation techniques. It is also about international partnerships where we, as responsible horticulturists, work with personnel from the host nation, sharing living material and information equally. This ethic has been key to the development of the Edinburgh *Vireya* collection throughout the years.

Equally so has been the development of a close relationship between scientific and horticultural staff, a working environment which has encouraged the growth of an expanding skill base and good staff continuity plus skill transferral. These factors combined have delivered a dedicated provision of service to the living collection, ensuring its success. All this would be impossible to obtain, however, without that sometimes most difficult to secure resource, finance. This is crucial particularly regarding the provision of adequate dedicated funding for field work and collecting, especially over such a long term project where the immediate output and benefits are at times difficult for the funding provider to visualise. To address this problem the importance of regular measurable outputs in the form of publications and lectures cannot be over-emphasised.

In regard to the establishment of the collection we are often asked about the most successful method of plant introduction from the wild. Analysis of the methods of introduction used for the Edinburgh collection indicates that 48% of it was established from living plants, 20% from cuttings, 10% from seedlings and finally 22% from seed. However, due to the success we have enjoyed from seed in recent years, I would expect this figure to rise in the future, as the introduction from seed is now the best and most efficient method of collection expansion. In addition, it provides a broad genetic base, which is increasingly important for research purposes. This method is reflected in Figure 5, which indicates the rise in the number of species currently held. The sharp rise at the end is due to two factors – improvements in seed germination technique as a direct result of the Rouse Case and to a lesser extent a slight increase in the frequency of expeditions.

The geographical makeup of the collection is shown in Figure 6. As can be seen, we now hold 178 species from throughout the range, with the total rising as living material from the collection produces flowers for the first time. In the coming years it is envisaged that the collection will expand further, through a focused collection programme, or

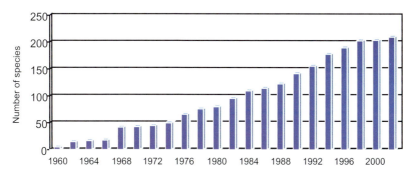

Figure 5 Growth of the RBGE living collection of *Rhododendron* section *Vireya*, 1960–2002.

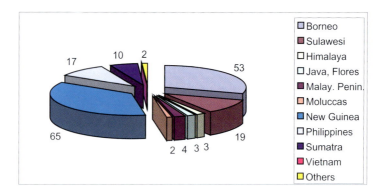

Figure 6 Geographical make-up of the RBGE living collection of *Rhododendron* section *Vireya* (species, subspecies and natural hybrids held, 2002).

the development of an Edinburgh range of hybrids. In addition, proposals are being developed to investigate flower inducement in difficult species, e.g. *R. maxwellii* Gibbs, as well as the creation of a computerised flowering database from our present paper-held records. Finally our aim is to prepare the definitive guide to the containerised cultivation and landscape display of this most interesting and attractive group of plants.

Acknowledgements

This work has been made possible only by the help of many colleagues and former staff members, who provided both advice and practical assistance both in Edinburgh and in the field. I am indebted to them for their encouragement and enthusiasm, especially Dr G. Argent, Mr J.D. Main, Dr D. Rae, Mr B. Burtt, Mrs M. Mendum, Mr P. Smith, Miss L. Galloway, Miss F. Inches, Mr A. Reid, Mr P. Woods, Mr R. Kerby, Mr I. Sinclair, Mr J. Sandham, Mr J. Fernie and finally Catherine Dawson from Melcourt Industries who assisted with the trials.

References

ARGENT, G. & HAMLET, E. (1985). Edinburgh's vireyas – *Vireya* rhododendrons cultivated at the Royal Botanic Garden Edinburgh. *Notes Roy. Bot. Gard. Edinburgh* 43(1): 185–186.

ARGENT, G., MITCHELL, D. & SMITH, P. (1993). Introducing vireyas. *The Garden* 118(11): 492–494.

MITCHELL, D. (1991). *Royal Geographical Society, Brunei Rainforest Expedition – Microclimate Studies*. Unpublished report, Royal Botanic Garden Edinburgh.

MITCHELL, D. (1992). *Trekforce Expedition, Irian Jaya – Microclimate Studies*. Unpublished report, Royal Botanic Garden Edinburgh.

MITCHELL, D. (1995). *Gunung Kinabalu Expedition, Sabah*. Unpublished report, Royal Botanic Garden Edinburgh.

ROUSE, J. L. (1985). The propagation of *Rhododendron* section *Vireya* from seed. *Notes Roy. Bot. Gard. Edinburgh* 43(1): 99–115.

ROYAL HORTICULTURAL SOCIETY (1990). *The Rhododendron Handbook*, 6th edition. London: The Royal Horticultural Society.

SINCLAIR, I. (1984). A new compost for *Vireya* rhododendrons. *The Plantsman* 6(2): 101–104.

SLEUMER, H. (1966). *Flora Malesiana*, Ericaceae [ser. I, vol. 6(1)], 469–668.

Rhododendron registration and the preparation of a new Register

Alan C. Leslie

RHS Garden Wisley, Woking, Surrey GU23 6QB, UK

The development of the International Rhododendron Register and the rules that govern its operation are reviewed, together with the progress made in preparing a new edition of the Register. The functioning of the registration process is explained and the importance of Standards emphasised. Credit is given to a small international team of individuals who have been assisting in the preparation of the new Register.

Introduction

We in the West think of ourselves as in the vanguard of those involved in the development and recording of rhododendron and azalea cultivars. In reality we were late starters. As far back as 1692, in Japan, there was sufficient interest and knowledge for Ito Ihei, 'a prosperous nurseryman of Old Japan', to publish a detailed and illustrated account of 332 named azalea cultivars. Many of these plants persist to this day and Ito's account too is still available, translated and re-published in 1984 under the title *A Brocade Pillow*. Ito was trying to standardise the names of the burgeoning number of azalea cultivars and to develop a terminology to describe them: a job the registration system is still trying to do! His work is merely an early example of a tradition of fine Japanese books on azaleas, continuing to the present day with the detailed and lavishly illustrated accounts published by the Japanese Satuski Azalea Society.

It was much later, well into the 19th century, before a similar number of rhododendron and azalea cultivars had been developed in Europe, later still in North America. But then they came pouring out of British nurseries such as Standish and Noble and the various Waterer establishments at Knap Hill and Bagshot. In continental Europe too the houses of Vervaene and van Houtte, for example, launched countless new azaleas whilst the Seidel family in Germany developed a steady stream of new rhododendron cultivars. With so much activity in different countries, using different languages and alphabets and involving such a wide range of firms, it is not difficult to see why names were soon duplicated and sometimes confusing. Without a single international authority to turn to, raisers had to cope on their own. One needs to remember too that in the 19th century *Rhododendron* and *Azalea* were treated as separate genera so the use of the same cultivar name in both was quite normal.

Here in the UK, for example, Standish and Noble listed an Indian azalea called 'Comet' in 1848, having already offered a hardy hybrid rhododendron of the same name in 1847. Later James Veitch also raised an Indian azalea to which he gave this name, whilst in 1862 Liebig in Germany was offering a 'Comet' derived from a *formosum/ edgeworthii* cross. Examples such as this are not uncommon.

Codes of nomenclature

Trying to introduce some order and method into this rather haphazard process of naming new cultivars is clearly impossible without an agreed set of rules. The first steps in this direction were taken by Alphonse de Candolle in 1862, and by 1867 the Botanical Congress in Paris accepted his *Lois de la Nomenclature Botanique*, which incorporated an article indicating that plants of horticultural origin should be given fancy names, that is, names in a common language as distant as possible from the Latin names of species.

This is not the place to go into the intricate and prolonged discussion, debate and argument that eventually, in 1953, resulted in the first edition of an independent *International Code of Nomenclature for Cultivated Plants*. This Code included a statement that adequate and accurate registration of cultivar names was of the first importance in their stabilisation. The aim of registration was described as 'to avoid duplication of names and the creation of names which are unnecessary or are likely to produce confusion and controversy'. To a large extent this still applies today.

Origin of the Rhododendron Register

After this events moved more swiftly. In 1955, at the 14th International Horticultural Congress in The Netherlands, the Royal Horticultural Society (RHS) was amongst the first bodies to accept responsibility for international registration, agreeing to deal with *Rhododendron*, *Narcissus* and perennial *Delphinium*. By 1958 the first Rhododendron Register, containing something like 8000–9000 entries, had been published by the Society. This work had a very strong Edinburgh connection since it was compiled by Harold Fletcher who had taken the work with him when he moved from his position as Director of the Garden at Wisley to be Keeper at Edinburgh.

Rhododendron growers in the UK had not been entirely idle in bringing their house into order prior to the appearance of the first Register, and indeed their efforts to a large extent shaped its form and content. *The Rhododendron Society Notes* for 1926 include a list of 'Rhododendron hybrids that have flowered and have been named, and of which the parentage can be traced back to species on both sides'. This consisted of only 100 names and was a very exclusive list for a very exclusive Society in which every member would no doubt have been just as conscious of their own lineage. No mention at all of vireyas or azaleas.

By 1929 the newly established and marginally more egalitarian Rhododendron Association produced a list of 600 names, 'compiled from the lists of principal nursery gardeners' and including some indication of flower colour and hardiness. An azalea list appeared for the first time in 1934. In 1945, after the end of the Second World War, the Association became the RHS Rhododendron Group, and this survives with various nomenclatural additions to its title to the present day; its early Handbooks continued to include cultivar lists up to 1969.

The registration process

Registration itself is merely the addition of an approved name to an existing list of approved names and is a process that confers no legal protection whatever over the plant or its name. The whole exercise relies on the voluntary co-operation of all concerned and in effect is a self-policing exercise conducted by the horticultural community. The development and increasing use of legally established protection for the plant material and the name under which it is sold is a more recent phenomenon which poses new and separate problems. An 'approved name' in this context is one that in the opinion of the registration authority is in accord with the *Cultivated Plant Code*, the most important element being that it should be novel. As soon as two plants have the same name that epithet can never be used again without some element of ambiguity, and this causes many problems in interpreting parentages or in other cases where the context does not help. To set the process of registration in motion an application form needs to be obtained, either directly from the International Registrar or from regional representatives based in North America, Japan, Australia and New Zealand. The form can also be downloaded from the RHS website (www.rhs.org.uk).

The form prompts for the full range of data we like to hold on each cultivar and which will be found in its Register entry. So we include parentage, plus the names and addresses of all those involved in the evolution of the new plant. Believe it or not, different people may be responsible for the initial hybridisation, for growing it on to flower, for its selection, naming, introduction and registration! The form also asks for a diagnostic description and we list the characters most likely to help in this regard, with colour references to the RHS Colour Chart preferred. We even ask for the edition of the chart to be stated since despite our best endeavours the four printings are not quite as identical as one would wish.

Increasing emphasis is placed, as part of the registration process, on asking for Standards. These are the equivalent of type specimens for cultivars. They are not as yet compulsory but at best are definitive, contemporary examples of a cultivar from the hand of the originator. Standards can be herbarium specimens, like most botanical types, but prints, slides or paintings are equally eligible if these would provide a more informative record.

When all is well and the name has passed all the tests and all the data gathered in, we issue a certificate. This whole service, certificate and all, is free, a small, but not inconsiderable, contribution to the Society's work as a charity.

It must be emphasised that it is not the registration authority's job to pontificate on the quality or distinctness of the plants whose names are submitted for registration. This very important and often neglected role is very much one for the originator to deal with.

As an International Cultivar Registration Authority (ICRA) the Society realises that some growers could not care less about this system and will never co-operate with it. Some just never hear about it. We recognise this and rather than stick our heads into the nomenclatural sand pretending their plants don't exist, we try to gather in data about them from whatever sources we can. Indeed where there is duplication of names it is all the more important that we can draw attention to them. For this reason we are now styling our publications as *Registers and Checklists* so that all names, registered or not, can appear logically in one list.

Register revision

In referring earlier to the *first* Register the heavy implication is that there was to be a second. The good news is that it is on its way. Since 1958 the Society has produced more-or-less regular annual supplements, usually adding several hundred names each time. These 40 lists are now an unwieldy reference tool. We need a new, fully revised and expanded Register. Work to provide this started formally on 5 April 1983.

Initially the plan was just to add in the new records and publish it all in hard copy. This later expanded to take into account some further significant new sources, most notably Galle's *magnum opus* on azaleas, but on the whole the work was not intended to involve a major historical trawl. We were then overtaken by the electronic revolution and the revision refocused itself on reworking the data to ensure it was consistently presented, updated and entered into a database. This meant reworking every single entry, looking at botanical names and names of the people involved, and standardising units of measurement and as far as possible colour coding. Altogether a massive task. For better or worse we are using an Access database, with specially written software, which, with very little effort, has been adapted for many of our other registers. At the present moment the Register is duplicated on the database and on over 27,000, 5 × 3″ index cards, housed in 24 box files. To this one needs to add c.3000 names of raisers, introducers and registrants: itself a significant set of data.

A great deal of time has been spent trying to iron out some of the inconsistent data and conflicting entries – or simply trying to decide if one has two uses to record or one use that has become garbled or misrepresented. As far as botanical names are concerned we have followed the revision undertaken in Edinburgh by James Cullen and David Chamberlain, with significant contributions from several other authors. For some this will be a painful memory as this revision was the first significant overhaul of the genus since the great influx of species in the first part of the 20th century, in response to which Balfour had devised a set of Series to accommodate all the new plants. Indeed the revision caused such a furore in horticultural circles that in possibly a unique exercise, the Society called a public open meeting to discuss the proposals. This was chaired by the then RHS President Lord Aberconway who was flanked on the platform by the two principal authors, together with Chris Brickell, then Director of the Garden at Wisley. There was a vigorous and at times impassioned debate from the floor: the Balfourian Series had clearly taken deep root amongst growers. More importantly it soon became apparent that many gardeners did not appreciate the potential for natural variation within species, believing that the one form of a species known to them was the one and only true plant!

The debate confirmed the Society's resolve to adopt the new proposals but at Chris Brickell's instigation many sunk taxa were retained as 'Groups' where it was felt they had horticultural value. This is what we still aim to do. A number of subsequent changes have been absorbed into the system, although it must be admitted that we 'stuck in our heels' about demoting *R. yakushimanum* and continue to use this at specific rank. It must surely be one of the most widely grown and hybridised species of all rhododendrons.

Advisory team

The overlong period of gestation for this new Register is in part a reflection of the size of the task but has as much to do with other demands on the Registrar's time. However, as entries under individual letters of the alphabet were finished the accounts were printed off and sent round to a small international team of individuals, to whom we would like to pay tribute. They have taken on this work for no reward, save the satisfaction of passing on their knowledge and experience, and have given unstintingly of their time over many years. They are: Alan Hardy (UK), John Bond (UK), Charles Puddle (UK), Hideo Suzuki (Japan), Graeme Eaton (Australia), Michael Cullinane (New Zealand), Albert de Raedt (Belgium), Walter Schmals-

cheidt (Germany) and Jay Murray (USA). Sadly both Alan Hardy and John Bond have died without seeing the finished product.

Together this team have produced so much new data that it has taken several years to get it all incorporated, but we are nearly there. We fervently hope and intend to complete the project in 2002 and to get the data made available during 2003. We certainly still intend to publish in hard copy and we will be putting the data on our website (where they will join both our orchid and daffodil registers, with clematis soon to join them). These web accounts are regularly updated.

As it currently stands the Register starts with the cultivar 'A. Abels' and ends with General Harrison's 'Zyxya'. The shortest entry must be the Lem/Barefield hybrid 'Oz' whilst the longest name is the old van Nes cultivar 'The Honourable Jean Marie de Montgue' (31 letters, 11 syllables). We wouldn't be able to allow that today!

Rhododendron powdery mildew – a continuing challenge to growers and pathologists

Stephan Helfer

Royal Botanic Garden Edinburgh, 20A Inverleith Row, Edinburgh EH3 5LR, UK

There appears to have been a resurgence of powdery mildew infection of rhododendrons in the recent past. This is probably mainly due to environmental factors, but at least in some cases a spread of the pathogens responsible for the disease can be assumed.

Reports from growers in Scotland indicate that powdery mildew has been on the increase during the past two years, mainly affecting known susceptible cultivars but also spreading out onto plants which have been considered resistant in the past. On the other hand, new pathogens have been discovered (one of the specimens collected 10 years ago has turned out to be a species new to science) and hitherto geographically confined species have been reported from new, previously unaffected areas. Thus the 'North American' species *Microsphaera azaleae* Braun (see Note 1) has recently been found in Britain and notably in the Royal Botanic Garden Edinburgh.

The common 'British' *Rhododendron* powdery mildew continues to be an enigma, as it is still known only in its anamorphic (asexual) condition. Rhododendrons in North America have now also been found infected with this pathogen. Possible control measures are discussed and suggestions for future research strategies are presented.

Introduction

Powdery mildews, diseases caused by members of the Erysiphales, are still causing major problems in *Rhododendron* plantings in many parts of the globe (Farr *et al.*, 1996). Problems with powdery mildew on this genus were first reported in 1955 (Anon., 1955). In the 1980s powdery mildew diseases became increasingly important to growers, and now 12 taxa of mildew pathogens have been described for this host (Table 1). Unfortunately it has not been possible to verify all of these, and Braun (1987) lists only five taxa with a possible sixth (our main pathogen). *Erysiphe rhododendri* Kapoor, a pathogen

which has been found only once in Sikkim, *Microsphaera azaleae* on North American deciduous azaleas and *M. izuensis* Y. Nomura from the Far East have been reported from wild-growing rhododendrons, with the latter two also occurring on cultivated species and hybrids. The other taxa have so far been found only on cultivated *Rhododendron*.

Recent research and findings

At the International Rhododendron Conference in Bad Zwischenahn in 1992, two papers were submitted on the subject (Helfer, 1994; Kenyon *et al.*, 1994), and these provided the baseline of the recent research

Table 1 Taxa of powdery mildew on *Rhododendron* spp.

Oidium sp. on *Rhododendron* sect. *Vireya* under glass
Oidium sp. on evergreen outdoor rhododendrons
Oidium sp. on deciduous outdoor azaleas
Oidium ericinum Erikss.
Erysiphe polygoni DC. reported from California and Virginia
Erysiphe sp. on outdoor rhododendrons (*cinnabarinum*)
Erysiphe rhododendri Kapoor
Microsphaera azaleae Braun (outdoor azaleas, USA, Germany)
Microsphaera digitata Inman & Braun (*R. mecogense* ex Belgium)
[*Microsphaera vaccinii* (Schw.) Cook & Peck]
[*Microsphaera izuensis* Nomura]
[*Phyllactinia guttata* (Wall.: Fr.) Lév.]

into these diseases. Most of the conclusions of the 1992 presentations are still valid, with some additional ideas on recommended control measures and pathogen taxa. A world survey on the problem was initiated, and results published in 1995 (Basden & Helfer, 1995, 1996). This highlighted the host range of the mildew pathogens as well as different environmental conditions favoured by some of the various pathogens involved. At that time 69 species and a further 123 hybrids or cultivars were found to be susceptible to mildew. The species most often mentioned were *R. cinnabarinum* Hook.f. and *R. thomsonii* Hook.f. However, no new light could be shed on where the current mildew epidemic might have originated. Further occurrence of pow-

dery mildews on rhododendrons in cultivation has also been reported since from other localities in Europe (Braun, 1997).

The biology of the main pathogen on British (outdoor) rhododendrons has been studied by Kenyon *et al.* (1994, 1995, 1997, 1998a,b, 2002) and others (Beales & Hall, 1994; Hall & Beales, 1998; Inman *et al.*, 2000). Kenyon *et al.* (1995) developed a method of sterile dual culture of the pathogen on micropropagated *Rhododendron* plantlets, which allowed a number of in-depth studies to be carried out (see Figures 1–6). It transpired that pathogen development is favoured by cool temperatures in high humidity and at low light levels; this was attributed to higher disease resistance in plants under higher light levels and longer photoperiod (Kenyon *et al.*, 2002). Higher light levels and a longer photoperiod also induces the mildew to form appressoria less frequently, leading to overall smaller colonies.

In a further study selected fungicides were applied to plantlets *in vitro*. Eight fungicides were evaluated for activity against *Erysiphe* sp. using *Rhododendron ponticum* L. microplantlets grown *in vitro*. Pathogen development changed with both the type of fungicidal compound and the concentrations applied. The most active materials were fenpropidin and penconazole, which showed high activity at the lowest concentrations. Six of the compounds performed more effectively than a mixture of bupirimate and triforine (Nimrod T), the then standard recommendation (see Note 2) for control of this pathogen on *Rhododendron*. All fungicides showed an effect on the sporulation of *Erysiphe* sp., with propiconazole, pyrazophos and triadimenol causing a significant increase in sporulation at the lowest concentrations (Tables 2 and 3). At higher concentrations, sporulation was

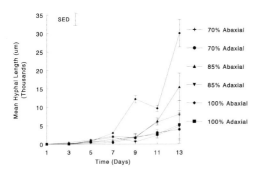

Figure 1 The effect of relative humidity on the total length of hyphae produced by *Erysiphe* sp. colonies on *Rhododendron*.

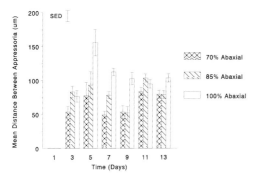

Figure 2 The effect of relative humidity on the frequency of appressoria produced by *Erysiphe* sp. colonies on *Rhododendron*.

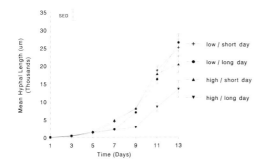

Figure 3 The effect of light intensity and photoperiod on the total length of hyphae produced by *Erysiphe* sp. colonies on *Rhododendron*.

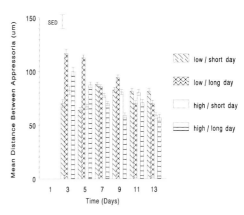

Figure 4 The effect of light intensity and photoperiod on the frequency of appressoria produced by *Erysiphe* sp. colonies on *Rhododendron*.

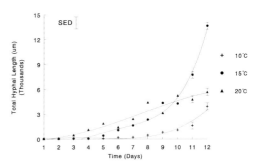

Figure 5 The effect of temperature on the total length of hyphae produced by *Erysiphe* sp. grown on *Rhododendron* cv. Elizabeth.

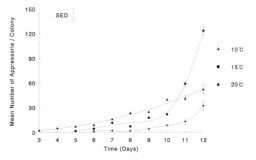

Figure 6 The effect of temperature on the number of appressoria produced by *Erysiphe* sp. grown on *Rhododendron* cv. Elizabeth.

Table 2 The effect of fungicide dose rate on the number of infected leaves on *Rhododendron ponticum* plantlets in dual culture.

	Fungicide dose rate (ai ppm)			
	0.1	1	10	100
Bupirimate and triforine	2.9	3.5	1.4 *a*	1.1 *a*
Fenarimol	3.6	1.7 *a*	0.1 *a*	0 *a*
Fenpropidin	1.8 *a*	1.1 *a*	0.1 *a*	0 *a*
Penconazole	3.3	0.7 *a*	0 *a*	0 *a*
Prochloraz	2.7 *a*	0.1 *a*	0 *a*	0.1 *a*
Propiconazole	3.3	0.7 *a*	0 *a*	0 *a*
Pyrazophos	3.9	2.7 *a*	1.6 *a*	0.9 *a*
Triadimenol	4.1	3.5	0 *a*	0 *a*
Control	3.6			

ai = active ingredient; ppm = parts per million; *a* = number of infected leaves is significantly lower than control.

Table 3 The effect of fungicide dose on sporulation of *Erysiphe* sp. infecting *Rhododendron ponticum* in dual culture. Figures are spore numbers = conidia per mg leaf tissue (dry weight).

	Fungicide dose rate (ai ppm)			
	0.1	1	10	100
Bupirimate and triforine	7095	5623	1296 *a*	1943 *a*
Fenarimol	**7478 *b***	3105	114 *a*	0 *a*
Fenpropidin	4509	3179	54 *a*	0 *a*
Penconazole	7335	0 *a*	0 *a*	0 *a*
Prochloraz	5140	2174 *a*	0 *a*	36 *a*
Propiconazole	**7693 *b***	**7513 *b***	0 *a*	0 *a*
Pyrazophos	**9557 *b***	**12515 *b***	5079	2390 *a*
Triadimenol	**7699 *b***	890 *a*	0 *a*	0 *a*
Control	5184			

ai = active ingredient; ppm = parts per million; *a* = sporulation significantly reduced; *b* = sporulation significantly increased.

significantly reduced by all treatments. No phytotoxic effects were detected with any fungicide at any concentration. The growth of plantlets in most treatments showed no significant difference from the untreated controls (Kenyon *et al.*, 1997).

Currently recommended control measures of the powdery mildew problems include

> **Box I** Advances in control measures. Refinement of and additions to earlier recommendations.
>
> *Earlier recommendations (still) holding true*
>
> • Where possible plant resistant cultivars
>
> • Apply good hygiene
>
> • Avoid overapplication of fertiliser (soft growth)
>
> • Be vigilant, reacting to first signs of disease
>
> • If necessary use sulfur-based fungicides
>
> *New recommendations*
>
> • If necessary use bicarbonate-based products
>
> • Propiconazole products have been shown to be effective
>
> • *Ampelomyces quisqualis* may be used for biological control

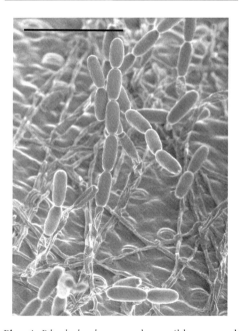

Plate 1 *Rhododendron* powdery mildew caused by *Erysiphe* sp.: chains of mildew conidia (SEM). Scale bar = 100 μm.

most of the older recommendations plus several additional observations (see Box 1). A number of rhododendrons have shown at least some lasting resistance. These are the dwarf rhododendrons, taxa of subsection *Triflora*, Asiatic rhododendrons such as *R. yakushimanum* Nakai (= *R. degronianum* Carrière ssp.) and relatives, and subsection *Arborea*. Further feedback from growers is needed to complement this list. Recently the use of bicarbonate-based products has shown promise, as has treatment with propiconazole (Robson, 2000; Elmhirst, 2001; Pscheidt, 2001) and the phylloplane fungus *Ampelomyces quisqualis* Ces. ex Schltdl. (Pscheidt, 2001).

Conclusions

Powdery mildews are continuing to cause extensive damage to rhododendrons. The nursery trade is particularly affected, as no economically viable control is possible at the moment. However, new control strategies are available to growers, and there are a number of resistant plants which can be exploited in breeding programmes. The taxonomy of the main pathogen (*Erysiphe* sp.) on outdoor rhododendrons in Britain is still unresolved, as no cleistothecia have been observed so far. *Microsphaera azaleae*, hitherto known mainly from North American deciduous rhododendrons, has now spread to Europe including Britain, and appears to be increasingly problematic.

References

ANONYMOUS (1955). *Agricultural Gazette NSW* 65: 100–103.

BASDEN, N. & HELFER, S. (1995). World survey of rhododendron powdery mildews. *J. Amer. Rhododendron Soc.* 49: 147–156.

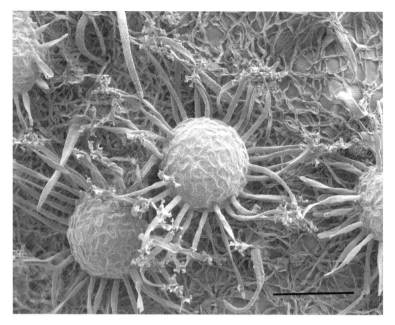

Plate 2 *Rhododendron* powdery mildew caused by *Erysiphe* (*Microsphaera*) *azaleae*: cleistothecia (SEM). Scale bar = 100 μm.

BASDEN, N. & HELFER, S. (1996). World survey of rhododendron powdery mildews. *Dunedin Rhododendron Group Bull*. 24: 18–43.

BEALES, P. & HALL, A. M. (1994). Over wintering and early disease development of *Rhododendron* Powdery Mildew. *Brighton Crop Protection Conference: Pests and Disease*, pp. 949–950.

BRAUN, U. (1987). A monograph of the Erysiphales (powdery mildews). *Beih. Nova Hedwigia* 89.

BRAUN, U. (1997). *Mycrosphaera azaleae* U. Braun (Erysiphales). In: TRIEBEL, D. (ed.) Microfungi Exsiccatae Fasc. 8–10 (nos. 176–250). *Arnoldia* 14: 12.

BRAUN, U. & TAKAMATSU, S. (2000). Phylogeny of *Erysiphe*, *Microsphaera*, *Uncinula* (*Erysipheae*) and *Cystotheca*, *Podosphaera*, *Sphaerotheca* (*Cystotheceae*) inferred from rDNA ITS sequences – some taxonomic consequences. *Schlechtendahlia* 4: 1–33.

ELMHIRST, J. (2001). *Rhododendron* Powdery Mildew. *BC Crop Protection Newsletter*, May 2001, p. 5. http://www.agf.gov.bc.ca/cropprot/cpnmay01.pdf

FARR, D. F., ESTEBAN, H. B. & PALM, M. E. (1996). *Fungi on Rhododendron: A World Reference.* Boone, NC: Parkway Publishers.

HALL, A. M. & BEALES, P. A. (1998). The epidemiology and control of rhododendron powdery mildew. ICPP 1998 abstracts.

HELFER, S. (1994). *Rhododendron* Powdery Mildews. *Acta Horticulturae* 364: 155–160.

INMAN, A. J., COOK, R. T. A. & BEALES, P. A. (2000). A contribution to the identity of rhododendron powdery mildew in Europe. *J. Phytopathol.* 148: 17–27.

KENYON, D. M., DIXON, G. R. & HELFER, S. (1994). Powdery mildew pathogens of rhododendron. *Acta Horticulturae* 364: 161.

KENYON, D. M., DIXON, G. R. & HELFER, S. (1995). Culture in vitro of *Rhododendron* and *Erysiphe* sp. *Plant Pathol.* 44: 350–354.

KENYON, D. M., DIXON, G. R. & HELFER, S. (1997). The repression and stimulation of growth of *Erysiphe* sp. on rhododendron by fungicidal compounds. *Plant Pathol.* 46: 425–431.

KENYON, D. M., DIXON, G. R. & HELFER, S. (1998a). The effect of temperature on colony growth by *Erysiphe* sp. infecting rhododendron. *Plant Pathol.* 47: 411–416.

KENYON, D. M., BRUNDIN, J., DIXON, G. R. & HELFER, S. (1998b). Investigation of the genetic diversity of *Erysiphe* sp. infecting *Rhododendron* spp. ICPP 1998 abstracts.

KENYON, D. M., DIXON, G. R. & HELFER, S. (2002). Effects of relative humidity, light intensity and photoperiod on the colony development of *Erysiphe* sp. on rhododendron. *Plant Pathol.* 51: 103–108.

PSCHEIDT, J. W. (2001). *Information on Rhododendron Powdery Mildew: An Online Guide to Plant Disease.* http://plant-disease.orst.edu/disease.cfm?recordID=969

ROBSON, M. (2000). *Rhododendron* Powdery Mildew. *Regional Garden Column*, June 2000. http://gardening.wsu.edu/column/06-11-00.htm

Notes

1   Molecular data published in February 2000 by Braun & Takamatsu suggest that *Microsphaera* (and *Uncinula*) can no longer be justified as separate genera of mildews. According to the rules of priority, all species have to be amalgamated with *Erysiphe*; all necessary new combinations and new names have been proposed and published but they have not been followed in this paper.

2   The registration of Nimrod T for amateur use is under phased revocation in the UK, with an expiry date of 31 March 2004.

'botanika' – a biodiversity project in the Botanic Garden and Rhododendron Park Bremen, Germany

Hartwig Schepker[1] & Michael Werbeck[2]

[1]Beratungs- und Planungsbüro für Gartenbau, Naturschutz und Pflanzenökologie,
Rampenstrasse 16, 30449 Hannover, Germany (e-mail: postbox@hartwig-schepker.de) and
[2]Senator für Bau und Umwelt, Ansgaritorstr. 2, 28195 Bremen, Germany (e-mail:
Michael.Werbeck@umwelt.bremen.de)

The 'botanika' project in the Botanic Garden and Rhododendron Park Bremen aims to illustrate biodiversity using the genus *Rhododendron* as the prime example. A visitor centre hosts a modern biodiversity exhibition, presenting many plant-related topics in an attractive way that is not directly instructive. Non-hardy rhododendron species from the Sino-Himalayan mountains and the South East Asian islands along with many of their associated plants are displayed in a conservatory to demonstrate diversity in this group of plants. The unique Bremen collection of Indian azaleas provides the setting for showing the potential uses of plants for mankind. 'botanika' is due to open in summer 2003.

Introduction

In June 1992 the first global agreement on biodiversity, the Convention on Biological Diversity (CBD), was passed during the United Nations Conference in Rio de Janeiro. It is based on an integrated and holistic approach acknowledging the human factor as an essential part of the complex of life on Earth. The main components of the CBD are the conservation of biological diversity, the sustainable use of its components and the fair and equitable sharing of the benefits.

A decade of many activities and projects originating around the term 'biodiversity' has followed. For several regions of the world, surveys of their biological diversity have been undertaken, but there is still an immense lack of information. The current state of global biodiversity has been summarised in the impressive *World Atlas of Biodiversity* (Groombridge & Jenkins, 2002).

Whereas 'biodiversity' is a well-used scientific term within modern ecology, it is surprisingly little known outside these circles. Despite an increased coverage in the public debate, the average citizen hardly has a clear picture of what 'biodiversity' means or what it represents in detail. Clearly Article 13 of the CBD, 'Public Education and Awareness', needs much more practical implementation. Article 13 urges all contracting parties to 'promote and encourage understanding of the importance of the conservation of biological diversity'.

In terms of biological diversity, there is hardly any other woody genus which serves as such a prime example as *Rhododendron*.

No matter which characteristics one looks at, the variation within the genus is extreme. Rhododendrons are distributed over almost all of the Northern Hemisphere, their vertical range stretching from sea level in Alaska and Greenland to 5800 m altitude on the Himalayan peaks. Rhododendrons grow along the icy coasts of Greenland but also in the tropical forests of Borneo. *Rhododendron pumilum* Hook.f. forms creeping shrubs just 10 cm tall, whereas *R. arboreum* Sm. grows into a tree of 30 m. Others species grow as epiphytes high up in the tree-tops. Some rhododendron flowers are just 1 cm long (*R. ericoides* Low ex Hook.f.) whilst others, such as *R. dalhousiae* Hook.f., reach a length of 20 cm. A similar range exists for the foliage: the leaves of *R. nivale* Hook.f. are less than 1 cm long, whilst those of *R. sinogrande* Balf.f & W.W.Sm. sometimes reach 90 cm. Last but not least, the variation of corolla colours and shapes within the genus is extraordinarily wide, making this one of the most eye-catching genera in temperate horticulture.

We feel the genus *Rhododendron* is ideal to illustrate to the public the meaning and importance of biodiversity in the spirit of the CBD, at all three levels of biological diversity:

1 It provides several examples of genetic diversity in a single species (e.g. *R. arboreum*).
2 The 1000 currently accepted *Rhododendron* species excellently demonstrate an enormous level of species variation.
3 The many different *Rhododendron* habitats in North America, Europe and Asia ideally demonstrate ecosystem diversity.

This idea was the starting point of the '*botanika*' project in Bremen.

The project

We aim to present the topic 'biological diversity' in an attractive and exciting way that is not directly instructive, for a broad range of visitors. Many *Rhododendron* examples will be used to illustrate the variation of species and habitats and to underline the necessity of preserving biological diversity. Thereby on the one hand '*botanika*' fulfils a national obligation to strengthen public awareness of conservation and management initiatives. On the other hand it demonstrates the ecological, social and economic significance of a genus which has been much criticised in the past due to the invasiveness of *R. ponticum* L. (see, e.g., Rotherham, 2003). Thus, '*botanika*' will hopefully open the eyes of its many visitors to the true role of *Rhododendron* in the global ecosystem, along with its broader environmental implications.

Location

The project is situated in the old Hanseatic city of Bremen in northern Germany. Bremen's Rhododendron Park is the largest one in the country. The park covers a total area of approximately 46 ha. It was founded in 1936 and has been enlarged several times. The park contains different sections: a botanic garden, a rose garden, a heath garden, an alpine garden, and a display area for the newest *Rhododendron* cultivars amongst others. The *Rhododendron* collection includes approximately 500 species, subspecies and varieties, about 2000 different cultivars and a unique historic assemblage of Indian azaleas. About 300,000 people visit the park each year, most of them during the main flowering period between April and June. But throughout the year the park is an important greenbelt recreational area for the citizens of Bremen.

'botanika' is located in the centre of the park. It consists of a renovated greenhouse used as entrance and foyer, a modern visitor centre hosting a biodiversity exhibition, a new 2500 m² conservatory for mostly non-hardy *Rhododendron* species, and a restored greenhouse for cultivated Indian azaleas. This complex of new and old buildings forms a thematic unit which can be walked through on a circular tour. There will be an admission fee for 'botanika', while admission to the Botanic Garden and the Rhododendron Park is free.

'botanika' will be built and managed by the municipal Rhododendron Park Company. The total budget of the project is about 17 million euros. The German Federal Agency for Nature Conservation is supporting this project with a share of 2.8 million euros. It considers the concept a very promising way to create more interest in and a better understanding of nature conservation issues.

The visitor centre

The visit to 'botanika' begins in the visitor centre. The biodiversity exhibition consists of different themed sections. It aims to address people of all ages and provides information at different levels. Many of the exhibits encourage young and older visitors to interact. The visitor centre is composed of two levels which are interlinked in the form of an ascending spiral (Plate 1). The spiral encircles a wall which symbolises a cup-like blossom, embracing on each level a secret room. The first mysterious room, called the 'origin', has an aesthetic, flowery atmosphere and provides an introduction to the exhibition. From there on, a winding path leads through the building, taking the visitor from the 'dark to the light', from the 'microcosm to the macrocosm', from the basic facts of life on Earth to global biological diversity as the overall result.

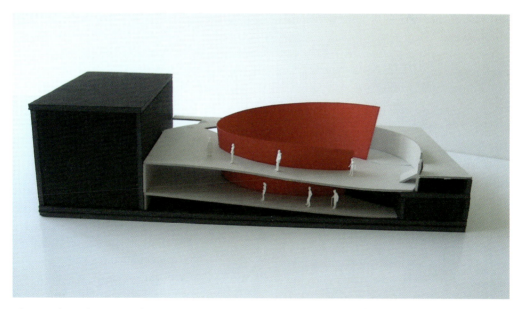

Plate 1 The architecture of the new visitor centre of 'botanika': on both levels the path encircles a 'mysterious' room in the centre of the building. The black box on the left side is the administrative building.

The path starts in the 'underworld', where environmental fundamentals like soil, water and climate are illustrated. Each themed section provides general information and *Rhododendron*-related facts. 'Discoverer stations' invite further exploration. Special children's exhibits keep younger visitors busy. The anatomy and morphology of plants is the next compartment, followed by one introducing the functions of plants. The 'darker' regions of the visitor centre are left behind when the visitor enters the section displaying life cycles of plants. On the upper floor, a second mysterious room, the 'kingdom of blossoms', offers a chance to relax and enjoy a visual and sensual experience. The following themed stations impart more complex knowledge about ecosystems, biological diversity and the contents and objectives of the CBD. The highest level of the visitor centre introduces the diversity of *Rhododendron* habitats. Monitors show short films about natural *Rhododendron* stands from Switzerland, Yunnan, Borneo and Georgia. Thereafter, a bridge disguised as a sluice marks the end of the 'theoretical' exhibition area and leads to the real plant world of the new conservatory.

Himalayan and Borneo greenhouse

The new, 17 m high conservatory is divided into two parts (Plate 2). One section hosts *Rhododendron* species from the lower altitudes of the Himalaya and neighbouring Chinese mountains. The other is dedicated to vireya species from South East Asia. All these tender rhododendrons have to be cultivated indoors due to the local climate conditions in Bremen, where hard frosts occasionally occur. The plants are displayed in natural mountain-like scenery with a stream, a lake, a waterfall and shady gorge-like sections.

The two sections can be explored on a main route with a few sidepaths. A lookout tower in the middle of the conservatory allows a closer look at the flowers of some of the large trees in the vicinity, such as *R. arboreum* and *R. protistum* Balf.f. & Forrest var. *giganteum* (Tagg) D.F.Chamb.

The visitor is taken on an imaginative round trip through the principal natural *Rhododendron* areas of mainland Asia. The countries and regions of this continent which have most of the temperate species are introduced. Facts about the flora of these areas are given as well as information about the history of plant collection in Asia. Different ethnic exhibits establish the connection of the plants to the people living in the natural *Rhododendron* territories. Along the main path the visitor can turn prayer wheels, speculate about stone carvings with Buddhist mantras, pass a Mani wall, marvel at a 4 m long Nirvana Buddha, said to be the largest outside Asia, and interpret the paintings on a Medicine Thangka.

Starting on a 4 m high gallery, the main path leads at first through a section representing the *Rhododendron* landscape of Nepal. Next to typical *Rhododendron* species of this region, such as *R. grande* Wight and *R. griffithianum* Wight, grow many of their associated plants, e.g. primulas or cobra-lilies. The journey proceeds, figuratively speaking, eastwards. The path leads through the Indian federal state of Sikkim and the Kingdom of Bhutan, another little-known Indian region, Arunachal Pradesh, and Myanmar (Burma). It continues into the Chinese provinces of Yunnan and Szechuan. Along the way representative plants of the associated native flora accompany typical *Rhododendron* species of these regions. The collection includes rare non-hardy *Quercus*

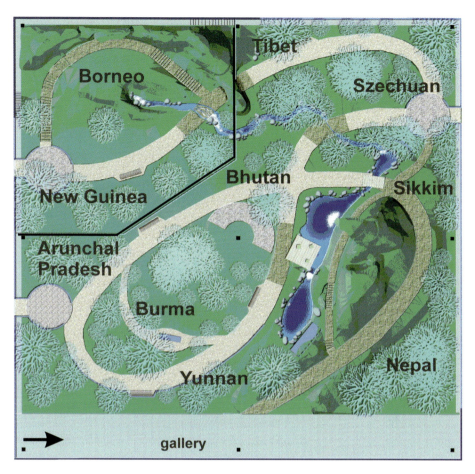

Plate 2 The new conservatory of '*botanika*' with its two sections: a 'Himalaya' section representing regions with natural rhododendron stands in the Himalaya and neighbouring Chinese mountains, and a 'Borneo' section hosting rhododendron species from South East Asian islands such as Borneo and New Guinea.

species and several *Lilium*, *Nomocharis* and *Meconopsis* species. The first part of the round trip ends in a section devoted to the rhododendrons of Tibet.

A grotto in another rock formation is the crossing point to the second part of the conservatory. Here, the vireya species from Borneo, New Guinea, Sulawesi, and other South East Asian islands are displayed. The main paths lead through plantings of tree ferns

and large *Elaeocarpus*, *Agathis* and *Eugenia* trees which are up to 10 m high. In this greenhouse section the rhododendrons grow both terrestrially and epiphytically. Small paths direct visitors into the rock formation where embedded planting pockets contain vireya species and natural hybrids from higher altitudes, such as *R. rarum* Schltr. and *R.* × *sheilae* (Sleumer) Argent. A partition divides the Borneo section from the Himalayan part of the greenhouse to enable higher mean temperature and humidity to be maintained for the more tropical plants. As in the Himalayan section, associated plantings are made from natural *Rhododendron* stands in Borneo or New Guinea, such as the ferns *Dipteris* and *Blechnum* or shrubs such as *Leptospermum* and *Medinella*, which accompany the vireyas in the wild.

The Japanese Garden

Whereas the *Rhododendron* species diversity in the new conservatory is displayed in a natural landscape, the Japanese Garden highlights the results of the cultivation of *Rhododendron*. Embedded in a formal Japanese Zen Garden with artificial water features are grove-like bamboo plantings and formal pine trees, amongst which the colourful result of the work of azalea breeders is displayed. A historic collection of Indian azaleas, with some cultivars introduced more than 150 years ago, provides the setting for demonstrating another important aspect of the CBD: the sustainable use of plants. Different exhibits and information terminals provide examples of the potential uses of plants. A special showcase displays different products made out of *Rhododendron*. The Japanese Garden is followed by a gift shop, which is the last stop on the round trip through '*botanika*'.

The future

'*botanika*' is due to open in summer 2003. Together with the surrounding Rhododendron Park, it will form a unit confirming the park's position as the largest and most varied rhododendron garden in Germany. '*botanika*' and the Rhododendron Park aim to serve as the German centre for rhododendron lovers and rhododendron breeders, creating a platform where information can be imparted and exchanged, new cultivars can be seen, and ideas for garden design with rhododendrons can be explored. Special courses for schools as well as temporary exhibitions are planned. In the future, it is also intended to support biodiversity-related projects in the home territories of the genus *Rhododendron*. It is hoped that '*botanika*' will serve to illustrate very successfully the beauty and importance of biological diversity, and that it will stimulate its visitors to act in a responsible way towards wildlife and the environment.

References

GROOMBRIDGE, B. & JENKINS, M. D. (2002). *World Atlas of Biodiversity: Earth's Living Resources in the 21st Century*. Berkeley: University of California Press.

ROTHERHAM, I. D. (2003). The ecology and history of *Rhododendron ponticum* as an invasive alien and neglected native, with impacts on fauna and flora in Britain. In: ARGENT, G. & McFARLANE, M. (eds) *Rhododendrons in Horticulture and Science*, pp. 233–246. Royal Botanic Garden Edinburgh.

Phylogenetic relationships and major clades of *Rhododendron* (Rhodoreae, Ericoideae, Ericaceae)

Kathleen A. Kron

Department of Biology, Wake Forest University, Winston-Salem, NC 27109-7325, USA

Evolutionary relationships within the large and diverse genus *Rhododendron* are not well understood despite the fact that rhododendrons and azaleas are among the showiest of garden plants. Parsimony analyses of DNA sequence data from the nuclear and chloroplast genomes indicate that *Rhododendron* belongs in the Ericoideae along with genera such as *Kalmia*, *Empetrum*, and *Erica*. *Rhododendron* falls within the tribe Rhodoreae, which also includes *Menziesia* and *Therorhodion*. Results of parsimony analyses indicate several unexpected relationships within *Rhododendron*. These include: *Menziesia* is likely to be derived from within *Rhododendron*; the lepidote rhododendrons may be more closely related to members of section *Pentanthera* than to other evergreen rhododendrons; the deciduous azaleas do not form a natural (monophyletic) group; and section *Sciadorhodion* (subgenus *Pentanthera*) is likely not to be monophyletic.

Introduction

Rhododendrons and azaleas are among the world's showiest shrubs and trees. Because of this they are familiar components of our landscape; however, their evolutionary history is not well known. Traditionally placed within the Ericaceae (blueberry family), rhododendrons and azaleas are most abundant in tropical and temperate Asia (Chamberlain *et al.*, 1996). Stevens (1971) included them in the subfamily Rhododendroideae and considered them closely related to *Kalmia*, *Bejaria*, and *Epigaea*. Recent molecular studies (Kron *et al.*, 2002) have shown that *Rhododendron* (rhododendrons and azaleas) belongs in a clade that also includes *Calluna* and *Ceratiola* (the latter formerly in Empetraceae) (Figure 1). This clade (Ericoideae) comprises one of the larger groups within

the blueberry family and contains two of the largest genera within the Ericaceae, *Rhododendron* and *Erica*. Based on the studies of Kron *et al.* (2002), potential morphological indicators of relationship (synapomorphies) include the septicidal dehiscence of the capsular fruit and erectly held corollas (as compared with the pendant corollas found in most of the rest of the Ericaceae). Many members of the Ericoideae also lack anther appendages and have viscin threads that help to disperse the pollen.

Parsimony analysis (Kron *et al.*, 2002) of molecular and morphological data from selected taxa within the Ericoideae indicates five clades (tribes) including: Bejarieae, Empetreae, Ericeae, Phyllodoceae, and Rhodoreae (Table 1). These groups of plants are diverse in their morphology and are geographically widespread. Within the Phyllodoceae are the

Table 1 Classification of Ericoideae Link (Kron *et al.*, 2002). Groups are listed alphabetically.

Tribe	Genus
Bejarieae Copeland	*Bejaria* Mutis ex L. (15 spp.)
	Bryanthus D.Don (1 sp., *B. gmelini*)
	Ledothamnus Meisn. (7 spp.)
Empetreae D.Don	*Ceratiola* Michx. (1 sp., *C. ericoides*)
	Corema D.Don (1 sp., *C. alba*)
	Empetrum L. (~ 6 spp.)
Ericeae DC. ex Duby	*Calluna* Salisb. (1 sp., *C. vulgaris*)
	Daboecia D.Don (1 sp., *D. cantabrica*)
	Erica L. (725 spp.)
Phyllodoceae Drude	*Elliottia* Muhl. (4 spp.)
	Epigaea L. (2 spp.)
	Kalmia L. (9 spp.)
	Kalmiopsis Rehder (1 sp., *K. leachiana*)
	Phyllodoce Salisb. (5 spp.)
	Rhodothamnus Reichb. (2 spp.)
Rhodoreae DC ex Duby	*Menziesia* Sm. (~ 4 spp.)
	Rhododendron L. (> 900 spp.)
	Therorhodion Small (2 spp.)

genera *Elliottia*, *Kalmia* (mountain laurel, lambkill), and *Epigaea* (trailing arbutus). The Bejarieae contain the tarflower (*Bejaria*), the rare *Ledothamnus*, restricted to tepuis in northern South America, and *Bryanthus*

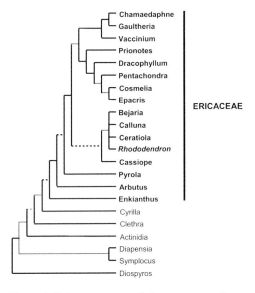

Figure 1 Strict consensus of three most parsimonious trees obtained in the analysis of 18S, *mat*K, and *rbc*L of selected members of the Ericales and Ericaceae (modified from Kron *et al.*, 2002). Narrow lines represent bootstrap support of less than 50%, dashed lines represent bootstrap support of 59–73%, thicker lines (dashed or solid) represent more strongly supported clades.

gmelini D.Don, endemic to northern Japan. The heathers (*Calluna*, *Daboecia*, and *Erica* spp.) form a group that is most diverse in the Cape Region of South Africa. These plants are characterised by their small, needle-like leaves, referred to as 'ericoid' (Stevens, 1971). Ericoid leaves generally have tightly inrolled margins and specialised anatomical features. The crowberries (*Empetrum*) and their relatives (*Ceratiola*, *Corema*) also have superficially similar leaves but have previously been considered a clade distinct from the Ericaceae because most species are wind pollinated (Anderberg, 1994). Recent studies have shown that *Empetrum*, *Corema*, and *Ceratiola* form a clade that is likely to be closely related to the Rhodoreae (Kron *et al.*, 2002; Kron, unpublished data). *Rhododendron*,

Therorhodion, and *Menziesia* are the three genera currently recognised within the tribe Rhodoreae. This tribe is characterised by specialised bracts, or perulae, that surround the inflorescence (Kron *et al.*, 2002). These bracts are green and leafy in *Therorhodion*, but brown and scarious in *Menziesia* and *Rhododendron*. The flowers in the Rhodoreae possess slightly zygomorphic corollas that often are marked with spots or blotches. In addition, most members of the Rhodoreae have ovoid to cylindric capsules (Kron *et al.*, 2002). Phylogenetic studies using parsimony analysis of nucleotide sequence data from the chloroplast *mat*K gene indicate that *Therorhodion* is sister to all other members of the Rhodoreae sampled (Kron, 1997; Kron *et al.*, 2002).

Method and results

DNA sequence data from the nuclear ribosomal internal transcribed spacer region (ITS) and two chloroplast genes, *rbc*L and *mat*K, were obtained from 14 species that are representative of the tribes Ericeae, Phyllodoceae, and Rhodoreae within the Ericoideae. These were included in a single data matrix (available from the author) along with *Cassiope mertensiana* (Borg.) G.Don, which was used to root the tree. Parsimony analysis (PAUP*4.0) (Swofford, 1999) was performed, using the following settings: all characters and states equally weighted, heuristic search, 100 random replicates, and MULPARS on. This analysis indicates that the Rhodoreae is a strongly supported group (Figure 2), as measured by bootstrap analysis (Felsenstein, 1985). Also indicated is the close relationship of *Menziesia* to *Rhododendron* (here represented by *R. kaempferi* Planch.). *Therorhodion* is sister to the remaining sam-

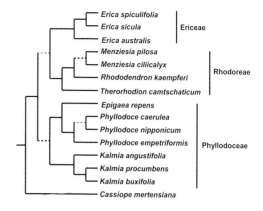

Figure 2 Single most parsimonious tree obtained in the analysis of ITS, *mat*K, and *rbc*L of selected members of the Ericoideae. Length (L) = 670, Consistency Index (C.I.) = 0.69, Retention Index (R.I.) = 0.76, autapomorphies removed. Dashed lines represent bootstrap support of 73–86%, solid lines represent bootstrap support of 96–100%, thicker lines (dashed or solid) represent more strongly supported clades.

pled members of the Rhodoreae, and this has also been found in other analyses using larger taxon sampling (Kron, 1997; Kron *et al.*, 2002). In the present analysis the Ericeae are more closely related to Rhodoreae than to Phyllodoceae (Figure 2). However, representatives of Empetreae and Bejarieae were not included in the analysis.

Because of the great morphological variability within *Rhododendron*, data from DNA sequences in the nuclear and chloroplast genomes can also be very useful for addressing evolutionary relationships within the genus. By using genes or non-coding regions of DNA the genetic relationships within *Rhododendron* can be compared with major groups that have traditionally been recognised (Table 2), such as the lepidotes (subgenus *Rhododendron*) and the deciduous azaleas (subgenus *Pentanthera*). Genetic information is particularly important for

Table 2 Classification of *Rhododendron* (Chamberlain *et al.*, 1996) excluding *Therorhodion* Small. Subgenera and sections are listed alphabetically.

Rhododendron L.

Subgenus *Azaleastrum* Planch.

Section *Azaleastrum* (Planch.) Maxim.

Section *Choniastrum* Franch.

Subgenus *Candidastrum* Franch.

Subgenus *Hymenanthes* (Blume) K.Koch

Section *Ponticum* G.Don

Subgenus *Mumeazalea* (Sleumer) W.R.Philipson & M.N.Philipson

Subgenus *Pentanthera* (G.Don) Poyarkova

Section *Pentanthera* G.Don

Section *Rhodora* (L.) G.Don

Section *Sciadorhodion* Rehder & Wilson

Section *Viscidula* Matsum. & Nakai

Subgenus *Rhododendron*

Section *Pogonanthum* Aitch. & Hemsl.

Section *Rhododendron*

Section *Vireya* (Blume) Copel.f.

Subgenus *Tsutsusi* (Sweet) Poyarkova

Section *Brachycalyx* Sweet

Section *Tsutsusi* Sweet

investigating biogeographic relationships and in the study of ecophysiology, and can also be important to breeders and hybridisers of rhododendrons and azaleas.

Despite their morphological diversity, rhododendrons and azaleas are likely very closely related because many regions of DNA are identical or extremely similar between species that are quite distinct morphologically (Kron, unpublished data). This might suggest that the diversity of species currently seen in the genus is a result of rapid speciation. However, relatively few genes have been investigated within *Rhododendron* and most of these are from the more conserved (i.e. less variable) chloroplast genome. A preliminary study of the relationships of major groups (clades) within *Rhododendron* used DNA sequence data from six regions and sampled 14 species of *Rhododendron*, *Menziesia pilosa* (Michx.) Juss. and *Therorhodion camtschaticum* (Pall.) Small. Three regions were obtained from the nuclear genome: the nuclear ribosomal internal transcribed spacer (ITS), the second intron of the homeotic gene *LEAFY*, and the ninth and tenth intron, and tenth exon, of the gene encoding granule-bound starch synthase (*WAXY*). Data from the chloroplast genome were also obtained and included complete sequence for the *matK* gene (ribosomal maturase K), the 5′ two-thirds of *ndh*F (nitrogen dehydrogenase subunit F), and the coding and non-coding flanking regions of *rps*4 (encodes protein 4 of the small chloroplast ribosomal subunit). Each of these data sets were analysed separately using the parsimony criterion as available in PAUP*4.0 (described above). *Therorhodion* was designated as the outgroup for rooting purposes based on previous analyses (see above, and Kron, 1997; Kron *et al.*, 2002). Trees resulting from the single gene or region did not exhibit conflicting relationships. Therefore the six regions were combined into one matrix and analysed simultaneously (as described above). This resulted in a single most parsimonious tree (Figure 3). Relationships within the tree were generally strongly supported as measured by bootstrap analysis (Felsenstein, 1985). The results indicate that some traditionally recognised groups are well-supported clades based on molecular data, but other traditionally recognised groups are fragmented in the tree.

The scaly, or lepidote, rhododendrons (subgenus *Rhododendron*) are strongly supported as a group by the data from the nuclear and chloroplast genomes (Figure 3). Also strongly supported as a group are the elepidote rhododendrons (subgenus *Hymenanthes*). However, in this analysis these two groups of evergreen rhododendrons are not closely related. The lepidotes (subgenus *Rhododendron*) are more closely related to members of section *Pentanthera*, which contains deciduous azaleas. The results also indicate that subgenus *Tsutsusi* (evergreen

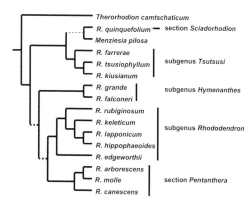

Figure 3 Single most parsimonious tree obtained in the analysis of ITS, *WAXY*, *LEAFY*, *mat*K, *ndh*F, and *rps*4 for selected members of *Rhododendron* and *Menziesia*. L = 715, C.I. = 0.65, R.I. = 0.66, autapomorphies removed. Dashed lines represent bootstrap support of 60–89%, solid lines represent bootstrap support of 90–100%, thicker lines (dashed or solid) represent more strongly supported clades.

azaleas) is likely closely related to the section *Sciadorhodion* (represented by *R. quinquefolium* Bisset & S.Moore) and to *Menziesia*. Based on these results the azaleas are not a natural group within *Rhododendron*. This means that some members of subgenus *Pentanthera* (including sections *Sciadorhodion* and *Pentanthera* that were included in this study) are more closely related to the evergreen rhododendrons than to other members of subgenus *Pentanthera*. This study also indicates that *Menziesia* is derived from within *Rhododendron* and is likely closely related to the evergreen azaleas (subgenus *Tsutsusi*). The placement of *Tsutsiophyllum tanakae* Maxim. within *Rhododendron* (= *R. tsusiophyllum* Sugim.) by Chamberlain & Rae (1990) is supported in this analysis, where it is strongly supported as sister to *R. kiusianum* Makino (Figure 3) and in the same clade as *R. farrerae* Tate, both in subgenus *Tsutsusi*.

Table 3 Classification of *Rhododendron* sub-genus *Pentanthera* (G.Don) Poyarkova (Judd & Kron, 1995; Kron & Creel, 2000). An asterisk indicates species sampled in the analysis of *mat*K and ITS (Figure 4).

Subgenus *Pentanthera*

    Section *Pentanthera* G.Don

        R. alabamense Rehder

        R. arborescens (Pursh) Torrey*

        R. atlanticum (Ashe) Rehder

        R. austrinum (Small) Rehder*

        R. calendulaceum (Michx.) Torrey

        R. canescens (Michx.) Sweet*

        R. cumberlandense E.L.Braun

        R. eastmanii Kron & Creel

        R. flammeum (Michx.) Sargent

        R. luteum Sweet

        R. molle (Blume) G.Don*

        R. occidentale (Torrey & A.Gray) A.Gray*

        R. periclymenoides (Michx.) Shinners*

        R. prinophyllum (Small) Millais

        R. prunifolium (Small) Millais

        R. viscosum (L.) Torrey

    Section *Rhodora* (L.) G.Don

        R. canadense (L.) Torrey

        R. vaseyi A. Gray

    Section *Sciadorhodion* Rehder & Wilson

        R. albrechtii Maxim.*

        R. pentaphyllum Maxim.

        R. quinquefolium Bisset & S.Moore*

        R. schlippenbachii Maxim.*

    Section *Viscidula* Maxim.

        R. nipponicum Matsum.*

In order to examine the relationships within subgenus *Pentanthera* in more detail, representatives of three of the four sections cur-

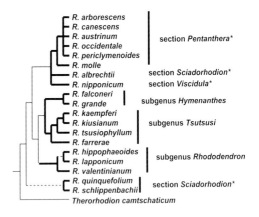

Figure 4 Strict consensus of eight most parsimonious trees obtained in the analysis of ITS and *mat*K for selected members of *Rhododendron*. L = 174, C.I. = 0.63, R.I. = 0.78, autapomorphies removed. Narrow lines represent bootstrap support of less than 50%, dashed lines represent bootstrap support of 65–69%, thick solid lines represent bootstrap support of 98–100%.

rently included within the subgenus (Table 3) were included in an analysis of *mat*K and ITS sequences. In addition, *mat*K and ITS were obtained for representatives of subgenera *Hymenanthes* (elepidotes), *Rhododendron* (lepidotes), and *Tsutsusi* (evergreen azaleas). Twenty species were analysed using parsimony (as described above). The tree was rooted with *Therorhodion*. The summary tree of all trees obtained (strict consensus) (Figure 4) indicates that *R. nipponicum* Matsum. (i.e. section *Viscidula*) and *R. albrechtii* Maxim. (section *Sciadorhodion*) are more closely related to the members of section *Pentanthera* than to other members of subgenus *Pentanthera* (i.e. *R. quinquefolium* and *R. schlippenbachii* Maxim., section *Sciadorhodion*). This analysis does not support the monophyly of either subgenus *Pentanthera* or section *Sciadorhodion*, making the deciduous azaleas, as currently recognised, an artificial group.

Conclusion

The results of this study challenge current ideas about relationships within *Rhododendron*. These include the position of the azaleas and *Menziesia* relative to the evergreen rhododendrons (subgenera *Rhododendron* and *Hymenanthes*), the likely close relationship of the deciduous section *Pentanthera* azaleas to the lepidotes (subgenus *Rhododendron*), and the potential derivation of *Menziesia* out of the subgenus *Tsutsusi* clade. However, this study also supports the recognition of the two traditionally defined evergreen rhododendron groups, the lepidotes (subgenus *Rhododendron*) and the elepidotes (subgenus *Hymenanthes*). The results also support the recognition of the section *Pentanthera*, but not the subgenus *Pentanthera* (which currently includes sections *Viscidula*, *Rhodora*, *Sciadorhodion*, and *Pentanthera*).

Owing to the size and complexity of the genus *Rhododendron* future analyses must include many more species. The relationships indicated in this study will be investigated in much greater detail using more sequence data from the nuclear and chloroplast genomes from many more representatives within *Rhododendron*. In addition, it will be important to investigate morphological characters within the group so that characteristics that are visible in the field may be used as reliable indicators of genetic relationships. The results of these studies will be important in the understanding of the evolution and diversification of *Rhododendron*.

Acknowledgements

I gratefully acknowledge the help and support of the following people and institutions: George Argent and David Chamberlain; Royal Botanic Garden Edinburgh; Rhododendron Species Foundation; Royal Botanic Gardens, Kew. The following agencies provided funding for this project: National Science Foundation; Wake Forest University Research Fund; American Rhododendron Society.

References

ANDERBERG, A. A. (1994). Phylogeny of the Empetraceae, with special emphasis on character evolution in the genus *Empetrum*. *Syst. Bot.* 19: 35–46.

CHAMBERLAIN, D. F. & RAE, S. J. (1990). A revision of *Rhododendron* IV. Subgenus *Tsutsusi*. *Edinb. J. Bot.* 47: 89–200.

CHAMBERLAIN, D., HYAM, R., ARGENT, G., FAIRWEATHER, G. & WALTER, K. S. (1996). *The Genus Rhododendron. Its classification & synonymy*. Royal Botanic Garden Edinburgh.

FELSENSTEIN, J. (1985). Confidence limits on phylogenies: An approach using the bootstrap. *Evol.* 39: 783–791.

JUDD, W. S. & KRON, K. A. (1995). A revision of *Rhododendron* VI. Subgenus *Pentanthera* (Sections *Sciadorhodion*, *Rhodora* and *Viscidula*). *Edinb. J. Bot.* 52: 1–54.

KRON, K. A. (1997). Phylogenetic relationships of Rhododendroideae (Ericaceae). *Amer. J. Bot.* 84: 973–980.

KRON, K. A. & CREEL, M. (2000). A new species of azalea (*Rhododendron* section *Pentanthera*: Ericaceae) from South Carolina. *Novon* 9: 377–380.

KRON, K. A., JUDD, W. S., STEVENS, P. F., CRAYN, D. M., ANDERBERG, A. A., GADEK, P. A., QUINN, C. J. & LUTEYN, J. L. (2002). Phylogenetic classification of Ericaceae: Molecular and morphological evidence. *Bot. Rev. (New York)* 68: 335–423.

STEVENS, P. F. (1971). A classification of the Ericaceae: Subfamilies and tribes. *J. Linn. Soc., Bot.* 64: 1–53.

SWOFFORD, D. (1999). *PAUP*4.0: Phylogenetic Analysis Using Parsimony (*and other methods)*. Sunderland, MA: Sinauer Associates.

Glenarn: a Scottish West Coast rhododendron garden

Michael Thornley

Glenarn, Rhu, Nr. Helensburgh, G84 8LL, UK

Glenarn is a West Coast garden and is representative of many other larger gardens in Scotland with significant collections of rhododendrons. Its history mirrors the social, economic and horticultural changes that have taken place over the last 150 years.

The house at Glenarn was built by Andrew Macgeorge, who was William Hooker's neighbour and lawyer when they both lived in Glasgow. Some of the venerable rhododendrons at Glenarn, notably *R. falconeri*, are said to come from Joseph Hooker's expedition to Sikkim. The informal layout of Victorian paths, through small, steep-sided glens, provided the framework for the woodland garden that the Gibson family developed from the late 1920s onwards, with species rhododendrons forming the backbone to a wonderful collection of ericaceous and other plants. In all of this they were encouraged by John Holms, an avid collector, who created one garden at Formakin and, later, another at Larachmhor.

At this time a vast amount of collected material had already arrived in this country from the expeditions of Forrest, Rock and Kingdon-Ward, later to be supplemented by that from the explorations of Ludlow & Sherriff. But the majority of rhododendrons at Glenarn were collected in weekend forays to major landowners in Scotland. The results of these plant-hunting trips were carefully documented in the garden's accession book, which provides an important record of this period.

The Second World War was a watershed in what may now be seen as a golden age of gardening. Up to 1939, owners of larger gardens, such as Lord Strathcona at Colonsay, were supported by small armies of gardeners and labourers. After the war the owner–gardeners were in the ascendancy. However, by the end of the 1970s many of the original owners had died, some of the gardens had been sold or passed into the hands of trusts, and others had become neglected and overgrown. These changes heralded a period of garden restoration. At the same time there has been a third influx of new rhododendrons which have been made available here in Scotland by Peter and Kenneth Cox and other modern-day plant-hunters. We now know much more about the habitats where rhododendrons grow, in countries which are far more accessible than previously, to the extent that the continuation of botanical collections in gardens has been seriously questioned. However, it is suggested that large gardens with important collections of rhododendrons still have a future role, namely to:

- Record the history of the gardens and their plants.
- Identify and preserve particularly good forms of rhododendrons.
- Provide destinations for new introductions.
- Conserve that sense of place which makes gardens such attractive places to visit.

Glenarn is a garden on the west coast of Scotland and is representative of many other larger gardens in Scotland with significant collections of rhododendrons. The history of the garden mirrors the social, economic and horticultural changes that have taken place over the last 150 years.

The house at Glenarn was built in the 1840s. At that time 15-acre plots, each with a mansion house, were being laid out across the hillside above the village of Rhu, overlooking the Gareloch. Glenarn was first occupied as a holiday home but, with improving communications by train and by steamer up the River Clyde to Glasgow, it soon became a permanent residence and was extended, both upwards and outwards, at least twice in the following 20 years. The first owner was Andrew Macgeorge, a lawyer and near neighbour to William Hooker who was Professor of Botany at Glasgow University and Curator of the Botanical Garden, then located in Anderston. William Hooker also had a holiday home in Helensburgh, close to Rhu, to which he dispatched his family during the summer. He would walk out on Friday night from Glasgow and return on foot late on Sunday in order to be able to lecture to his students on Monday morning.

Andrew Macgeorge was not only an acquaintance of the family but acted as William Hooker's lawyer before and after his move to Kew in 1843. At the side of the house at Glenarn there is a huge plant of *Rhododendron falconeri* Hook.f., dubbed, by Sir Islay Campbell, as one of the Giants of the West and reputed to have been grown from seed collected by Joseph Hooker's expedition to India and Sikkim in 1848–51. To mark this rhododendron's 150th year we wrote to Kew to ask if there was any information that might corroborate the story.

Plate 1 A general view of Glenarn, with *Magnolia campbellii*.

Staff at Kew kindly provided a photocopy of the herbarium material, originally collected by Joseph Hooker in the Tonglo mountains. But while there is information on seed that arrived in London (the second batch of which had arrived over land by post from India) as well as details of organisations and individuals to whom the seed had been distributed, no direct connection to Glenarn could be traced. Perhaps this famous *Rhododendron* had been purchased from one of the local nurseries, of which there were four in Helensburgh at that time.

In any event, it appears that Andrew Macgeorge was not interested in gardens. In a letter that he wrote following a dinner party at the Hooker's family home, there is no mention of plants or horticulture, which seems improbable in that company. He laid out large hybrid rhododendrons to define the front lawn and had a vegetable patch at the back of the house to feed the household. The main legacy, apart from the ubiquitous *Sequoiadendron giganteum* (Lindl.) Buchholz, which towers over the garden like a lighthouse, is the network of Victorian paths that spread out from the small quarry, now the Rock Garden, which supplied the stone to build the house and for the boundary walls of the property. These paths wind their way around the garden, up small glens, under a canopy of oak, sycamore and birch, with lime trees marking the western boundary.

The curtains close and re-open again only in the mid-1920s when James Bogle Gibson of Portencross on the Ayrshire coast purchased Glenarn (Campbell, 1983/84). He died before the sale was completed but, subsequently, grandmother Gibson, her daughter and her three children, along with various dogs and hens, moved into Glenarn. The two sons, Archie and Sandy Gibson, were in their late teens or early twenties and training to be chartered accountants, travelling into Glasgow every day. Archie subsequently married Betty Graham and these three created the garden that we see today. In 1928 a huge storm swept down the west coast of Scotland, knocking over 60 trees in the garden. It was in the aftermath of this disaster that rhododendrons were planted in clearings in the *R. ponticum* L. and laurel that had been smashed flat by the falling timber, the Gibsons being encouraged in all of this by John Holms of Formakin (Gibson, 1967).

John Holms' father had made his money in the mills of Paisley and his son took it upon himself to lose the family fortune on two occasions. He was a friend of William Burrell (whose fabulous collection of antiques, historic artefacts and paintings is preserved in Glasgow) and competed against him in the auction houses. The third member of this triumvirate was Scotland's premier architect of the time, Robert Lorimer, who designed a Scottish tower house at Formakin for Holms. Even before the building work commenced, Holms laid out a many compartmented walled garden, in the Arts and Crafts style, divided by yew and beech hedges, with herbaceous borders and roses on trellises, all looked after by 23 gardeners. By the end of the First World War Holms was in deep financial trouble and his army of gardeners much depleted. But being an insatiable collector, and not a man to give up a fight easily, he turned his attention to rhododendrons, which were beginning to flood into the country, notably from Forrest's plant-hunting expeditions. Realising that Formakin did not have suitable conditions he scoured the West Coast for a more favourable site and started to create another garden or, more accurately, planting at Larachmhor, near Arisaig (Hedge, 1967).

Plate 2 A general view of Glenarn, with *Rhododendron pingianum* in the foreground.

John Holms introduced the Gibson brothers to other major landowners in Scotland who were also avid collectors of rhododendrons. The first was John Stirling Maxwell of Pollok in Glasgow, who supplied *R. orbiculare* Decne. and one of the 'top ten' plants, *R. caloxanthum* Balf.f. & Farrer (= *R. campylocarpum* Hook.f. ssp.) which is to be wondered at not only for its beautiful pale lemon flowers but also for the glaucous blue new foliage which follows a month later. The next port of call was Lochinch and Lord Stair who provided the tall white *R. arboreum* Sm. that stand liked pillars in the lower garden as well as *R. crassum* Franch. (= *R. maddenii* Hook.f. ssp.) which grows in thickets about the place. Balfour at Dawyck swapped choice plants for Chinese pheasants with the Gibsons before directing them to the Royal Botanic Garden Edinburgh which is the source of *R. strigillosum* Franch., 'the best form', with its black red flowers of early spring. Thomas North Christie of Blackhills, near Elgin, handed out packets of seed including *R. augustinii* Hemsl., as well as a plant of *R. eximium* Nutt. (= *R. falconeri* Hook.f. ssp.) which covers itself in flower down to the ground, a bride in the woodland going to her wedding.

Thus the first influx of species rhododendrons to Glenarn came not from plant-hunting expeditions abroad, but from weekend forays in Scotland when, presumably, the two young men were welcomed with open arms by collectors of rhododendrons who were being almost overwhelmed by their ever-expanding beds of seedlings. By this time, in the early 1930s, the Forrest introductions must have been coming into flower for the first time and it is for this reason, we believe, that the garden has particularly good forms which had been hand picked from the profusion of material that was available. However, this leaves us with a problem of now finding equally good replacements as it is an unfortunate fact that a 70-year-old woodland garden is a kind of horticultural eventide home, with a distressingly high death rate.

Other sources of rhododendrons were further afield, in England, with Magor sending *R. haematodes* Franch. and the extraordinary *R. quinquefolium* Bisset & S.Moore which, if it is the original 1933 plant, is a true bonsai, standing less than a metre high and hardly putting on any growth each year. Some plants came from nurseries such as Gills and White, including *Rhododendron* 'Loderi' which Archie Gibson purchased, appropriately, to celebrate his honeymoon. That we have such detailed information is due to the fact that the Gibsons maintained an accession book, 'The Bible', that gives the dates, size and source of all new rhododendrons. From time to time the records were passed between the two brothers, who lived in either side of the house, for comments and queries as to where they had been planted, and whether they had survived, all of which helps to build up an overall picture of the garden.

However, by far the greatest contributor in this period was John Holms himself who, for instance, provided *R. valentianum* Forrest ex Hutch., grown on stumps, for good drainage and acting as navigation marks on the paths. The records indicate that some of these were 'in lieu of interest' plants which suggests that the Gibsons were helping Holms financially. Others are designated as 'RIP JAH', i.e. they were obtained after his death in 1938, when the banks requisitioned his properties and sold off his collection of antiques. Likewise his creditors arranged for his *Rhododendron* collection at Larachmhor to be catalogued for an intended sale in Glasgow. But being greedy men, they instructed Drennan, the gardener, to load the largest rhododendrons on to the goods train. It became stuck in the first tunnel and the rhododendrons were returned to Larachmhor where they reside to this day. In the circumstances Betty Gibson's slightly guilty comment that 'John Holms always meant us to have this' is understandable.

What is more difficult to comprehend is the huge amount of work put in by these three young people to develop the garden in its first 10 years, assisted by only one full-time gardener who was largely confined to the vegetable patch. The first planting took place in the sheltered lower glen, which had been hit hardest by the storm, but the clearing process continued throughout the garden in areas colloquially known as: 'Melrose' (named after relations from the Borders who came to help); 'Betty's clearing' (at the top of the drive); 'Germany' (where two prisoners of war felled the trees); and 'Granny's Hens' (where granny kept her hens).

At this time the Gibsons were also starting to hybridise. With time on their side and aided and abetted by John Holms, they 'cocked a snook' at the Establishment and concentrated their efforts on big leaf crosses. Notably they crossed *Rhododendron* 'Ronald' with *R. hodgsonii* Hook.f., with its vibrant, tennis ball size truss of flowers, and blew it up using additionally *R. sinogrande* Balf.f. & W.W.Sm., into a strawberry pink confection which, unfortunately, exhibits the same failings as *R. hodgsonii* and quickly fades away to white. An even better, but unregistered, cross is *R. falconeri* × *R. macabeanum* Watt ex Balf.f., where the primrose yellow of the latter is introduced losing neither the parliamentary green leather foliage, nor the shapely truss of the former.

The Second World War marked a change of direction. Having established the backbone of the rhododendron collection, the Gibsons now started to obtain rarer or new introductions, notably from Ludlow & Sherriff and, later, Kingdon-Ward expeditions. For instance, *R. viscidifolium* Davidian came as seed of *R. thomsonii* Hook.f. var. *pallidum* Cowan (= *R.* × *candelabrum* Hook.f.) from the Ludlow & Sherriff 1938 expedition and, after correspondence with the Royal Botanic Garden Edinburgh, was subsequently confirmed as a new species. *Rhododendron luteiflorum* (Davidian) Cullen, with its acid green and moderately tender flower, was one of Kingdon-Ward's introductions from upper Burma. Another relative rarity is the frothy yellow-flowered *R. zaleucum* Balf.f. & W.W.Sm. var. *flaviflorum* Davidian (KW 20837) with the white underside of the leaf evenly scaled with tiny pointillist dots. However, the best-known introduction from this period is *R. lindleyi* T.Moore, which thrives outside and sometimes sports 12 flowers in its crown-like trusses that hover in the dusk, scenting the oncoming darkness. The Gibsons named the pink striped *R. lindleyi* 'Geordie Sherriff' in honour of their friend and a great plant collector.

In the late 1950s and early 1960s, the garden must have reached some kind of peak of perfection with burgeoning plants exhibiting all the beauty of youth, with plenty of sunlit space in which to grow gracefully and erect. The Gibsons themselves must have been in their physical and intellectual prime, with Archie contributing acerbic and witty notes to the *Rhododendron Year Book* and Sandy chairing the National Trust for Scotland's Garden Committee. Peter Clough, then a trainee gardener at Achamore House on the

Plate 3 *Rhododendron falconeri.*

island of Gigha, recalls a visit to Glenarn at that time. Having attended to his business with Sandy, Peter was dispatched off to Archie and Betty's side of the house where he was liberally entertained. Next morning he woke with an acute hangover and, looking out into the rain at 7.00 a.m., was appalled to see all three of his hosts already hard at work in the Rock Garden.

New plants continued to arrive, including an interesting collection of straggly *R. yakushimanum* Nakai (= *R. degronianum* Carrière var.) from seed collected by Doleshy and distributed by the Royal Horticultural Society. *Rhododendron* 'Avalanche' came from Gigha and carpets the ground in a snowfall of petals sending its scent sliding away down the garden on the cool evening air. Later the Gibsons started to obtain rhododendrons from Peter Cox at Glendoick,

who has continued to be the main source of new introductions to this day, including the foxy-eared *R. bureavii* Franch. and *R. pachysanthum* Hayata, neither of which had been previously represented in the garden. However, it should not be forgotten that rhododendrons and many other interesting and unusual plants arrive as gifts to the garden from individuals from all over the world. John Basford, for instance, when on his occasional forays from his island fastness at Brodick Castle, would leave boxes of plants on our doorstep for us to discover when we returned home from work, and this generosity has been repeated on countless occasions by many other visitors to the garden.

Betty Gibson died in May 1975, predeceasing Archie by six months. Sandy Gibson soldiered on, suffering from Parkinson's disease, and died in 1982. Towards the end of his life he would leave the nursing home where he was staying in Helensburgh and return to his bedroom at Glenarn to watch the garden he had created come into flower once again in the spring. It was on such a day, at the end of March in the following year, after a bitter winter, that we first saw Glenarn and were amazed to find such a remarkable collection in its sheltered garden so close to Glasgow. By the time we had reached the top of the drive we had decided that we would try to restore Glenarn, a resolve that was sorely tested by the sight of the derelict house that came with the garden.

It was Peter Cox who advised that, as we were going to spend most of our time outside, we should immediately take action to make the house more comfortable and manageable. This was achieved by the simple expedient of demolishing half of it, and constructing a two-storey conservatory that re-connected the previously inward-looking house to the garden. There then commenced a seven-year period during which we gradually cleared our way through the overgrown garden, moving from one area to another until, finally, the jigsaw pieces started to interlock (Thornley, 2000). While we were clearing, lifting the canopy, felling sycamores and making huge bonfires, we mapped the areas where we were working, recording the fallen tallies and the labels on the plants. We would spend the evenings putting this information together with the accessions book and slowly began to identify the collection of rhododendrons. Even now, 20 years later, new ones come to light.

Rhododendron enthusiasts quickly arrived to check that we were not destroying rare species and warned us of the threat of honey fungus which we took seriously enough to purchase a winch that could lift 5 tonnes.

The problem was how to deal with 5 tonnes of stump when it had been prised from the ground. Between the demolition material from the house and the removal of fallen and rotten trees in the garden more than 100 skips went down the drive. Out of necessity we re-cut all the paths, to gain access to the furthest reaches of the garden, and replaced all the step fronts. (Seventeen years later I have repeated the operation ordering the new step fronts from the original invoice and, in more philosophical moments, wonder whether I will be replacing them again 17 years hence.)

By far the greatest problem was the Rock Garden, which had disappeared beneath invading birch and brambles. A remarkable number of far from dwarf rhododendrons had survived under this rampant regeneration and the attention of people seeking to liberate the plants from the hands of the developers, who were threatening to purchase the garden and top field, where they planned to build 120 houses. We decided that we would not start on the Rock Garden until everywhere else had been cleared as we knew that, to do otherwise, would have been like entering the Labyrinth, with no hope of escape. Nevertheless, the time-consuming Rock Garden still causes frustration, and remains a rather overblown and colourful version of its original neat self, a kind of fading music hall artiste who has seen better days.

Fortunately Glenarn is more than a collection of rhododendrons. Archie Gibson planted thousands of daffodils: small, elegant old hybrids rather than the buxom versions that are popular today. There are also many species *Narcissus* that seed themselves in the short turf which is one of the benefits of taking off all the grass from the garden in the late summer, although an unpleasant and fly ridden job. The major impact, however, comes from the tree magnolias which are like huge beacons of light in the spring, drawing the eye down through the garden and out into the landscape beyond. The Gibsons were rhododendron enthusiasts of the highest order but their greatest contribution was to create a truly wondrous sense of place, by immense physical work and an encyclopaedic knowledge of plants, that is impossible for us to emulate.

The purpose of seeking to restore Glenarn is to:

- Record the history of the garden, both social history and that of the collections of plants.
- Identify the particularly important or good forms as well as rhododendrons under threat, and seek to propagate these wherever possible.
- Continue to provide a destination for new introductions (a task that is not easy to achieve when space is at a premium, particularly in an ageing collection).
- Maintain and conserve that sense of place

which so often creates such strong and emotional responses from visitors. Many people come to gardens because they provide the security of continuity, proof that the world continues from season to season. Paradoxically everything they see that engenders these feelings lies in the past.

If we understand and achieve this last objective we may unlock the interest and enthusiasm that will be so necessary to achieve the first three objectives.

References

CAMPBELL, SIR I. (1983/84). Glenarn and the Gibson family. In: *Rhododendrons with Magnolias and Camellias*. London: The Royal Horticultural Society.

GIBSON, J. F. A. (1967). The garden at Glenarn. *J. Royal Hort. Soc.* 92(8): 341–347.

HEDGE, I. C. (1967). The garden of Larachmhor, Arisaig. In: *The Rhododendron and Camellia Year Book*. London: The Royal Horticultural Society.

THORNLEY, M. G. (2000). Gardening against time: the maintenance of Glenarn. In: *The World of the Rhododendron*, No. 3. The Scottish Rhododendron Society.

Vireya rhododendrons: an insight into their relationships

Gillian K. Brown

Centre for Plant Biodiversity Research, CSIRO Plant Industry, GPO Box 1600, Canberra, ACT 2601, and School of Botany, The University of Melbourne, Vic. 3010, Australia

For the first time a phylogeny of *Rhododendron* section *Vireya* has been produced using chloroplast DNA data. Section *Vireya* is the largest section within the genus *Rhododendron*, consisting of approximately 300 species. The species are predominantly montane plants of the tropics, found throughout Malesia and in neighbouring countries such as Australia, China, Nepal and Taiwan.

Section *Vireya* is defined by the possession of seeds with a tail at each end; in addition, they are the only lepidote rhododendrons found growing in the Malesian region. At present, seven subsections are recognised based on leaf scale type and corolla shape.

Two chloroplast DNA (cpDNA) regions have been sequenced, the *psb*A-*trn*H spacer and the *trn*T-*trn*L spacer, and used in a combined phylogenetic analysis. The resultant phylogeny does not support the majority of traditionally recognised subgroups within section *Vireya*; many of the supported clades correlate strongly with the geographical areas in which they grow, rather than with the subsections identified in the current classification. The geographical regions repeated throughout the six major clades are: western and middle Malesia; New Guinea, Australia and the Solomon Islands; and mainland Asia.

Introduction

Vireya rhododendrons are a group of the lepidote or scaly-leaved rhododendrons classified in subgenus *Rhododendron* L. as section *Vireya* (Blume) Copel.f. They are one of the largest groups in the genus, consisting of around 300 species (Chamberlain *et al.*, 1996). Species of section *Vireya* are predominantly distributed throughout Malesia, but extend to India and Nepal in the west, China and Taiwan in the north, the Solomon Islands in the east and north Queensland, Australia in the south. Only 12 non-*Vireya* species of *Rhododendron* are found in Malesia (Sleumer, 1966).

Seeds of *Vireya* rhododendrons are distinct, having appendages, generally tail-like and long, at each end (Hedegaard, 1980). They are small, 3–10 mm long, light and presumably dispersed by wind. Other distinguishing characters of section *Vireya* have been identified through embryological studies. Palser *et al.* (1991) found that the ovules of *Vireya* rhododendrons are shortly stalked and tilted downwards, unlike other rhododendrons that have virtually sessile ovules. Another, yet not solely defining, character of vireyas is the lack of visible zygomorphic markings or blotches on the corolla (Rouse *et al.*, 1986). Even though visible marks are absent, UV markings have been detected for some species (Rouse *et al.*, 1986).

Rhododendron section *Vireya* is currently divided into seven subsections – *Albovireya* Sleumer, *Euvireya* Copel.f., *Malayovireya* Sleumer, *Phaeovireya* Sleumer, *Pseudovireya* (Clarke) Sleumer, *Siphonovireya* Sleumer and *Solenovireya* Copel.f. – on the basis of leaf scale type and corolla shape (Sleumer, 1966; Table 1). Subsection *Phaeovireya* has very distinctive leaf scales that are deeply lobed and raised on stalks. The corollas of species in subsection *Phaeovireya* are generally short tubular, but they can also be funnel-shaped or salver-shaped. Subsection *Malayovireya* has two distinct size classes of scales on the underside of the leaf; they are sessile, with a moderately lobed margin and generally large, dark centre. The corollas of subsection *Malayovireya* can be campanulate, tubular or funnel-shaped (Sleumer, 1966).

Subsections *Albovireya*, *Euvireya* and *Solenovireya* share the same type of leaf scales. Like *Malayovireya* they are sessile and moderately lobed; however, the centres of the scales are smaller and all of the one size class. Subsection *Albovireya* is differentiated from the other two by the leaf scales being very dense and overlapping, while in subsections *Euvireya* and *Solenovireya* they are more widely spaced. Corolla shape is used to distinguish between subsections *Solenovireya* and *Euvireya*: corollas of species in subsection *Solenovireya* are salver-shaped, while for species in subsection *Euvireya* the corolla can be tubular, campanulate or funnel-shaped, but they are never salver-shaped (Sleumer, 1966).

The last two subsections, *Pseudovireya* and *Siphonovireya*, share 'disc'-shaped leaf scales. The margin is entire, not lobed, and the centre is often swollen. Corolla shape

again distinguishes these subsections, with species of subsection *Siphonovireya* having salver-shaped corollas, and species of subsection *Pseudovireya* having short tubular flowers (Sleumer, 1966).

Since the 1966 *Flora Malesiana* treatment of *Rhododendron*, which forms the basis of the presently accepted classification of section *Vireya*, numerous new species have been described and investigations into grafting compatibilities, cross pollinations, embryology, and flower types have contributed useful taxonomic information (Stevens, 1976; Palser *et al.*, 1991; Rouse *et al.*, 1993; Rouse & Williams, 1997). Results from these studies support some taxonomic groupings within *Vireya* but raise questions as to the naturalness of others.

Little is known of the evolutionary relationships of *Vireya* rhododendrons; the present classification is practical for identification purposes, although the naturalness of the vireyas has not been tested. Are each of the subsections of section *Vireya* monophyletic? Which species are closely related and which are more distantly related? How did section *Vireya* come to have its present distribution? This paper addresses the first of these issues using evidence from chloroplast DNA (cpDNA) and preliminary morphological measurements. Investigations into the other questions are continuing.

Materials and methods

Plant materials

Species were sampled from six of the seven subsections across the distribution range of *Vireya* (Table 2). Subsection *Siphonovireya* is the only subsection not represented here, because of a lack of leaf material. Leaf material for molecular work came from the pri-

Table 1 Traditional distinguishing characters for the seven subsections of *Rhododendron* section *Vireya*. The two defining characters are flower shape and leaf scale type (adapted from Sleumer, 1966).

Subsection	Leaf scales	Flower shape
Albovireya	Sessile; centre small, variously coloured; moderately lobed; equal in size; dense and overlapping	Various shapes
Euvireya	Sessile; centre small, variously coloured; moderately lobed; equal in size; lax to sub-dense	Tubular, campanulate, funnel-shaped
Malayovireya	Sessile; large dark centre; margin moderately lobed; two distinct sizes; dense to overlapping	Campanulate, funnel-shaped, tubular
Phaeovireya	Stalked; margin deeply lobed	Mostly short tubular, rarely funnel- or salver-shaped
Pseudovireya	Sessile; marginal zone narrow, entire or almost so ('disc-shaped'); centre often thick or swollen	Short tubular
Siphonovireya	Sessile; marginal zone narrow, entire or almost so ('disc-shaped'); centre often thick or swollen	Salver-shaped
Solenovireya	Sessile; centre small, variously coloured; moderately lobed; equal in size; lax to sub-dense	Salver-shaped

vate *Rhododendron* collections of Lyn Craven, David Binney and the late John Rouse. Herbarium specimens from A, CANB, L and NY were used for collection of morphological data and supplemented by information from the literature when specimens were not available.

DNA isolation and amplification

Genomic DNA was isolated from fresh, silica dried or CTAB preserved leaves following a CTAB protocol (Thomson & Henry, 1993) and then cleaned with the QIAquick PCR purification kit (QIAGEN). DNA was quantified with a fluorometer (Hoefer® DyNA Quant™ 200) and two cpDNA regions, the *psb*A-*trn*H spacer and the *trn*T-*trn*L spacer,

were amplified by polymerase chain reaction (PCR).

The *psb*A-*trn*H spacer was amplified using the primers *psb*AF (5′ GTTATGCAT-GAACGTAATGCTC 3′) and *trn*HR (5′ CGCGCATGGTGGATTCACAAATC 3′) (Sang *et al*., 1997). The *trn*T-*trn*L spacer was amplified using the primers a (5′ CATTACAAATGCGATGCTCT 3′) and b (5′ TCTACCGATTTCGCCATATC 3′) (Taberlet *et al*., 1991). PCR reactions were 50 µl in volume and contained 0.2 mM dNTPs, 0.5 µM of each primer, 1.25 units of Taq polymerase, 3 mM $MgCl_2$, and c.50–100 ng of genomic DNA. Amplifications for both cpDNA regions were performed on either a Hybaid PCR Express Thermal Cycler or a Hybaid Touchdown Thermal Cycler. The

Table 2 Species list. Subsections of *Vireya* following Chamberlain *et al.* (1996), except for *Rhododendron rousei*, which was described after the Chamberlain *et al.* (1996) publication – this subsection has been noted from the original species description (Argent & Madulid, 1998). Distribution of taxa (total range). PNG = Papua New Guinea, IJ = Irian Jaya, MP = Malay Peninsula, LSI = Lesser Sunda Islands. L.C. = Lyn Craven, D.B. = David Binney. Morphology data source is the collector and collector number, and herbarium symbol and accession number; there were no accession numbers for the NY or A specimens. Australian National Herbarium (CANB); Nationaal Herbarium Nederland, Universiteit Leiden branch (L); The New York Botanical Garden (NY); Arnold Arboretum, Harvard University (A). Literature = specimens not yet seen, all measurements as description (Argent & Madulid, 1998). N/A = morphology not recorded.

Taxon	Subsection	Distribution	DNA leaf material	Morphology data
R. acuminatum Hook.f.	*Malayovireya*	Borneo (Sabah)	D.B.	G. Argent 1617, L 442108
R. alborugosum Argent & J.Dransf.	*Euvireya*	Borneo (Kalimantan)	D.B.	Gen Murata, Masahiro Kato & Yohanis P. Mogea B-3443, L 442118
R. album Blume	*Albovireya*	Java	D.B.	Junghuhn 76, L 442121
R. carringtoniae F.Muell.	*Solenovireya*	PNG	L.C.	Craven & Brown 10396, CANB 639658
R. christi Foerste	*Euvireya*	PNG	L.C.	R.D. Hoogland 9472, CANB 146166
R. commonae Foerste	*Euvireya*	PNG	L.C.	Schodde 1738, CANB 107192
R. culminicolum F.Muell. var. *angiense* (J.J.Sm.) Sleumer	*Euvireya*	IJ	L.C.	Craven & Brown 10385, CANB 639659R
R. ericoides Low ex Hook.f.	*Pseudovireya*	Borneo (Sabah)	L.C.	S.H. Collenette 21500, CANB 205339
R. gracilentum F.Muell.	*Euvireya*	PNG	L.C.	A.N. Miller NGF 22778, CANB 178950
R. hyacinthosmum Sleumer	*Euvireya*	PNG	L.C.	Paul J. Kores WEI 1429, L 442208
R. inconspicuum J.J.Sm.	*Euvireya*	IJ, PNG	L.C.	R.D. Hoogland & R. Schodde 7399, CANB 85594
R. jasminiflorum Hook.	*Solenovireya*	Borneo, MP, Philippines, Sumatra	L.C.	W. de Wilde & B. de Wilde-Duyfjes 18592, L 442216
R. kawakamii Hayata	*Pseudovireya*	Taiwan	L.C.	Craven & Brown 10381, CANB 639660
R. konori Becc.	*Phaeovireya*	IJ, PNG	L.C.	R.D. Hoogland 9208, CANB 143862

R. laetum J.J.Sm.	Euvireya	IJ	L.C.	H. Sleumer & W. Vink 4416, A; H. Sleumer & W. Vink 4389, L 442237/8
R. lagunculicarpum J.J.Sm.	Albovireya	Sulawesi	D.B.	Eyma 649, A
R. leucogigas Sleumer	Euvireya	IJ, PNG	L.C.	L.A. Craven s.n. Aug/Sept 1974, clone collected by L.A. Craven, CANB 285420
R. lindleyi T.Moore	Outgroup – sect. Rhododendron	Bhutan, China, India, Nepal	L.C.	N/A
R. loranthiflorum Sleumer	Solenovireya	Bismark Archipelago, Solomon Islands	L.C.	K. Paijmans 508, CANB 215716
R. luraluense Sleumer	Euvireya	Solomon Islands	L.C.	L. Craven & R. Schodde 346, CANB 148926
R. notiale Craven	Euvireya	Australia (Qld)	L.C.	L.A. Craven & R. Elliot 9105, CANB 451845 (holotype)
R. pneumonanthum Sleumer	Solenovireya	Borneo (Sarawak, Sabah)	D.B.	H.P. Nooteboom & P. Chai 2179, CANB 251119
R. quadrasianum Vidal var. rosmarinifolium Vidal	Pseudovireya	Philippines	D.B.	G. Banlugan et al. Field No. 589, Herb No. 72578, L 442310
R. rarilepidotum J.J.Sm.	Euvireya	Sumatra	L.C.	Lorzing 13480, L 442311
R. rarum Schltr.	Phaeovireya	PNG	L.C.	T.M. Reeve 6205, CANB 370593
R. retusum (Blume) Benn.	Pseudovireya	Java, Sumatra	L.C.	J.R. Flenley & C.D. Oliver ANU 2901, CANB 272300
R. rousei Argent & Madulid	Unplaced	Philippines	D.B.	Literature
R. rubineiflorum Craven	Euvireya	PNG	L.C.	R. Pullen 227, CANB 42862
R. ruttenii J.J.Sm.	Solenovireya	Moluccas	L.C.	G. Argent C8786, CANB 494446
R. santapaui Sastry et al.	Pseudovireya	India	L.C.	Craven & Brown 10384, CANB 369661
R. saxifragoides J.J.Sm.	Euvireya	IJ, PNG	D.B.	J.M. Wheeler ANU 6388, CANB 178271
R. superbum Sleumer	Phaeovireya	PNG	L.C.	R. Pullen 6037, CANB 155052

Table 2 Continued

R. tuba Sleumer	*Solenovireya*	PNG	L.C.	J.R. Croft et al. LAE 65054, CANB 255065
R. vaccinioides Hook.f.	*Pseudovireya*	Bhutan, China, India, Nepal	L.C.	Craven & Brown 10372, CANB 639662
R. williamsii Merr. ex Copel.f.	*Euvireya*	Philippines	L.C.	R.S. Williams 990, NY
R. zoelleri Warb.	*Euvireya*	IJ, Moluccas, PNG	L.C.	R. Pullen 5979, CANB 154931
R. zollingeri J.J.Sm.	*Albovireya*	Java, LSI, Philippines, Sulawesi	D.B.	C. Monod de Froideville 271, L 442354

cycling parameters for the *psb*A-*trn*H spacer region were one ramp of 94°C for 3 minutes, 30 cycles of 94°C for 1 minute, 50 or 55°C for 1 minute and 72°C for 2 minutes, and a final extension step of 72°C for 7 minutes. The conditions for the *trn*T-*trn*L spacer were the same as for the *psb*A-*trn*H spacer region, but with an annealing temperature of 55 or 58°C. The PCR product was cleaned using the QIAquick PCR purification kit (QIAGEN) and cleaned products were quantified with the fluorometer (Hoefer® DyNA Quant™ 200) for sequencing.

DNA sequencing, alignment and analysis

Direct sequencing was completed as per the BIG Dye sequencing kit, with approximately 100 ng of template DNA. Sequencing reactions were analysed on an ABI–Perkin Elmer 377XL sequencer. Contiguous sequences were edited in Sequencher v.3.0 (Gene Codes) and manually aligned to each other in BioEdit sequence alignment editor v.4.8.6 (Tom Hall, North Carolina State University – available from www.mbio.ncsu.edu/BioEdit/bioedit.html). Sequences are lodged with GenBank, accession numbers AY196045-AY196081 for the *trn*T-*trn*L sequences, and AY196008-AY196044 for the *psb*A-*trn*H

sequences. The sequences of each region were aligned separately and the partition homogeneity test (Farris *et al.*, 1995) was carried out using PAUP, v.4.0b10 (Swofford, 1998), to test if the regions were compatible. They were then combined into one data matrix and analysed using PAUP.

Individual base positions were coded as unordered multistates and gaps were treated as missing data. Seventeen indels were identified across the two regions and coded as single binary characters at the end of the data matrix; nine of these were informative. A heuristic search was performed on the combined cpDNA data set with random sequence addition replicated 1000 times. Uninformative characters were excluded from the analysis. A strict consensus tree was calculated for the equally parsimonious trees and these trees were rooted to the outgroup *Rhododendron lindleyi* T.Moore. Bootstrap analysis (Felsenstein, 1988) was performed on the combined data matrix to test the support for nodes; this analysis was replicated 100 times.

Morphology

Fifty-five morphological characters, both floral and vegetative, were measured for the

36 *Vireya* species from the cpDNA analysis; only the 13 characters that were informative for the cpDNA clades are used in this paper. Morphological data were plotted on the cpDNA phylogeny rather than analysed separately. Fragrance was assessed from comments in the literature and also from discussions with collectors and growers of vireyas.

Results

Sequencing of the two non-coding chloroplast DNA spacer regions, the *psb*A-*trn*H and the *trn*T-*trn*L spacers, was completed for 36 species from *Rhododendron* section *Vireya* and one species from *Rhododendron* section *Rhododendron* as an outgroup (Table 2). Sequence lengths for the *psb*A-*trn*H spacer region ranged between 346 and 357 base pairs (bp) for species of section *Vireya*. The *trn*T-*trn*L spacer region had slightly more sequence length variation, with section *Vireya* sequences ranging between 656 and 703 bp long. The outgroup taxon *R. lindleyi* fell within these ranges with a sequence length of 336 bp for the *psb*A-*trn*H spacer region and 748 bp for the *trn*T-*trn*L spacer region. The partition homogeneity test indicated there was no conflicting signal between the two cpDNA regions, therefore they were able to be analysed together.

A heuristic search revealed four equally parsimonious trees of 135 steps. The consistency index (C.I.) for these trees was 0.63 and the retention index (R.I.) was 0.81 (Figure 1). The strict consensus of these four trees is shown in Figure 1 with bootstrap values > 50% shown above the nodes; there are six main clades supported by the cpDNA data (Figure 1). Given the molecular tree, morphological characters for individual clades are mostly homoplasious (Figure 2).

At least four of the currently accepted *Vireya* subsections are polyphyletic. Representatives from subsections *Albovireya* (A), *Euvireya* (Eu), *Phaeovireya* (Ph) and *Solenovireya* (So) are intermingled across the phylogeny (Figure 1). Species of subsection *Pseudovireya* (Ps) are grouped into two clades (5 and 6) that are strongly supported (99%) as sister to the rest of the species from section *Vireya* (clade 1), forming an unresolved trichotomy (Figure 1). Resolution of this trichotomy may show *Pseudovireya* to be monophyletic. Within clade 1 there are three major clades with support: 2, 3 and 4 (Figure 1). Clade 4 consists of two species from different geographical areas. Clade 3 consists of species from New Guinea, Australia and the Solomon Islands, while clade 2 contains the species from western and middle Malesia (Figure 1). The only representative of subsection *Malayovireya*, *Rhododendron acuminatum* Hook.f. from Borneo, is weakly supported (51%) as sister to clades 2 and 3 (Figure 1).

Clade 1: Non-*Pseudovireya* species of section *Vireya*

The strongly supported (99%) clade 1 consists of all species sampled from subsections *Albovireya*, *Euvireya*, *Malayovireya*, *Phaeovireya* and *Solenovireya* (Figure 1). Morphologically these species share only a few characters: the margin of their leaf scales are lobed (Table 1) and the perulae margins are not ciliate (Figure 2; Sleumer, 1966).

Clade 2: Western and middle Malesia

The species in clade 2 – *Rhododendron alborugosum* Argent & Dransf., *R. jasminiflorum* Hook., *R. pneumonanthum* Sleumer, *R. rarilepidotum* J.J.Sm., *R. lagunculicarpum* J.J.Sm., *R. zollingeri* J.J.Sm., *R. williamsii* Merr. ex Copel.f., *R. rousei* Argent

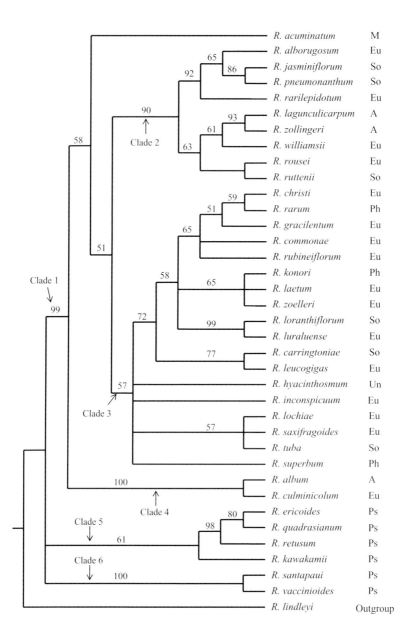

Figure 1 Strict consensus of four trees of length 135 for the combined chloroplast DNA data set (*psbA-trn*H spacer and *trn*T-*trn*L spacer). Bootstrap numbers are shown above the nodes. The most parsimonious trees have a C.I. of 0.6 and an R.I. of 0.8. Major clades discussed in the text are labelled, and the subsection each species belongs to is indicated. A = *Albovireya*, Eu = *Euvireya*, M = *Malayovireya*, Ph = *Phaeovireya*, Ps = *Pseudovireya*, So = *Solenovireya*.

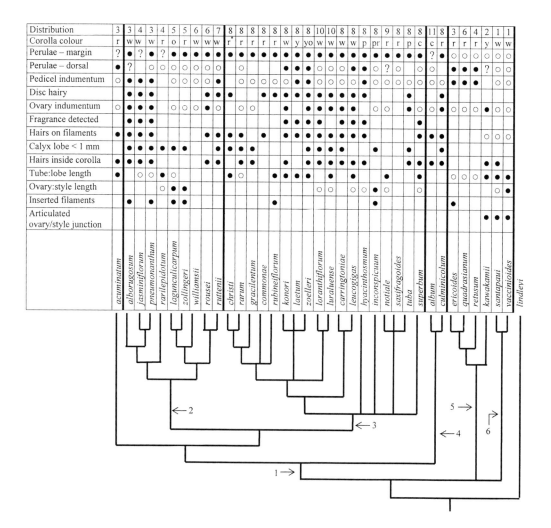

Figure 2 Morphological characters, fragrance and geographical distribution plotted against the strict consensus tree (Figure 1). A '?' shows that the character state could not be identified from available herbarium specimens and has not been commented on in the literature. **Character states**: *Distribution*: 1 = mainland Asia, 2 = Taiwan, 3 = Borneo, 4 = Sumatra, 5 = Sulawesi, 6 = Philippines, 7 = Moluccas, 8 = New Guinea, 9 = Australia, 10 = Solomon Islands, 11 = Java. *Corolla colour* (only the predominant corolla colour is listed): r = red (r* yellow tube, red lobes), w = white, o = orange, y = yellow, yo = yellow-orange, p = pink, pr = pink-red, c = cream. *Perulae – margin* (indumentum on margin of perulae): ● = scales, ○ = cilia. *Perulae – dorsal* (indumentum on dorsal surface of perulae): ● = scales and hairs, ○ = scales only. *Pedicel indumentum*: ● = scales and hairs, ○ = scales. *Disc hairy*: ● = yes. *Ovary indumentum*: ● = scales and hairs, ○ = scales. *Fragrance*: ● = detectable. *Hairs on filaments*: ● = present from the base, ○ = present in the middle 1/3 only. *Calyx lobe < 1 mm* (in length): ● = yes. *Hairs inside corolla*: ● = yes. *Tube:lobe length* (ratio of corolla tube length to corolla lobe length): ● = 1:1, ○ = 1:3. *Ovary: style length* (ratio of ovary length to style length): ● = 1:1, ○ = 1:4. *Inserted filaments* (not exceeding the mouth of the corolla tube): ● = yes. *Articulated ovary/style junction*: ● = yes.

& Madulid and *R. ruttenii* J.J.Sm. – are strongly supported as a group by a bootstrap value of 90% (Figure 1); they are all found growing on islands to the west of New Guinea (Figure 2). Within this clade there are two main groups: western Malesia and middle Malesia.

The western Malesia group consists of *R. alborugosum*, *R. jasminiflorum*, *R. pneumonanthum* and *R. rarilepidotum* and is strongly supported with a bootstrap value of 92% (Figure 1). These four species are found in either Sumatra or Borneo and they have calyx lobes that are less than 1 mm in length (Figure 2). *Rhododendron alborugosum*, *R. jasminiflorum* and *R. pneumonanthum* share several other characters that are not possessed by *R. rarilepidotum*: they have white, fragrant corollas, the indumentum on the pedicel, ovary and style consists of scales and hairs, and their staminal filaments and discs have hairs (Figure 2).

The middle Malesia clade is supported by a 63% bootstrap value and includes the species *R. lagunculicarpum*, *R. zollingeri*, *R. williamsii*, *R. rousei* and *R. ruttenii* (Figure 1). These five species are found on the islands of Sulawesi, the Philippines or the southern Moluccas at altitudes greater than 1500 m (Sleumer, 1966; Argent & Madulid, 1998); the margin and the dorsal surface of their perulae both possess scales (Figure 2). Within this middle Malesian group *R. lagunculicarpum*, *R. zollingeri* and *R. williamsii* (61%, Figure 1) have glabrous staminal filaments, ovaries and pedicels covered with scales, and their discs are glabrous (Figure 2). The clade of *R. lagunculicarpum* and *R. zollingeri* is strongly supported with a bootstrap value of 93% (Figure 1). They are the only representatives in this study from Sulawesi and have many morphological characters in common, including a 1:1 ovary: style length and filaments inserted, not protruding past the mouth of the corolla tube (Figure 2).

Clade 3: New Guinea, Australia and the Solomon Islands

Clade 3 is sister to clade 2 and contains all representatives in this study from the islands of New Guinea, Australia and the Solomon Islands: *R. christi* Foerste, *R. rarum* Schltr., *R. gracilentum* F.Muell., *R. commonae* Foerste, *R. rubineiflorum* Craven, *R. konori* Becc., *R. laetum* J.J.Sm., *R. zoelleri* Warb., *R. loranthiflorum* Sleumer, *R. luraluense* Sleumer, *R. carringtoniae* F.Muell., *R. leucogigas* Sleumer, *R. hyacinthosmum* Sleumer, *R. inconspicuum* J.J.Sm., *R. lochiae* F.Muell., *R. saxifragoides* J.J.Sm., *R. tuba* Sleumer and *R. superbum* Sleumer (Figure 1; Table 2). The perulae margins of these species have scales only (Figure 2). At present I have recognised only three character states for the perulae margin – scales, cilia or smooth – but this is probably an oversimplification, as scanning electron microscopy (SEM) work on *R. rushforthii* Argent & D.F.Chamb., a *Vireya* rhododendron not included in this study, shows it in fact possesses both scales and hairs on the perulae margin (Argent & Chamberlain, 1996).

Within clade 3 there are several smaller groups of species supported by both bootstrap values and morphology. *Rhododendron christi*, *R. rarum*, *R. gracilentum*, *R. commonae* and *R. rubineiflorum* are from the island of New Guinea and group together with a bootstrap value of 65% (Figures 1 and 2). They are all predominantly red-flowered taxa, with *R. christi* having a yellow tube but red lobes. None of these species has been noted as having fragrant flowers (Figure 2).

New Guinean endemics *R. konori*, *R. laetum* and *R. zoelleri* are grouped together with 65% bootstrap support; they share several morphological characters and a fragrance has been detected (Figures 1 and 2). These three species have both scales and hairs on the dorsal surface of their perulae, their corolla tube length is equal to that of the corolla lobe length, and they have hairs present on the disc and staminal filaments (Figure 2).

Rhododendron loranthiflorum and *R. luraluense* are grouped together with a bootstrap value of 99% (Figure 1). They are the only two representatives in this study from the Solomon Islands and have a number of morphological characters in common (Figure 2). The indumentum on the dorsal surfaces of their perulae consists of scales and both species have white flowers, scaly pedicels, hairs on the staminal filaments and disc, a style approximately four times the length of the ovary, hairs present on the inside of the corolla tube and calyx lobes < 1 mm in length (Figure 2).

The last noteworthy grouping within clade 3 is *R. carringtoniae* and *R. leucogigas* with a bootstrap value of 77% (Figure 1). They are both from New Guinea, with *R. carringtoniae* occurring in the southeast and *R. leucogigas* in the central north of the island. *Rhododendron carringtoniae* and *R. leucogigas* have white flowers, hairs present on top of the disc, and both hairs and scales present on the ovary and style (Figure 2). The relationships of the remaining species in clade 3 are unresolved or have weak bootstrap support and there are no obvious morphological characters in common.

Clade 4: *Rhododendron album* and *R. culminicolum*

Rhododendron album Blume, from Java, and *R. culminicolum* F.Muell., from West Papua (New Guinea), form the strongly supported (100%) clade 4 (Figures 1 and 2). These species have a few morphological characters in common: their pedicels are scaly and they possess hairs on their staminal filaments and also the inside of the corolla (Figure 2).

Clades 5 and 6: Subsection *Pseudovireya*

The relationships between clades 5 and 6 are unresolved and this section may be monophyletic or paraphyletic. As well as the traditional morphological characters grouping subsection *Pseudovireya*, i.e. tubular corolla and disc-shaped leaf scales (Table 1), they all have ciliate margins on their perulae (Figure 2). A ciliate perula margin is not unique to this subsection though, as in the living collection at the Royal Botanic Garden Edinburgh I have observed this in a number of species from subsection *Malayovireya*.

Clade 5 is supported by a bootstrap value of 61% (Figure 1). Within this clade *R. ericoides* Low ex Hook.f., *R. quadrasianum* Vidal and *R. retusum* (Blume) Benn. are strongly supported as a group with a bootstrap value of 98% (Figure 1); these are the only Malesian representatives of subsection *Pseudovireya* included in this study, found on the islands of Sumatra, the Philippines and Borneo (Figure 2). These three species share a number of morphological traits: the dorsal surfaces of their perulae possess both hairs and scales, they have glabrous staminal filaments, their pedicels are both scaly and hairy, their ovaries are covered in scales, and their flowers are all red or red-orange in colour (Figure 2). Also, their corolla lobes are approximately 1/3 of the length of their corolla tubes (Figure 2). Sister to these three species is *R. kawakamii* Hayata from Taiwan.

Clade 6 is strongly supported with a bootstrap value of 100% (Figure 2) and it includes the only representatives in this study from mainland Asia: *R. santapaui* Sastry et al. and *R. vaccinioides* Hook.f. (Figure 2). They both have white flowers, scales on the dorsal perulae surface and scaly pedicels and ovaries (Figure 2). These two species also share a number of morphological characters with *R. kawakamii* from clade 5: an articulated ovary/style junction, staminal filaments with hairs in the middle 1/3, and the corolla tubes equal in length to the corolla lobes (Figure 2).

Discussion

This is the first phylogenetic work to concentrate on *Rhododendron* section *Vireya*. Previous studies have been aimed at higher level problems within the genus, and at tribal and subfamily level (Kron & Judd, 1990; Kron, 1997; Kurashige *et al.*, 1998, 2001), or they have concentrated on specific groups within the genus such as subgenus *Hymenanthes* (Denton & Hall, in press). Kurashige *et al.* (2001) investigated sectional relationships in the genus *Rhododendron* and found subgenus *Rhododendron* to be monophyletic, while two of its sections, *Rhododendron* and *Vireya*, were shown to be polyphyletic. From the present study further comments cannot be made as to the monophyly of section *Vireya* because only one outgroup taxon was included. The results presented here do, however, provide preliminary insights into the subsectional relationships of *Rhododendron* section *Vireya*.

Four of the seven currently recognised subsections of section *Vireya* appear to be polyphyletic: *Albovireya*, *Euvireya*, *Phaeovireya* and *Solenovireya* (Figure 1). The monophyly of the other three subsections cannot be commented on because representatives of subsection *Siphonovireya* have not yet been sequenced, only one taxon from subsection *Malayovireya* has been sequenced and the relationships of the two clades identified within subsection *Pseudovireya* are unresolved. The polyphyly indicated by this analysis is not surprising, though, as the subsections are differentiated by only two characters: leaf scale type and corolla shape (Sleumer, 1966). Further, one of these traditionally defining characters, leaf scale type, can be difficult to interpret (Stevens, 1985).

The combined cpDNA phylogeny presented here reveals that the relationships between species in *Rhododendron* section *Vireya* correlate strongly with the geographical areas in which they grow (Figure 2). Clade 2 has a western and middle Malesian distribution; the distribution of taxa in clade 3 is eastern Malesian; clade 5 has a western Malesian assemblage of taxa but also includes a representative from the island of Taiwan; clade 6 includes the two representatives in this study from mainland Asia (Figure 2). Similar separation into eastern and western geographical areas has been found in other groups of flowering plants with a centre of diversity in Malesia, for example *Aeschynanthus* (Gesneriaceae; Denduangboripant *et al.*, 2001) and *Nepenthes* (Nepenthaceae; Meimberg *et al.*, 2001).

Clade 2: Western and middle Malesia

Clade 2 shows a relationship between islands of western Malesia (Borneo, Sumatra) and middle Malesia (Philippines, Sulawesi and the Moluccas). A similar pattern was found in several lepidopteran genera, where the islands of Sumatra and Borneo are sister to Sulawesi, the Philippines and the Moluccas (Vane-Wright, 1991). Even though clade 2 is strongly supported (99% bootstrap) from the cpDNA evidence (Figure 1), there do not appear to be any morphological characters shared by these taxa. Some of the subgroupings within clade 2, however, are supported by morphological evidence.

Rhododendron alborugosum, *R. jasminiflorum* and *R. pneumonanthum* have morphological similarities that are in agreement with their grouping in the cpDNA phylogeny. They are an interesting group possessing a number of characters in common, but *R. alborugosum* (subsection *Euvireya*) is not placed in the same subsection as *R. jasminiflorum* and *R. pneumonanthum* (subsection *Solenovireya*) (Chamberlain *et al.*, 1996). When *R. alborugosum* was described, the authors noted that it keyed out to the New Guinean solenovireyas, and its fragrance was similar to *R. jasminiflorum* (Argent & Dransfield, 1989). Its subsequent placement in subsection *Euvireya* was an oversight (G. Argent, personal communication).

Clade 3: New Guinea, Australia and the Solomon Islands

Within clade 3 there are several smaller clades, and some taxa where their position is unresolved (Figure 1). *Rhododendron christi*, *R. rarum*, *R. gracilentum*, *R. commonae* and *R. rubineiflorum* are predominantly red-flowered taxa with no detectable fragrance, restricted to the Papua New Guinea Highlands. Perhaps this is because red-flowered rhododendrons in Papuasia are most abundant between 3000 and 4000 m above sea level (Stevens, 1976), and there are many mountain peaks within this altitudinal range in the Papua New Guinea Highlands.

This high endemicity in the highlands of Papua New Guinea has been found for a number of other taxa (see van Welzen, 1997; Heads, 2001), and may be a result of orogeny, caused by the accretion of terranes onto the Australian craton, the southern part of New Guinea (van Welzen, 1997).

Relationships of distributions are not always readily explained, for example the clade of *R. carringtoniae* and *R. leucogigas*. This clade is reasonably supported (77%) by the cpDNA evidence and the two taxa share a number of floral characters (Figure 2). The taxa occur in two different regions of New Guinea: the east Peninsula (Mt. Dayman and Mt. Suckling) and the central north coast (Cyclops Mountains and Mt. Hunstein). This disjunction may be the result of a previously widespread ancestral taxon that has been restricted or fragmented to leave relic species (van Welzen, 1997). An alternative explanation for such a disjunction is given by De Boer & Duffels (1996) for several cicada genera. They suggest that the cicadas in these areas evolved on sister island arc terranes and that their present distributions reflect these historic areas because cicadas have poor dispersal abilities.

Clade 4: *Rhododendron album* and *R. culminicolum*

The position of these two taxa, *R. album* and *R. culminicolum*, clustering with all non-subsection *Pseudovireya* species of *Vireya*, is reasonable; however, their strongly supported grouping (100%) was not expected as they are from different islands: Java and New Guinea. Superficially these two species look dissimilar although they do share several morphological characters (Figures 1 and 2). Perhaps this relationship is a result of lack of sampling, and maybe the western and eastern groupings of clades 2 and 3 are being repeated here. The inclusion of more taxa and also a phylogeny produced using a nuclear marker may help to elucidate the relationships of these species.

Clades 5 and 6: Subsection *Pseudovireya*

Since the first attempt to divide up the vireyas by Charles Baron Clarke (1882), subsection *Pseudovireya* has been identified as unique. Clarke split the vireyas into two subgroups on the basis of capsule valves twisting (the true vireyas) or not (*Pseudovireya*) at dehiscence (Philipson & Philipson, 1996). Sleumer (1966) maintained the species of *Pseudovireya* as a subsection based on their 'disc-shaped' scales and generally short tubular flowers, and recent pollination and grafting studies have suggested an incompatibility exists between subsection *Pseudovireya* and the rest of the section (Rouse *et al.*, 1993; Rouse & Williams, 1997). The distinctiveness of subsection *Pseudovireya* (clades 5 and 6) from the rest of section *Vireya* (clade 1) is further supported by the cpDNA evidence (Figure 1).

Rhododendron ericoides, *R. quadrasianum* and *R. retusum* share a number of morphological features unique to the *Pseudovireya* clades, but sister to them is *R. kawakamii* from Taiwan, which in fact shares more morphological characters with the clade 6 species – *R. santapaui* and *R. vaccinioides* – than it does with the other members of clade 5 (Figures 1 and 2). Interspecific pollinations support the morphological data, showing a breeding barrier between the temperate *Pseudovireya* species (*R. kawakamii*, *R. santapaui* and *R. vaccinioides*) and all other species in the genus, including the tropical members of subsection *Pseudovireya* (Rouse *et al.*, 1993). Perhaps a

nuclear DNA marker will help to determine the relationships between *R. kawakamii* and its relatives.

Conclusion and further work

The currently accepted subsections of *Rhododendron* section *Vireya* are not all monophyletic and further sampling is required to test the monophyly of the section. The groups supported by the cpDNA phylogeny presented here reflect geographical regions rather than traditional taxonomic groupings. Several morphological traits have been identified as potentially informative for delineating groupings within section *Vireya*, particularly the indumentum of the perula margin and dorsal surface, but further investigations of these characters are required to define better the variation between species. Work in progress to further elucidate the natural groups and relationships within section *Vireya* includes broadening the sample size, analysing morphological data, and sequencing a nuclear DNA region.

Acknowledgements

I thank my supervisors Mr Lyn Craven, Professor Pauline Ladiges, Dr Randy Bayer and Dr Frank Udovicic. This project would not be possible without the financial support of The Baker Foundation, CSIRO Plant Industry and The University of Melbourne. My travel to Edinburgh for the Rhodo '02 conference was possible thanks to funding from the Australian Rhododendron Society Victorian Branch Inc., the Royal Horticultural Society and The School of Botany (Nicholas Travel Fellowship) and Scholarships Office (M.A.T.S. and M.A. Bartlett Research Scholarship Fund), The University of Melbourne. I also thank Mr Lyn Craven, Dr David Binney and the late Dr John Rouse for supplying leaf material for molecular studies, and the curators of A, CANB, L and NY for access to *Rhododendron* specimens for morphological studies.

References

ARGENT, G. C. G. & DRANSFIELD, J. (1989). *Rhododendron alborugosum*: a new species of vireya rhododendron from Borneo. *Notes Roy. Bot. Gard. Edinburgh* 46: 27–31.

ARGENT, G. & CHAMBERLAIN, D. (1996). *Rhododendron rushforthii*: a new species from Vietnam. *The New Plantsman* 3: 195–200.

ARGENT, G. & MADULID, D. (1998). *Rhododendron rousei* (Ericaceae): a beautiful new species from the Philippines. *The New Plantsman* 5: 25–31.

CHAMBERLAIN, D., HYAM, R., ARGENT, G., FAIRWEATHER, G. & WALTER, K. S. (1996). *The Genus Rhododendron. Its classification & synonymy*. Royal Botanic Garden Edinburgh.

CLARKE, C. B. (1882). In: HOOKER, J. D., *The Flora of British India* 3: 456–477. Ashford, Kent: L. Reeve & Co.

DE BOER, A. J. & DUFFELS, J. P. (1996). Historical biogeography of the cicadas of Wallacea, New Guinea and the West Pacific: a geotectonic explanation. *Palaeogeogr. Palaeoclimatol. Palaeoecol.* 124: 153–177.

DENDUANGBORIPANT, J., MENDUM, M. & CRONK, Q. C. B. (2001). Evolution in *Aeschynanthus* (Gesneriaceae) inferred from its sequences. *Pl. Syst. Evol.* 228: 181–197.

DENTON, A. L. & HALL, B. D. (in press). Diversification of *Rhododendron* subgenus *Hymenanthes* (Ericaceae) in the Neogene Himalaya. *Amer. J. Bot.*

FARRIS, J. S., KÄLLERSJO, N., KLUGE, A. G. & BULT, C. (1995). Testing significance of incongruence. *Cladistics* 10: 315–319.

FELSENSTEIN, J. (1988). Phylogenies from molecular sequences: inference and reliability. *Annu. Rev. Genet.* 22: 521–565.

HEADS, M. (2001). Birds of paradise, biogeography and ecology in New Guinea: a review. *J. Biogeog.* 28: 893–925.

HEDEGAARD, J. (1980). Morphological studies in the genus *Rhododendron*, dealing with seeds, fruits and seedlings and their associated hairs. In: LUTEYN, J. L. & O'BRIEN, M. E. (eds) *Contributions Toward a Classification of Rhododendron*, pp. 117–185. New York: The New York Botanical Garden.

KRON, K. A. (1997). Phylogenetic relationships of Rhododendroideae (Ericaceae). *Amer. J. Bot.* 84: 973–980.

KRON, K. A. & JUDD, W. S. (1990). Phylogenetic relationships within the Rhodoreae (Ericaceae) with specific comments on the placement of *Ledum*. *Syst. Bot.* 15: 57–68.

KURASHIGE, Y., MINE, M., KOBAYASHI, N., HANADA, T., TAKATANAGI, K. & YUKAWA, T. (1998). Investigation of sectional relationships in the genus *Rhododendron* (Ericaceae) based on *mat*K sequences. *J. Jap. Bot.* 73: 143–154.

KURASHIGE, Y., ETOH, J.-I., HANDA, T., TAKAYANAI, K. & YUKAWA, T. (2001). Sectional relationships in the genus *Rhododendron* (Ericaceae): evidence from *mat*K and *trn*K intron sequences. *Pl. Syst. Evol.* 228: 1–14.

MEIMBERG, H., WISTUBA, A., DITTRICH, P. & HEUBL, G. (2001). Molecular phylogeny of Nepenthaceae based on cladistic analysis of plastid *trn*K intron sequence data. *Plant Biol.* 3: 164–175.

PALSER, B., PHILIPSON, W. R. & PHILIPSON, M. N. (1991). Characteristics of ovary, ovule and mature megagametophyte in *Rhododendron* L. (Ericaceae) and their taxonomic significance. *Bot. J. Linn. Soc.* 105: 289–390.

PHILIPSON, M. N. & PHILIPSON, W. R. (1996). The taxonomy of the genus. In: POSTAN, C. (ed.) *The Rhododendron Story: 200 Years of Plant Hunting and Garden Cultivation*, pp. 22–37. London: The Royal Horticultural Society.

ROUSE, J. L. & WILLIAMS, E. G. (1997). Sexual and grafting compatibilities within section *Vireya* and between this section and other lepidote rhododendrons. In: JORDAN, N. (ed.) *Proceedings of the 1994 Pacific Region International Rhododendron Conference, Burnie, Tasmania, Australia, 28th–31st October 1994*, pp. 27–51. Burnie, Tasmania: Australian Rhododendron Society, North West Tasmania Branch.

Rouse, J. L., Williams, E. G. & Knox, R. B. (1986). Floral features related to pollination ecology in *Rhododendron*. In: Williams, E. G., Knox, R. B. & Irvine, D. (eds) *Pollination '86*, pp. 65–69. Plant Cell Biology Research Centre, School of Botany, University of Melbourne, Parkville, Victoria 3052, Australia.

Rouse, J. L., Knox, B. R. & Williams, E. G. (1993). Inter- and intraspecific pollination involving *Rhododendron* species. *J. Amer. Rhododendron Soc.* 47: 23–28, 40–45.

Sang, T., Crawford, D. J. & Stuessy, T. F. (1997). Chloroplast DNA phylogeny, reticulate evolution, and biogeography of *Paeonia* (Paeoniaceae). *Amer. J. Bot.* 84: 1120–1136.

Sleumer, H. (1966). *Rhododendron*. In: *Flora Malesiana*, ser. I, 6(4): 469–656. Groningen, The Netherlands: Wolters-Noordhoff Publishing.

Stevens, P. F. (1976). The altitudinal and geographical distributions of flower types in *Rhododendron* section *Vireya*, especially the Papuasian species, and their significance. *Bot. J. Linn. Soc.* 72: 1–33.

Stevens, P. F. (1985). Malesian vireya rhododendrons – towards an understanding of their evolution. *Notes Roy. Bot. Gard. Edinburgh* 43: 63–80.

Swofford, D. L. (1998). *PAUP*: Phylogenetic Analysis Using Parsimony (*and other methods)*. Sunderland, MA: Sinauer Associates.

Taberlet, P., Gielly, L., Pautou, G. & Bouvet, J. (1991). Universal primers for amplification of three non-coding regions of chloroplast DNA. *Plant Mol. Biol.* 17: 1105–1109.

Thomson, D. & Henry, R. (1993). Use of DNA from dry leaves for PCR and RAPD analysis. *Plant Mol. Biol. Rep.* 11: 202–206.

Vane-Wright, R. I. (1991). Transcending the Wallace line: do the western edges of the Australian region and the Australian plate coincide? *Aust. Syst. Bot.* 4: 183–197.

van Welzen, P. C. (1997). Increased speciation in New Guinea: tectonic causes? In: Dransfield, J., Coode, M. J. E. & Simpson, D. A. (eds) *Plant Diversity in Malesia III. Proceedings of the Third International Flora Malesiana Symposium 1995*, pp. 363–386. Royal Botanic Gardens, Kew.

Rhododendron collecting in Sulawesi, Indonesia

David Binney

85 Castles Road, Oropi, RD 3 Tauranga, New Zealand

As a keen collector of *Vireya* species I have travelled to SE Asia five times since 1995 looking to introduce new species to cultivation. Visits to Sabah, Sarawak and Sumatra have been followed by two expeditions to Sulawesi in 1998 and 2000. Over the last four or five years several collectors have visited Sulawesi but until recently only a few of the vireyas found here had been introduced. Only a small percentage of growers will venture out to climb these mountains so I hope to give you an idea of how these plants grow in their natural habitat. Plants are only part of the experience of climbing mountains in tropical countries and I would like to share some of the varied experiences involved in these visits. In New Zealand we have strict regulations covering the importation of plant material, effectively meaning only seed can be brought back. This has greatly restricted the number of species I have been able to introduce. The cultivation of these seedlings has produced some interesting results and I would like to present my experiences with the cultivation of the species collected on my various expeditions.

I am an enthusiastic hobbyist and have been growing *Vireya* species for the last 10 years. I come from the northern part of New Zealand, where the climate is fairly kind to vireyas. The hybrids grow easily and vigorously outdoors; the species are more easily grown under cover as this provides wind shelter and better control of watering. The original species in New Zealand have come mainly from the collections of Oz Blumhardt and Keith Adams in Sabah, Sarawak and West Malaysia and the Pukeiti collections, many from Graham Smith, from Papua New Guinea.

My interest in collecting vireyas arose after hearing George Argent speak at a conference in Burnie, Australia. His talk on collecting in Irian Jaya (West Papua New Guinea) got me enthused and led to my own collecting on Mt. Kinabalu and Mt. Trus Madi in Sabah

in 1995, Mt. Penrissen, Mt. Berumput and Mt. Santubong in Southern Sarawak in 1996, and Mt. Sibayak and Mt. Kemiri in Northern Sumatra in 1997. Mt. Rantemario in SW Sulawesi followed in 1998 and Mt. Sojol in Northern Sulawesi in 2000.

My first experience of the rhododendrons of Sulawesi was from material collected from the summit of Mt. Sesean in SW Sulawesi in 1996 by Keith Adams and John Farbarak. The species, *R. rhodopus* Sleumer, *R. zollingeri* J.J.Sm. and *R. quadrasianum* Vidal, that were collected have grown vigorously and have now all flowered in cultivation. John Farbarak and his friend Hank Helm returned in 1997 from climbing Mt. Rantemario and I have included some of his observations in this paper together with those of my own visit in 1998.

Mt. Rantemario (3440 m) is most easily climbed from the SW via the village of Karangan. From here a good ridge track can be followed with a two-day trek to the summit. We approached the mountain from the west, in the valley between Mt. Rantemario and Mt. Sinaji to the north, and we ended up taking five days to get close to the summit. It did allow us to find *R. vanvuurenii* J.J.Sm. at 750 m altitude in pine forest remnants on the foothills of Rantemario. Buffalo and goats grazed the surrounding areas and the local residents actively cut out the rhododendron as they said it was poisonous.

Leaving these lowland hills we ascended through river valley rice paddy and onto the mountain proper. We ascended through mountain gardens with some coffee plantations, but largely of shifting cultivation of subsistence crops. At about 1500 m pine forest gave way to a large terraced area of regenerating broad-leaved forest. Our guides told us that the local people had inhabited this area during the Japanese occupation in the Second World War. It was in these more open areas that we found *R. zollingeri* growing and seeding terrestrially. This is the same species Adams and Farbarak had collected on Mt. Sesean and it turned out to have a wide distribution on Mt. Rantemario up to an altitude of 2500 m.

From here we had to traverse a small headland called Mt. Tirowal towards the south to resume our ascent. The canopy was high above us, with light levels low at ground level. Flowers of *R. celebicum* (Blume) DC. covered the ground but the plants were largely out of sight high in the canopy above. It wasn't until we regained the more open ridges that light levels increased and rhododendrons became common at ground level. We found *R. quadrasianum* and large areas of *R. malayanum*

Jack, in both a nice crimson form and also a rather spectacular white variety. This species became so thick on one part of the ridge that we had to crawl on our hands and knees through the twisted trunks of the 5 m high plants.

The mountain weather in the tropics is usually fairly predictable. Mornings are mostly fine but as the day heats up, convection over the hills causes the air to rise and cloud forms at higher altitudes. This leads to mid-afternoon and evening rain, which clears in the late evening. The cloud can make navigation difficult and at one stage we did get lost for half a day, descending on the wrong track. With clearing weather the next morning we found our way back to the correct track and found the first of the white species, *R. impositum* J.J.Sm. This grew both terrestrially and epiphytically, tending to grow in moister areas just off the main track. The larger-leaved plants tended to grow in moister areas just off the ridges while the smaller-leaved vireyas of subsection *Albovireya* – *R. zollingeri*, *R. lagunculicarpum* J.J.Sm. and *R. arenicolum* Sleumer – were more common in the drier, more exposed ridges. As we climbed we found further plants of *R. impositum*, mostly with white flowers but some were pink, and some had larger, more campanulate blooms. It was difficult to tell whether this represents species variation or some element of hybridisation with the next species we found, the spectacular *R. pudorinum* Sleumer. We found only one plant in flower and this was just at the end of its flowering, but the foliage was as spectacular as the malayovireyas previously seen on Mt. Kinabalu. Plants with intermediate foliage almost certainly represented hybrids of *R. pudorinum* and *R. impositum*. It certainly can be very difficult to determine whether some of these plants are variations within these species or part of the populations of natural hybrids found on the mountain.

This problem is further compounded when seed is collected and grown on. New Zealand agricultural regulations prevent the return of wild-collected cutting material, so seed is the only practical way of introducing these species to New Zealand. I have grown on seed from all my collections and have found at times large numbers of natural hybrids result. The low altitude species have tended to come true from seed, with 100% of *R. fallacinum* Sleumer and 99% of *R. polyanthemum* Sleumer collected on Mt. Kinabalu coming 'true'. The higher altitude plants, where they are more likely to grow in close proximity to other species, are much more likely to produce hybrid seedlings. Only about 10% of *R. lowii* Hook.f. and none of the *R. acuminatum* Hook.f. or *R. maxwelli* Gibbs seedlings have turned out to be true species. Although there are numerous obvious natural hybrids on the mountains it does surprise me they don't dominate as they are certainly vigorous in cultivation. I presume the true species do tend to be the best adapted for their natural environment. These natural hybrids do also offer insight into how new species can develop in response to changes in habitat or climate.

The flora of the high altitude ridges is always spectacular, especially the reds, purples and oranges of the new growth of *Vaccinium* species. As we climbed, *R. zollingeri* gave way to *R. lagunculicarpum*. This species was also variable, with the leaves tending to become more oval and the corolla tubes much shorter at higher altitude. Unfortunately it was at this stage we had to turn back due to lack of water. No rain for the previous two days meant we had to manage on water

squeezed from sphagnum moss. The three species seen higher up the mountain were *R. pseudobuxifolium* Sleumer, *R. nanophytum* Sleumer, and *R. eymae* Sleumer.

I returned to Sulawesi to climb Mt. Sojol (3000 m) in the year 2000. This is a much more isolated mountain than Mt. Rantemario. The area has been less influenced by 'the West' or even the Indonesian government, and the hill people lead a subsistence existence living mainly off taro, rice and maize supplemented by a small amount of protein. While in the forest they hunted with blowpipes and slingshots and would supplement their rice and taro with the birds they caught.

We did find vireyas, *R. javanicum* (Blume) Benn., *R. quadrasianum* and *R. radians* J.J.Sm., growing at around 1000 m in the coastal hills. For the first few days, most of the time was spent in hot dry areas of disused cultivation. One day we managed to walk only about 5 km 'as the crow flies' as it unfortunately involved ascending and descending multiple spurs and ravines. It was a relief to reach the cool shelter of the forest, and as we ascended we found more *R. radians* and, for the first time, *R. leptobrachion* Sleumer. This grew terrestrially on the main ridge at about 2000 m and appeared to have longer corollas than material seen on Mt. Rantemario. Mt. Sojol is only 100 km north of the equator but the climate is strongly influenced by its altitude and the narrowness of the peninsula, which means all sides are close to the sea. Above 2000 m we were constantly in cloud, with frequent rain. It was cold and our local guides suffered in their wet cotton clothing. The rain also meant there were none of the dry ridges, such as those we had experienced on Mt. Rantemario, where vireyas and the albovireyas especially were dominant. Even with higher altitude the canopy never became shorter than 5–6 m, and vireyas tended to be high up in the trees and much less abundant than on Mt. Rantemario.

The night before climbing to the summit we pitched our tent on a ridge. The ground was covered in tree roots and rose and fell about 6 inches as the trees swayed in the breeze. In the final 300 m to the summit we found two new vireyas. The red-flowered *R. pseudobuxifolium* grew as an epiphyte in relatively low light areas off the ridge, and a spectacular large-leaved, white-flowered species *R. bloembergenii* Sleumer was growing terrestrially to a height of 4–5 m.

Our guides were cold and wet and keen to get home. We had planned to take two days to get down the mountain; in the end we struggled to keep up with them on the descent and walked a total of 17 hours that day. We were glad to find a bamboo hut in which to spend the night, but not so happy the next morning to find the place crawling with cockroaches.

The rhododendrons of Sulawesi are still very poorly known, and many of its mountains have not been visited by collectors. The problem of the apparent frequency of hybrids, indicated here, often makes species identification difficult, but recent visits have brought several new and beautiful vireyas into cultivation.

Late flowering elepidote hybrids with reference to Exbury and other gardens in Southern England

Michael Robinson

Hindleap Lodge, Priory Road, Forest Row, East Sussex RH18 5JF, UK

It is a sad fact that a 'late flowering' rhododendron is often thought of as flowering in June, and that almost the only widely known varieties flowering later than this are 'Europa', 'Polar Bear' and *R. auriculatum*. However, a visitor to many of the great rhododendron collections in Southern England will see rhododendrons in full flower in July, August and even September. The purpose of this paper is to make such taxa better known.

Hybridisation for late flowering seems to have taken place sporadically at The Grange in Benenden, Leonardslee, South Lodge, Wakehurst, and the Earl of Limerick's garden at Chiddinglye, but a systematic effort to lengthen the flowering season was undertaken at Exbury before the Second World War. It is well documented that Lionel de Rothschild was especially interested in extending the rhododendron season at Exbury throughout the year, and he is quoted as having plans for a 'summer' garden there. These plans were probably thwarted by the Second World War, and many crosses for lateness, both named and unnamed, seem to have disappeared. However, some remain today, largely uncatalogued and forgotten until recently. Even the beautiful hybrid 'Leonore' (*R. kyawi* × *R. auriculatum*) lies outside the area surveyed so ably by Michael Lear a few years ago, and there are at least a dozen late flowering taxa in the area surrounding the five remaining clones of 'Leonore'.

The mention of late flowering in rhododendron almost always provokes two responses: that late means June, and really late means 'Polar Bear'. Although there are a few July and August flowering cultivars in commerce only 'Polar Bear' is well known. Such excellent hybrids as 'Europa', 'Ungerio', 'Red Cap', and the excellent 'Arthur Osborn' are available only through one supplier in the RHS Plantfinder (2002–3; www.rhs.org.uk/plantfinder/plantfinder.asp/). Other taxa flowering after the end of June are not generally known, with the exception of scented whites such as 'Albatross', 'Argosy', 'Antonio' and 'Angelo', which occasionally may flower as late as this.

When visiting the great gardens in the south of England in the height of summer it is therefore something of a surprise to encounter rhododendrons flowering in July, August and even September. There is the July flowering 'Isabella' at The Grange in Benenden, and there are a number of red and orange shaded hybrids derived from 'Tally Ho', *R. sanguineum* Franch. ssp. *didymum* (Balf.f. & Forrest) Cowan and

R. auriculatum Hemsl., at Leonardslee. Many more can be found at Exbury.

It is only at Exbury that the systematic hybridising for lateness and colour has been well documented. It is well known that the late Lionel de Rothschild wished to have rhododendrons in flower every month of the year, and that he used *R. auriculatum*, *R. ungernii* Trautv. and the '*parishia*' series, especially *R. kyawi* Lace & W.W.Sm., to achieve this end. Table 1, an excerpt from his stud book, shows some of the hybrids he produced with *R. kyawi* as a parent. The most striking thing about these crosses is how few of them are known today, with only the 'Penelope' group and 'Europa' appearing in the 2002–3 RHS Plantfinder. It was therefore a considerable

surprise to the author that there are so many July, August and September flowering rhododendrons extant at Exbury.

A few of the surviving late hybrids can be found within the area of Exbury open to the public. These include a cross made, not by Lionel de Rothschild, but by his son Edmund. Flowers of this cross labelled ER30 were in their main flowering period in August 1999. The Exbury tag numbers are 4494 (pale cream), 4495 (pink) and 4499 (salmon). With the exception of the 2002 season these plants flower in late July or August consistently. The cross is recorded as 'Barclayi' × 'Leonore' in the Exbury stud book. 'Barclayi' is *R. thomsonii* Hook.f. × *R. arboreum* Sm. It seems unlikely that the parentage of the cream

Table 1 An excerpt from the Exbury stud book – Lionel de Rothschild (LR) hybrids.

LR436	ungernii × griersonianum			
LR437	ungernii × kyawi	Europa	deep pink	
LR475	ungernii × diaprepes			
LR590	Mrs. Leopold de Rothschild × kyawi			also LR1146
LR595	blancmange (Corona × auriculatum) × kyawi			
LR596	maximum × kyawi			also LR1055
LR600	(discolor × Doncaster) × kyawi			also LR636
LR615	griersonianum × kyawi	Romarez	dark red	
LR620	trial 206 × kyawi	Ironside	crimson	
LR627	discolor × kyawi			
LR638	B. de Bruin × kyawi	Ivan	bright red	
LR640	Mrs. R.S. Holford × kyawi	Hypatia	red	
LR641	(discolor × auriculatum) × kyawi			
LR728	Romany Chai × kyawi			
LR757	Azor × kyawi			
LR995	(Mrs. J. Kelke × auriculatum) × kyawi			
LR1096	(ungernii × facetum) × kyawi			
LR1127	griersonianum × Dragonfly	Penelope	spotted red	

flowered taxon can be correctly stated, and if the other two are correct then the conclusion must be that 'Leonore' has great potential for inducing late flowering. These plants have recently been named 'Beverly Lear'.

A plant with the tag number 4440 which had flowers in August 2001 was identified on the Exbury computer database as *Azalea* 'Aladdin', which is clearly an error, but the plant does not match *Rhododendron* 'Aladdin' either. The foliage has a mixture of ramiform and stellate hairs, but there is no sign of the simple hairs which are a prominent feature of the foliage of *R. auriculatum*, and while this cannot be ruled out as a parent, it appears more likely to include *R. griersonianum* Balf.f. & Forrest, *R. elliottii* Watt ex Brandis or *R. facetum* Balf.f. & Kingdon-Ward, and *R. kyawi* (to give the lateness in flowering). The appearance of the new growth reinforces this view.

An August flowering red, tag number 4444, has a glabrous corolla which is speckled on all five lobes, and has nectar pouches. The stamens are glabrous, but the style, which is slightly shorter than the corolla, has short glands to its tip. The mature foliage has vestiges of stellate hairs only and the lamina is slightly decurrent to the petiole. This plant looks close to *R. kyawi*, but must be hardier, as it is over 4 m high and must have survived the 1962–63 winter. In fact all the plants discussed here must have done so. *Rhododendron kyawi*, which is not at all hardy, no longer survives at Exbury although there is a plant identified as *R. kyawi* on the Exbury database. This is pink, and resembles a plant that used to exist at Wakehurst of which the author has a graft. This is not only pink but fragrant, but it is not *R. kyawi*. A graft of *R. kyawi* is surviving winters in a sheltered part of the author's garden. The Exbury 'R.

kyawi' is therefore almost certainly a hybrid. It might be LR638 ('Bas de Bruin' (*R. catawbiense* Michx. hybrid) × *R. kyawi*) or, less likely, LR596 (*R. maximum* L. × *R. kyawi*).

A batch of 'Leonore' (LR580 *R. auriculatum* × *R. kyawi* AM1948), Plate 1, planted in a group at Exbury outside the public area of the garden, was in flower on 14 September 2001. 'Leonore' has a reputation for being tender, but of eight plants in the group five have survived for over 50 years. The surviving taxa are very similar to each other, but one is darker in flower and another is later, with a few flowers present in September even in 2002. The normal flowering time at Exbury is late August or early September, with the young and vigorous growth ripening very quickly.

There is a very late scarlet rhododendron that flowered in the last week of August 2001. Its normal flowering time appears to be early September. The corolla has no speckling. The upper leaf lamina shines noticeably, and has stellate and a few ramiform hairs left at maturity on the underside. There is no sign of *R. auriculatum* in its appearance. Again it looks close to *R. kyawi* but must be hardier. It is likely to be a cross between *R. kyawi* and an *R. griersonianum* hybrid. It might be LR728 ((*R. griersonianum* × 'Moser's Maroon') × *R. kyawi*), or possibly LR638 again.

This plant is the only red in the middle of two straight rows of large (4 m+ high) rhododendrons planted about 3 m apart looking like a set planted out for evaluation. The rest are late flowering scented pinks.

A smaller plant in bloom on 14 September 2001 was also flowering on 4 September 2002, with more flowers still to open. It reliably flowers every September and has the additional advantage of being of relatively small

Plate 1 *Rhododendron* 'Leonore' – a form – at Exbury, September 2001.

stature. The corolla is deep red, with dorsal speckles, five lobes, and six stamens. The style is glabrous and the stamens are glandular along half their length. The upper leaf surface is matt with vestiges of stellate hairs, and the lower surface is densely covered with hairs of dendroid or ramiform type, with perhaps an under layer of rosulate hairs (my microscope makes me uncertain about this) and some stellate hairs on the midrib. The growth buds are strongly reminiscent of *R. griersonianum*. It might be LR1096, but the size of the bush (it is not a young plant) indicates that it is more likely to be an *R. didymum* hybrid.

Two nursery rows some way beyond the group of 'Leonore' appear to be hybrids of 'Polar Bear' or *R. auriculatum*. All are fragrant, late flowering, large growing plants of good quality. Another pink flowering plant very similar to *R. auriculatum*, but later flowering, reveals nothing in its parentage from flower and foliage other than the obvious *R. auriculatum*. Its presence in a row of hybrids indicates it is unlikely to be a pure form of this species. The flowers are unfortunately not very scented. The latest flowering of this group was in bloom on 14 September 2001 when it was the only plant in this association in full flower, and there were still a few flowers unopened. It is very similar to *R.* 'Polar Bear' in flower and foliage, but the flowers are smaller.

In this general area of Exbury there are also upwards of 50 plants which appear to be hybrids of *R. ungernii* and *R. auriculatum*, or, more likely, 'Leonore', and 'Europa' (*R. ungernii* × *R. kyawi*). A typical example of these was in flower on 14 September 2001. While the flower colour varies from mid- to deep pink, tending towards red in the darkest examples, all are very vigorous large plants with *R. auriculatum*-like foliage. However, persistent stellate simple and dendroid hairs all occur on the lower leaf surface. These appear to be seedlings which are the result of crossing *R. ungernii* with 'Leonore', but may well be from LR1096 ((*R. ungernii* × *R. facetum* Balf.f. & Kingdon-Ward) × *R. kyawi*). The young foliage is outstanding, and the flowering period stretches from late July to mid-September. No plants of moderate growth were found.

Finally at Exbury, mention should be made of a rhododendron in the main garden (tag number 3411) which keys out to be, and almost certainly is, a cream form of the species *R. auriculatum*. This plant was flowering in late July 2000 (Plate 2). There is a similar plant of a deeper colour at Muncaster Castle, so it appears that the range of colour in the published descriptions of this species needs amending. The late pinks at Chiddinglye and South Lodge both appear to be hybrids *of R. facetum* or *R. elliottii*, *R. auriculatum* and *R. kyawi*.

It seems likely that the plants lined out in the far reaches of Exbury are a remnant of Lionel de Rothschild's plans for a summer garden in Exbury. It is a great pity that these were interrupted in 1939 by the outbreak of war which postponed the development of the garden there for many years.

However, Lionel de Rothschild has left us all yet another valuable legacy. The plants described here make it clear that it is not necessary for those seeking to extend the flowering season in their spring gardens to

Plate 2 A cream form of *Rhododendron auriculatum* at Exbury, August 2001.

start their own programme of hybridisation. Rhododendrons of high quality flowering from July to September already exist. It is only necessary to make them generally available and to have plenty of room to grow them. However, it does appear that 'Leonore' is very useful in producing late hybrids, and that the cream form of *R. auriculatum* may make it possible to achieve a really late flowering yellow. Producing late hybrids suitable for the small garden is a quest still to be pursued.

Acknowledgements

I am indebted to Edmund de Rothschild, the late Paul Martin, and Rachel Foster for permitting free access to Exbury and for their enthusiasm, and to Everard Daniel for his companionship, scepticism, humour, and expertise.

The importance of living collections in the classification of rhododendrons

James Cullen

Director, Stanley Smith (UK) Horticultural Trust, Cory Lodge, Cambridge CB2 1HR, UK

A survey of rhododendrons in cultivation is given, based largely on new information on these plants in botanic gardens throughout the world. How these collections have been used in, and have affected, the history of the classification of the genus is detailed. Some discussion is given concerning the numerous hybrids that have been raised, with special emphasis on the identification of their parents and their nomenclature. The value of properly recorded and curated living specimens is illustrated with examples of how they can be used to cast new light on the identification and relationships of the species.

Introduction

It is reasonable to suppose that most rhododendrons cultivated in gardens are grown for their aesthetic appeal, their freely produced, variously coloured flowers, their form, their leaves, their indumentum, etc. This is borne out by the existence of a vast range of garden hybrids and cultivars which extend and enhance the features that make these plants attractive to gardeners and landscape architects. This paper will concentrate on living *Rhododendron* collections which have a further set of purposes than just the aesthetic: that is, collections used for research, conservation and education. These are essentially what are called 'scientific' or 'botanical' collections, and are to be found almost exclusively in botanic gardens throughout the world.

Research on living collections

This is concerned mainly with the classification of the genus, providing characteristics of the plant that are additional to those available from the herbarium collections which form the basis of such research (Cullen, 1996). Such characteristics include posture, form and structure of buds (both vegetative and flowering), form and structure of quickly deciduous organs (bracts, bracteoles, prophylls) and the occurrence of chemical constituents of various kinds (including DNA). However, research on living collections can take other forms – physiological, horticultural and even ecological – so there is a wide range of potential uses. To give credibility to such research, the cultivated materials used must meet certain standards: the plants must be properly named, adequately grown and suitable for use as representatives of the wild populations from which they originally derived.

Conservation

We know a certain amount about the conservation status of some rhododendrons in the wild, especially concerning those species from North America, Europe and the Black Sea region. This is a very small proportion of the whole genus, and as we consider species from further east in temperate Asia, relevant information becomes more and more sparse (though the situation in subtropical and tropical South East Asia is perhaps somewhat better). Hence our living collections may well turn out to be a vastly important resource as our knowledge of the conservation status of the Himalayan and Chinese species of the genus improves. Again, it is important that our collections be properly verified taxonomically and suitable for use for this purpose. It is pointless and demoralising to provide special cultivation and horticultural techniques to a plant, thought to represent a species that is threatened or extinct in the wild, only to find out at some later date that the plant is wrongly identified and its species not at all threatened.

Education

The use of scientific collections for educational purposes is part of a wider use of plant collections generally, for both informal and formal educational activities.

The Botanic Gardens Conservation International (BGCI) database of plants in cultivation in botanic gardens

The collections to be considered here are those in botanic gardens. Now, for the first time, we have an opportunity to survey botanic garden collections across the world and to assess how representative they are in the coverage of the vascular plants as a whole (which, of course, includes *Rhododendron*).

The charitable organisation BGCI, of which I have been a Trustee for the last 10 years, exists to co-ordinate the efforts of *ex situ* plant conservation in botanic gardens (interpreted widely) throughout the world. It maintains and updates the original IUCN/WCMC database of threatened plants. It operates a service to member gardens with computerised plant records. On receipt of the total record in electronic form of a contributing garden, it can compare these with its own threatened plants database and mark in the garden's record those records of plants which are extinct, threatened or vulnerable in the wild. The garden can make sure, when the modified record is returned, that these plants are looked after in the best possible way so that they are not lost, and can be propagated so that they can be more widely distributed. This service is widely used by the member gardens and is highly appreciated by them. BGCI has over 500 members from all over the world, so the service it provides is worldwide in scope.

Because each member garden submits its total record for scanning for threatened plants, it occurred to the staff of BGCI that this process offered an opportunity to put together the records of all the gardens to produce a list of plants cultivated in botanic gardens throughout the world. With the aid of grants from the Stanley Smith UK and US Horticultural Trusts, Dianne Wyse-Jackson of BGCI has developed the software to carry out this process. Now, for the first time, we have at least a preliminary list of plants for consideration, covering the contents of 448 botanic gardens from 81 countries. This whole process has been possible only because of the existence of *The International Transfer Format for Botanic Gardens Plant Records*, version 1 (1987) and version 2 (1998) (BGCI, London) – known as the ITF. This sets out minimum standards for harmonising the content of such records, allowing direct comparisons to be made between records from different gardens (all of whom have given their permission for this process to be carried out).

This list now exists in a preliminary form: it consists of a list of plant names with, for each, an indication of the number of gardens which claim to grow the plant in question (that is, which have the name in their own record). The list is enormous; Table 1 gives some of the details. As it stands, the list is very definitely preliminary, needing a considerable amount of editing, because, even with the ITF, different gardens go about collecting and maintaining their plant record data in slightly different ways. I have carried out such an edit on the content of the list as far

Table 1 Summary statistics of the BGCI Botanic Gardens Database.

- Current number of gardens: 448 (including Kew and Edinburgh and many other major gardens from both temperate and tropical areas)

- Current number of countries represented: 81

- Current number of plant records included: over 79,000 (different species, subspecies, varieties, formae or cultivars)

Table 2 BGCI Botanic Gardens Database, *Rhododendron* records (partially edited): statistics.

Number of records: 1872, of which:

- 803 (43%) refer to hybrids, mostly under cultivar names

- 329 (17%) are synonyms, misspellings or mispresentations of names

- 740 (40%) are names of botanical taxa (species, subspecies, varieties or formae)

---

as the genus *Rhododendron* is concerned, which helps to give a clearer idea of what we have. Table 2 shows some basic statistics for the genus. Incidentally to all this, the most widely grown species, by a long way, is *R. luteum* (L.) C.K.Schneid. *Rhododendron schlippenbachii* Maxim. and *R. smirnowii* Trautv. follow somewhat behind.

The most striking feature of these figures is the very high proportion of hybrids represented. The total number of rhododendron hybrids and cultivars is vast (see Leslie, 2003), but the interest of such plants is mainly (though not totally) aesthetic, and it is perhaps surprising that botanic garden collections should include so many. Various reasons can be proposed for this: the fact that the plants are long-lived is probably very significant – it takes an act of horticultural will to remove a hybrid once it is growing. I would like such acts of will to be more frequent in botanic gardens.

Of the synonyms, misspellings and mispresentation of names, little need be said. Synonyms and, indeed, all taxonomic uses here are defined as such according to the various recent Edinburgh-based revisions (Cullen, 1980; Chamberlain, 1982; Philipson & Philipson, 1986; Chamberlain & Rae, 1990; Kron, 1993; Judd & Kron, 1995; Argent *et al.*, 1996 – see also Argent *et al.* (1998). Misspellings just happen – 'lochae' for 'lochiae', 'kyawii' for 'kyawi', etc. – and are easily cleaned up. Mispresentations of names are slightly more complex and arise from different gardens' interpretations of the ITF; Table 3 shows an example.

I now turn to the records of botanical taxa, of which there are 740. These cover 493 species, including 84 subspecies, 98 varieties and 13 formae. As the whole genus is considered to contain between 800 and 900 species,

Table 3 BGCI Botanic Gardens Database, *Rhododendron* records (partially edited): an example of mispresentation of names.

The same hybrid occurs under four different forms of its name:

R. × *cilpinense*

R. cilpinense

R. 'Cilpinense'

R. cv. Cilpinense

These are treated as four separate entities by the computer. Only the first of them is correct; the other three are mispresentations (typographic synonyms) of the first.

---

it looks as though the names of more than half of these (between 54% and 62%) are in cultivation. I use the word 'names' here quite deliberately. More than half the genus is a very sizeable representation, showing that we really do have a considerable resource available for research and conservation.

In the Edinburgh system of classification there are 71 basic supraspecific units (whether subgenera, sections or subsections). Of these, 68 are represented among the total 493 species; only subsections *Monantha*, *Fragariflora* and *Griersoniana* are not represented at all, even though species from all of these groups have been in cultivation in the past. They may, indeed, still be in cultivation in gardens but not included in the BGCI database. Table 4 shows the details, Table 5 provides a summary. The equality in numbers between 'lepidotes' and 'elepidotes' is very striking. These figures reinforce the idea that the material we have in cultivation represents the genus quite reasonably as a scientific resource. This present result and list is just a preliminary first attempt: the whole process needs to be taken further and greatly developed.

Taxonomic verification and the status of the actual plant material in relation to material in the wild

In editing the *Rhododendron* section of the database, I have been very aware that we are dealing with very raw data, and the reason I used the word 'names' earlier, rather than 'species' or 'taxa', was that, at present, all we have is a list of names. What the plants behind these names really are is a matter of some doubt unless we know what degree of confidence we can put in the names we have; that is, in the accuracy of the taxonomic identifications which have provided the names.

Even in the records as they stand, I know that there are misidentifications. For example, the name *R. shweliense* Balf.f. & Forrest occurs in several garden records. This species is known from only two collections from the subtropical part of Yunnan adjacent to the Burmese border, both made by George Forrest in 1919. The species is distinctive among its allies in its definitely subtropical provenance (which would make it difficult to grow out of doors in northern Europe) and in its style which is puberulent along its whole length. I have seen much material in gardens over the last 30 years labelled '*R. shweliense*', but not one specimen has shown the distinctive puberulent style; all of them have been hybrids involving *R. glaucophyllum* Rehder. I have concluded, after considerable searching, that *R. shweliense* is not in cultivation. It may have been in the past, though its origin makes that unlikely, and it may survive in some garden somewhere in the mildest parts of Europe, but I doubt it. Secondly, *R. excellens* Hemsl. & E.H.Wilson. This species was described on the basis of a single specimen collected

Table 4 The Edinburgh *Rhododendron* classification (with numbers of species in the database).

Subgenus Hymenanthes

    Sect. Ponticum

        Subsect. **Fortunea** 15

        Subsect. **Auriculata** 1

        Subsect. **Grandia** 10

        Subsect. **Falconera** 10

        Subsect. **Williamsiana** 1

        Subsect. **Campylocarpa** 4

        Subsect. **Maculifera** 8

        Subsect. **Selensia** 4

        Subsect. **Glischra** 7

        Subsect. **Venatora** 1

        Subsect. **Irrorata** 9

        Subsect. **Pontica** 12

        Subsect. **Argyrophylla** 11

        Subsect. **Arborea** 3

        Subsect. **Taliensia** 29

        Subsect. **Fulva** 2

        Subsect. **Lanata** 3

        Subsect. **Campanulata** 2

        Subsect. **Griersoniana** 0

        Subsect. **Parishia** 3

        Subsect. **Barbata** 4

        Subsect. **Neriiflora** 22

        Subsect. **Fulgensia** 1

        Subsect. **Thomsonia** 13

Subgenus Azaleastrum

    Sect. **Azaleastrum** 4

    Sect. **Choniastrum** 5

Subgenus Tsutsusi

    Sect. **Tsutsusi** 24

    Sect. **Brachycalyx** 12

Subgenus Pentanthera

    Sect. Pentanthera

        Subsect. **Pentanthera** 13

        Subsect. **Sinensia** 1

    Sect. **Rhodora** 2

    Sect. **Viscidula** 1

    Sect. **Sciadorhodion** 4

Subgenus **Mumeazalea** 1

Subgenus **Candidastrum** 1

Subgenus **Therorhodion** 2

Subgenus Rhododendron

    Sect. Rhododendron

        Subsect. **Edgeworthia** 3

        Subsect. **Maddenia** 29

        Subsect. **Moupinensia** 1

        Subsect. **Monantha** 0

        Subsect. **Triflora** 18

        Subsect. **Scabrifolia** 5

        Subsect. **Heliolepida** 3

        Subsect. **Caroliniana** 1

        Subsect. **Lapponica** 22

        Subsect. **Rhododendron** 3

        Subsect. **Rhodorastra** 2

        Subsect. **Saluenensia** 2

        Subsect. **Fragariflora** 0

        Subsect. **Uniflora** 4

        Subsect. **Cinnabarina** 2

        Subsect. **Tephropepla** 4

        Subsect. **Virgata** 1

        Subsect. **Micrantha** 1

        Subsect. **Boothia** 4

        Subsect. **Camelliiflora** 1

        Subsect. **Glauca** 5

        Subsect. **Campylogyna** 1

        Subsect. **Genestieriana** 1

        Subsect. **Lepidota** 3

        Subsect. **Baileya** 1

        Subsect. **Trichoclada** 4

        Subsect. **Afghanica** 1

    Sect. **Pogonanthum** 7

    Sect. Vireya

        Subsect. **Pseudovireya** 15

        Subsect. **Siphonovireya** 1

        Subsect. **Phaeovireya** 15

        Subsect. **Malayovireya** 6

        Subsect. **Albovireya** 3

        Subsect. **Solenovireya** 18

        Subsect. **Vireya** 59

Table 5 BGCI Botanic Gardens Database, *Rhododendron* records (partially edited): botanical taxa.

Total supraspecific units: **68** included (out of 71)

Total species: **493** (ssp. 84; vars. 98; formae 13)

Lepidotes: species 246 (ssp. 26; vars. 46; formae 6)

   Vireyas: species 127 (ssp. 0; vars. 9; formae 6)

Elepidotes: species 247 (ssp. 58; vars. 52; formae 7)

   '*Azaleas*': species 50 (ssp. 4; vars. 4; formae: 0)

---

by Augustine Henry in the late 19th century in the extreme south of Yunnan (Mengtze). This (herbarium) specimen, the type of the name, is at Kew, and there is a photograph of it in the herbarium at Edinburgh. The species was distinguished from *R. nuttallii* Booth by having 15 (not 10) stamens, and by flowering somewhat later. A recent collector has identified and distributed three or four collections as *R. excellens*, in spite of the fact that none of them has more than 10 stamens: these plants are all *R. nuttallii*. *Rhododendron excellens* has not been seen since Henry discovered it, and is perhaps an oddity or chance hybrid of *R. maddenii* Hook.f. or *R. nuttallii*. Further examples of such doubtful plants are *R. ciliicalyx* Franch. and *R. seingkhuense* Kingdon-Ward.

The ITF has produced some standards and rules on taxonomic verification, and these are shown in Table 6. At present the data in the database are of very mixed value from this point of view. Clearly, most of the living material at Edinburgh will qualify as level 3, as will some of that at Kew and other gardens visited by the Edinburgh group of taxonomists. In these cases we can have real confidence that these plants are really what they purport to be (errors in labelling

are not being considered here). In other collections, however, the level of verification is likely to be '0', with the plants being grown under the names they arrived with, or, more rarely, '1'; these names have to be viewed with a certain degree of scepticism, as they are open to change when checked. Presumably the specimens labelled '*R. shweliense*' and '*R. excellens*' mentioned above fall into the '0' category. It is worth bearing in mind here that in about 1978, when Edinburgh was collecting species for its new plantings, a vast number of plants of subsection *Lapponica* were acquired from many sources. On growing these on and checking them against wild-collected herbarium specimens, it was found that only about 40% were correctly named. It is a matter of some priority, therefore, to attempt to bring the verification levels for all these plants in botanic gardens up to at least level 2, and, if possible, to level 3. This is really only a matter of general living plant collection curation, but it needs some international co-operation to get it working properly. At least the BGCI list shows up the problem.

Over the past 30 years (and indeed before that) botanic gardens have stressed the importance of their 'wild-origin material'. This is generally understood to be material which can be traced directly to a source in the wild uncontaminated by garden conditions. Generally the place and date of origin of such plants are documented, and frequently the collector as well, with his/her collecting number when possible. Such material is clearly of greater value for research and conservation than material whose origin is not known. In fact, when one looks into the matter, it is a great deal more complicated than it first appears, and a number of issues need to be looked at in assessing the 'wild-origin' status of individual plants: not just the docu-

Table 6 Verification standards from ITF version 2 (1998).

[Code 'U']	It is not known if the name of the plant has been checked by an authority.
[Code '0']	The name of the plant has not been determined by any authority.
[Code '1']	The name of the plant has been determined by comparison with other named plants.
[Code '2']	The name of the plant has been determined by a taxonomist or other competent person using the facilities of a library and/or herbarium, or other documented living material.
[Code '3']	The name of the plant has been determined by a taxonomist who is currently or has been recently involved in a revision of the family or genus.
[Code '4']	The plant represents all or part of the type material on which the name was based, or the plant has been derived therefrom by asexual propagation.

mentation, which is vital, but also the type of material that was first introduced (seed, seedling, plant, scions of various kinds) and the propagation history of the material, as far as this can be determined, have a bearing on the matter.

The absolute classic *Rhododendron* collections are those made by George Forrest on his nine expeditions to western China and adjacent Burma between 1904 and 1931. Each Forrest expedition lasted over several seasons; he collected herbarium material of plants in flower, and marked the individual plants collected for a return visit by either himself or one of his trained Chinese helpers during seed production. At this return visit, another herbarium specimen was collected, as was as much seed from the plant as possible. Each herbarium specimen was numbered and an index was drawn up so the two numbers (one flowering, one fruiting) referring to a single plant could be collated. The seed was introduced to gardens under the number of the fruiting specimen. Other collectors who used this rather elaborate technique were Frank Kingdon-Ward (sometimes) and Joseph Rock. Unfortunately, the wild-collected herbarium specimens, though widely

distributed as duplicates to various herbaria, are not readily available to most gardens.

Because the material was introduced as seed, the resulting living material probably represented a different selection of seed success from that which would have occurred in the wild. Propagators will have employed an unconscious selection process among these seedlings that will have involved those that had made the strongest start (given that more than a few actually germinated). Hence the importance of the wild-collected herbarium material which is a persistent, unchanging record of what was actually there in the wild, and which can be compared with living plants bearing the same collector's number to ensure that they are broadly similar and taxonomically the same. If this is the case, then the living plant can be assumed to stand for the wild population it came from. If it is not the case, then the plant, or at least its wild-origin status and collector's number (if any), should be discarded; but how often has this been done?

This is complicated enough. But the perils of confusion in propagation and cultivation make the problem much worse. In many gardens, seedlings are pricked out (or cuttings or grafts lined out) in individual pots which, however, are not individually labelled; instead, some kind of positional convention is used to indicate where material from one collection begins and ends. In Edinburgh in the 1970s the convention was 'front to back, left to right'. This meant that the first pot of a batch was labelled, then those following it towards the back of the frame, and those in the next column, were all considered the same until a new labelled plant occurred. The possibilities for things going wrong in such a set-up are manifold. Later in the process, the planting out and labelling of a selection of these plants produces further possibilities for incorrect labelling. In dealing with such material it is essential that every propagule and young plant be individually labelled. Again, the wild herbarium material provides a proper check and allows for the correction of gross mistakes.

More recent collections have had to be made under more and more stringent rules about what can be collected, and under financial constraints on the length of expeditions. This means that the high standards of Forrest and his colleagues cannot be met. Seed collection is often not accompanied by the collection of herbarium specimens of the seeding plants, and collection of the same plants in flower is usually not possible. Thus material from more modern collections requires meticulous curation at all stages, from the packeting of the seed to the labelling of the plant. All gardens do what they can in this respect, but life is short and the art of curation is long, and things certainly do go wrong in even the best-run establishments.

As if this were not enough, there are further complications arising from propagation. Seed from garden rhododendrons should never be distributed (regrettably, much was in the 1950s and 1960s), except as potential ornamental material; it should certainly never bear the collector's number of its female parent. If propagation is done by asexual means (cuttings, layers, grafts, etc.) the problems of the variability of seed is obviated. However, it must be borne in mind that these methods involve the production of callus – disorganised tissue that can be cytologically diverse and give rise to morphologically disturbed plants. In most cases, and as far as we know, this is not very significant, but in some cases it clearly is and it is important that we are aware of it. Many of the classic collections have been through several generations of propagation: the chances of something going wrong increase with time and each propagation, so that the mistakes accumulate.

When we label a garden plant *Forrest* 19404 or *CLD* 935 we must understand the act of faith we are making in assuming that the plant so labelled is a reasonable representative of the original wild population. The ITF deals with some of these problems under the headings of 'provenance type', 'propagation history', 'wild provenance' and 'accession lineage'. However, in most gardens and with most existing collections, such information is hard to come by and time-consuming (and therefore costly) to enter into the records.

In order to improve the situation, botanic gardens will need to improve their curatorial procedures quite drastically. This is difficult to do when every garden is trying to grow the same range of plant material – the similarity in collection content between different botanic gardens has often been commented on. We perhaps need to develop a structure for designating particular collections as International Scientific Collections, which would hold new material of the particular genus or family involved, and would distribute duplicate material for safe keeping as necessary. Such a scheme would be very difficult to implement, but it would free individual gardens from the need to try to grow everything, giving more time for the proper curation of what was actually there. A step towards this would be for each garden to make good quality herbarium collections of its scientific material now, so that at least a base-line of permanent record would exist. This may not be possible for many gardens, which do not have herbarium facilities available, but for those that do, it is an important first step. Researchers must be aware of the importance of their collecting herbarium voucher material of the specimens on which they are working; without such vouchers, the value of their work is considerably reduced.

The problems of maintaining living collections are illustrated here by *Rhododendron*, but similar problems occur throughout gardens for every genus. At least the *Rhododendron* world is reasonably defined, and meets regularly to talk over problems, so perhaps it could show the way.

References

ARGENT, G., FAIRWEATHER, G. & WALTER, K. (1996). *Accepted Names in Rhododendron Section Vireya*. Royal Botanic Garden Edinburgh.

ARGENT, G., BOND, J., CHAMBERLAIN, D., COX, P. & HARDY, A. (1998). *The Rhododendron Handbook*. London: The Royal Horticultural Society.

CHAMBERLAIN, D. F. (1982). A revision of *Rhododendron* II. Subgenus *Hymenanthes*. *Notes Roy. Bot. Gard. Edinburgh* 39(2): 209–486.

CHAMBERLAIN, D. F. & RAE, S. J. (1990). A revision of *Rhododendron* IV. Subgenus *Tsutsusi*. *Edinb. J. Bot.* 47: 89–200.

CULLEN, J. (1980). A revision of *Rhododendron* I. Subgenus *Rhododendron* Sections *Rhododendron* and *Pogonanthum*. *Notes Roy. Bot. Gard. Edinburgh* 39: 1–208.

CULLEN, J. (1996). The importance of the herbarium. In: POSTAN, C. (ed.) *The Rhododendron Story*, pp. 38–48. London: The Royal Horticultural Society.

JUDD, W. S. & KRON, K. A. (1995). A revision of *Rhododendron* VI. Subgenus *Pentanthera* (Sections *Sciadorhodion*, *Rhodora* and *Viscidula*). *Edinb. J. Bot.* 52: 1–54.

KRON, K. A. (1993). A revision of *Rhododendron* [V]. Sect. *Pentanthera*. *Edinb. J. Bot.* 50: 249–364.

LESLIE, A. C. (2003). Rhododendron registration and the preparation of a new Register. In: ARGENT, G. & McFARLANE, M. (eds) *Rhododendrons in Horticulture and Science*, pp. 61–65. Royal Botanic Garden Edinburgh.

PHILIPSON, W. R. & PHILIPSON, M. N. (1986). A revision of *Rhododendron* III. Subgenera *Azaleastrum*, *Mumeazalea*, *Candidastrum* and *Therorhodion*. *Notes Roy. Bot. Gard. Edinburgh* 44: 1–24.

Leaf wax chemistry and *Rhododendron* taxonomy: light in a dark corner

David W. H. Rankin

Department of Chemistry, University of Edinburgh, West Mains Road, Edinburgh EH9 3JJ, UK

Analysis of the hydrocarbon fraction of leaf waxes from plants in *Rhododendron* subsection *Taliensia* shows that it consists mainly of *n*-alkanes with odd numbers of carbon atoms. In most taxa the distribution is centred at $C_{31}H_{64}$, but in some the centre is $C_{27}H_{56}$. More precise measurements allow a partial key to the subsection to be constructed. The data indicate that some varieties of *Rhododendron phaeochrysum, R. alutaceum, R. roxieanum* and *R. aganniphum* should be given higher status, while others are probably not supportable at all. Hybrids between taxa in the two major groups have wider distributions of alkanes. Some specimens appearing on morphological grounds to belong in one taxon are shown to be hybrids. In other cases, parents of hybrids can be identified.

Introduction

One could imagine an ideal world, in which the problems of taxonomy were solved by taxonomy machines. Into one end we would feed bits of plants, and at the other end we would be presented with a few Latin words, as well as a small amount of compost. In practice, if we could make our machines conform as far as possible to the behaviour of real taxonomists, the output would be a lot less decisive. 'Well, it could be *Rhododendron alutaceum* Balf.f. & W.W.Sm., perhaps var. *iodes* (Balf.f. & Forrest) D.F.Chamb., or maybe var. *russotinctum* (Balf.f. & Forrest) D.F.Chamb. ... or possibly a variety of *R. roxieanum* Forrest ... or it could be a hybrid.' Why are we so indecisive?

Ideally, every species (taxon) has one or more definitive characteristics (or a definitive set of characteristics), which in traditional taxonomy would be morphological. These characteristics may be prominent and unequivocal, such as the number of petals; they may depend on measurements, and thus could include ranges of dimensions for lengths and widths of leaves; or they may require detailed study in a laboratory, literally down to the level of splitting hairs. Similarly, each taxon ideally has one or more definitive chemical constituents (or a definitive set of constituents), which in principle would allow unequivocal identification by a suitably sophisticated analytical instrument.

The reality, of course, is much more complex. First, there can be immense variability within one taxon. Measurable parameters have statistical distributions, so that there will be specimens that fall outside any defined range, and concentrations of useful indicator chemicals can vary similarly, so that sometimes they may appear to be absent. Secondly, there are abundant hybrids, so that even those that appear on morphological grounds to belong to one taxon may in fact have properties of two or more taxa. Indeed, in the case of rhododendrons, even some type specimens may actually represent hybrids. Thirdly, we must recognise that the concepts of species, subspecies and variety are conveniences created for the benefit of communication, and not fundamental realities in the world of plants. Entities to which we give names, essential if we are to be able to communicate with one another, may be part of a continuum. What we regard as distinct taxa may overlap, and the situation is further confused when we recognise that it is not static, but in a state of continuing change, as plants interbreed and populations become isolated. The problems of interbreeding and incomplete speciation are severe with the genus *Rhododendron*, and particularly so in subsection *Taliensia* Sleumer.

There is therefore a need to focus on some specific properties of these plants, if we are to understand the complex relationships between them. Useful work has been done using morphological details, for example making use of characteristics of leaf hairs, and of course DNA analysis can make a substantial contribution, although it is by no means a universal panacea for taxonomic problems. In this paper, we focus on the chemistry of the wax that is present on the leaves of all rhododendrons. It turns out that the simplest components of these waxes can make a significant contribution to a tricky area of *Rhododendron* taxonomy. Using analytical information, we can develop a partial key to the *Taliensia* subsection, and in so doing provide insight into taxa that have so far proved difficult to classify. Finally, wax data help to identify some specimens

as hybrids, and provide pointers to possible parents.

Rhododendron leaf waxes

All *Rhododendron* leaves (and those of many other plants) are protected by a layer of wax. Early work (Evans *et al.*, 1980) showed that these waxes are extremely complex, containing many components. It is possible that many of these constituent compounds may be useful for taxonomic purposes, although it would require a huge amount of analytical work, on large numbers of wild and cultivated specimens, to develop a practical application for taxonomic purposes. However, the simplest compounds present in the waxes, *n*-alkanes, prove to be of considerable taxonomic utility. These compounds, which are the major constituents of paraffin wax and beeswax, are so-called 'straight-chain' hydrocarbons, in fact having zigzag chains of carbon atoms, and completely saturated with hydrogen.

The fraction of the leaf wax obtained by extraction with chloroform and passing over alumina consists almost entirely of hydrocarbons, and by far the largest part of this fraction is a mixture of *n*-alkanes (Chadwick *et al.*, 2000). These can be separated by gas chromatography, so that the distribution of chain length can be studied. It is found that the alkanes with odd numbers of carbon atoms are more abundant than their even-numbered analogues by a factor of at least ten to one. This study focuses entirely on these odd-numbered alkanes, although there may well be further information obtainable by study of the even-numbered alkanes and of the small quantities of other compounds, including unsaturated hydrocarbons, that are present.

131

In the large majority of *Rhododendron* species, including some members of subsection *Taliensia*, the compound with 31 carbon atoms, $C_{31}H_{64}$, predominates, with smaller amounts of the alkanes with 33, 29 and 27 carbons, $C_{33}H_{68}$, $C_{29}H_{60}$ and $C_{27}H_{56}$, and very small amounts of some others. On the other hand, in other members of subsection *Taliensia* it is $C_{27}H_{56}$ that is dominant and the whole distribution of chain length is shifted down. We therefore have a very simple test, which divides subsection *Taliensia* into two groups. Thus, for example, *R. taliense* Franch. and *R. traillianum* Forrest & W.W.Sm. belong in the $C_{27}H_{56}$ group, while *R. adenogynum* Diels, *R. beesianum* Diels and *R. bureavii* Franch. are members of the $C_{31}H_{64}$ group. This does not in itself provide a particularly useful diagnostic test, but it does illustrate the beginnings of a taxonomic key, based purely on this one fraction of the leaf wax.

A chemical key

Further information is available from the distributions of *n*-alkanes in the hydrocarbon fractions, but if this is to be used it is essential to study the extent to which the distributions for a single taxon can vary under different conditions. The consistency of the data was therefore studied, to see what precision could be achieved, and thus to assess the fineness of the resolution distinguishing different taxa. First, repeated chromatographic analyses of the same extract were checked for internal consistency. If the intensity of the chromatography peak for the most abundant compound is labelled I_n and those of the compounds with two more and two fewer carbon atoms are labelled I_{n+2} and I_{n-2}, then the values of the percentage ratios I_n/I_{n+2} and I_n/I_{n-2} were found to have a standard devia-

tion of 2%. Secondly, comparison was made of specimens from the same plant, including leaves taken from different parts of the plants, collections made at different times of the year and in different years. This also included young and mature leaves, as well as some that had been stored in the herbarium for years. The standard deviation for the various values of the ratios I_n/I_{n+2} and I_n/I_{n-2} was 10%. Finally, these ratios were compared for specimens of different plants, for a total of six taxa, and it was found that use of different plants contributed a further 5% to the standard deviation. Combining all these errors together yields a total uncertainty of 11 percentage units in the relative intensities of the peaks adjacent to the maximum, and any taxonomic key should therefore not depend on the accuracy of these measurements to any greater precision than this.

With the knowledge that the variations in wax distributions within any one taxon are not large, it is possible to define the position of the maximum of the distribution much more precisely than merely at C_{27} or C_{31}. Indeed, if a smooth curve is fitted to the intensities, the maximum of the curve does not have to fall at an odd number, nor even at an integer, but it can be at any value. In practice, the large majority of sets of data could be fitted with a Gaussian curve, with a full width at half height (FWHH) of about 3.3 carbon units. Typical distributions (Figure 1) easily allow measurement of the maximum to one or two tenths of a unit, and data for different specimens of the same taxon would typically yield maxima consistent to about 0.2 carbon units. Such data (Table 1) can easily form the basis of a taxonomic key. Such a key, in which the members of subsection *Taliensia* are divided primarily on the basis of the carbon number integer closest to

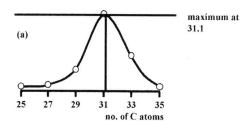

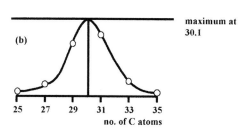

Figure 1 Distribution of alkanes in leaf waxes for two specimens of rhododendrons in subsection *Taliensia*, showing how the maximum of the distribution can be measured to within about 0.1 carbon atom units.

Table 1 Positions of maxima in distributions of alkanes in subsection *Taliensia*.

Taxon	Maximum carbon number
R. alutaceum var. *iodes* (Balf.f. & Forrest) D.F.Chamb.	27.0
R. phaeochrysum var. *levistratum* (Balf.f. & Forrest) D.F.Chamb.	27.0
R. traillianum var. *traillianum* Forrest & W.W.Sm.	27.0
R. taliense Franch.	27.7
R. alutaceum var. *russotinctum* (Balf.f. & Forrest) D.F.Chamb.	28.5
R. aganniphum var. *flavorufum* (Balf.f. & Forrest) D.F.Chamb.	30.1
R. adenogynum Diels	30.8
R. clementinae Forrest	31.0
R. balfourianum Diels	31.2
R. bureavii Franch.	31.4

the maximum in the distribution, has been published (Chadwick *et al.*, 2000). Maxima lie between 27 and 32 carbon units.

However, such a simple key is unsatisfactory in some ways. For example, one problem is that some taxa lie near to boundaries between two groups. But there are other problems of different kinds. One is the fact that for many taxa, few distinct clones are available. This is exacerbated by the inaccurate labelling of many specimens, some of which have been in cultivation for nearly 100 years. Getting unequivocal and consistent names for specimens is also difficult – but that illustrates exactly why there is a need for this kind of study. Accurate names cannot be applied to objects for which a clearly defined concept does not exist. Several of the taxa in subsection *Taliensia* are probably no longer

supportable, and it is to this issue that we now turn.

Problem taxa

Rhododendron phaeochrysum

Three varieties of *R. phaeochrysum* Balf.f. & Forrest were described by Chamberlain (1982). Analysis of leaf waxes shows all specimens of var. *levistratum* (Balf.f. & Forrest) D.F.Chamb. to have the maximum of their distribution at carbon number 27, averaging 27.0 for the fitted curves, while the maxima for var. *phaeochrysum* are all at 31, with the average maxima of the fitted curves at 31.1. This leads to the conclusion that there is a fundamental and unequivocal distinction between these two taxa. Given that there are also morphological differences, which were the basis of their separation into varieties, there is a case for separation at a higher level, at least subspecies, and probably species. In this paper it is not appropriate to make firm proposals for taxonomic changes. However, it is appropriate to present pointers to possible changes. Further study should be undertaken in such cases, as a matter of priority.

The third variety of *R. phaeochrysum* is var. *agglutinatum* (Balf.f. & Forrest) D.F.Chamb. Specimens of this variety include some with maxima at carbon number 27, some at 31, and some in between. This last type is unusual, and there is evidence (see below) that it is characteristic of a hybrid between C_{27} and C_{31} parents. It seems, therefore, that the agglutinated indumentum that defines this variety is a property that can occur in several distinct taxa, and that it should not be used to define a separate variety. Again, further study is needed, specifically to see whether characteristics other than agglutinated indumentum suffice to place these specimens in one or other of the varieties of *R. phaeochrysum*. The specimens with intermediate wax distribution could be hybrids between these two other varieties.

Rhododendron alutaceum

The situation with *R. alutaceum* is similar to that with *R. phaeochrysum*. The variety *alutaceum* consistently has the maximum in its wax alkane distribution at carbon number 31, var. *iodes* is equally consistently a C_{27} taxon, and var. *russotinctum* includes specimens with C_{27} maxima, others with C_{31} maxima, and some lying in between, suggesting that they are hybrids. There is a widespread belief (Cox & Cox, 1997) that var. *russotinctum* is not a valid taxon, and it is satisfying to find the evidence from the wax analyses consistent with other evidence. The other two

varieties are clearly distinct, and the data are probably consistent with the idea that *R. alutaceum* var. *alutaceum* should be included in *R. roxieanum* (see below). The position of var. *iodes* is less clear, and again it is important to undertake further study, preferably with a wider range of wild material.

Rhododendron roxieanum

The problem with *R. roxieanum* is that, at least in var. *oreonastes* (Balf.f.) T.L.Ming, it is so distinctive. The fine, narrow leaves immediately stand out, and I know from my own experience that a single specimen growing amongst other species is likely to be noticed. But such a specimen is very likely to be a hybrid, and any plants grown from seed collected from it are almost certain to be hybrids. As these hybrids may be morphologically very like one of the parents but chemically more like the other parent (see the section on hybrids below), working with waxes from the material in cultivation is fraught with difficulties. The single available specimen of *R. roxieanum* var. *roxieanum* had a C_{27} maximum, as did most of the specimens of var. *oreonastes*, although several had substantial C_{29} components, indicative of hybrid origin. There were just three specimens of var. *cucullatum* (Hand.-Mazz.) D.F.Chamb. available, and of these one had its maximum at C_{27}, and the other two at C_{31}. The status of this last variety is therefore questionable, but the shortage of material makes it difficult to make definitive statements. It is particularly important to resolve the position of *R. roxieanum* and its varieties, because of the close links to other species and varieties whose status is uncertain. The acquisition of material for study from pure populations of each of the three varieties (if such populations exist) is a matter of high priority.

Rhododendron aganniphum

The situation with the two varieties of *R. aganniphum* Balf.f. & Kingdon-Ward is perhaps the least satisfactory of all. All of the specimens of *R. aganniphum* var. *flavorufum* (Balf.f. & Forrest) D.F.Chamb. that we have been able to access have turned out to be of the same clone. The wax distribution in these is unusual, in that there are almost equal quantities of $C_{29}H_{60}$ and $C_{31}H_{64}$. It may be that the maximum of the distribution in this species, or at least this variety, is consistently lower than in others; the alternative interpretation is that this clone of var. *flavorufum* is in fact a hybrid. However, there is no doubt that there are extensive populations of this variety in the wild, and it is probable that it is a good taxon. Rather more specimens of *R. aganniphum* var. *aganniphum* were available. Some of these had C_{31} maxima, but others had the broader distribution characteristic of hybrids. Once again it is necessary to acquire many more specimens, preferably from populations that appear to consist solely of one variety of this species.

Hybrids

In the course of this work it has been found that the waxes for the large majority of specimens have maxima that are close to either $C_{31}H_{64}$ or $C_{27}H_{56}$. However, there are some that have broader distributions, typically with FWHH around six carbon units instead of the usual 3.3 and with the maximum around $C_{29}H_{60}$, or have two maxima, at $C_{27}H_{56}$ and $C_{31}H_{64}$ (Figure 2). This pattern has never been observed for specimens of species that are unequivocally pure, and

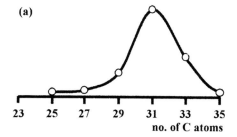

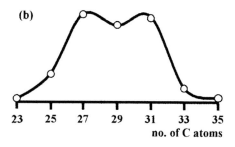

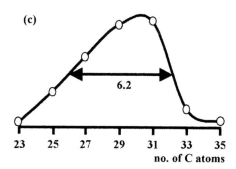

Figure 2 Distributions of alkanes in leaf waxes for rhododendrons in subsection *Taliensia*. (a) Typical distribution for a species. (b) Distribution with a double maximum, as observed for hybrids between C_{27} and C_{31} taxa. (c) Broad distribution, as observed for hybrids between C_{27} and C_{31} taxa.

has been found often in populations that are known to be hybrids between C_{27} and C_{31} taxa. We have therefore concluded that the observation of such a broad distribution, or of one centred at $C_{29}H_{60}$, is a very strong indicator of a hybrid. Of course, it is not essential that such hybrids have distributions of this kind; they may also follow the pattern of either parent. Three examples illustrate the information that can be obtained and show how wax analyses can help to identify the parents of hybrids.

Rhododendron beesianum × R. roxieanum var. oreonastes

The first example involves two of the most distinctive taxa within subsection *Taliensia*. Analyses were made of leaves from nine plants from a population growing on Bai Ma Shan, in Yunnan Province in China, that clearly consisted of hybrids. Three of these specimens are illustrated in Plate 1. The first of these (Plate 3a) looks like *R. beesianum*, and distribution of alkanes in its wax had a maximum at carbon number 31, which is consistent with it being pure *R. beesianum*. The second specimen (Plate 3b) definitely has the appearance of a hybrid, with several characters being intermediate between those of the two putative parents. However, the maximum in the wax distribution was at carbon number 27 and the distribution was narrow, as would be expected for pure *R. roxieanum* var. *oreonastes*. This emphasises the need to remember that although broad distributions indicate the probable presence of a $C_{27} \times C_{31}$ hybrid, it is quite possible for such plants not to give any hint of their hybrid origin in their wax.

The third specimen (Plate 3c) would undoubtedly be identified from its appearance as *R. roxieanum* var. *oreonastes*, but its wax contains substantial amounts of both $C_{27}H_{56}$ and $C_{31}H_{64}$, and thus clearly indicates that this plant is in fact a hybrid. This illustrates well the problem discussed in the section on problem taxa above, whereby many

cultivated plants purporting to be *R. roxieanum* var. *oreonastes* are in reality hybrids with C_{31} species. This in no way reduces their horticultural merit, but they clearly are of no use in scientific study of the species.

Rhododendron przewalskii × R. phaeochrysum

A population of plants from the Zheduo Pass, near Kanding in Sichuan Province in China, included some that had the appearance of *R. przewalskii* Maxim., others that looked like *R. phaeochrysum*, and a few that had characteristics inherited from both of these. The leaf wax distributions of seven plants that appeared to be hybrids were studied. They included some with maxima at carbon number 27, others at 31, and two with the broad distributions characteristic of hybrids between C_{27} and C_{31} taxa. *Rhododendron przewalskii* is a C_{31} species, but *R. phaeochrysum* has varieties that are C_{27} (var. *levistratum*) and C_{31} (var. *phaeochrysum*). In this case, therefore, the wax analysis confirms that the population does indeed include hybrids, but also, more importantly, identifies one of the parents as *R. phaeochrysum* var. *levistratum*.

Rhododendron proteoides × R. aganniphum

Finally, we consider a population of plants from Mei Li Shan in NW Yunnan in China. These included some that had the appearance of *R. proteoides* Balf.f. & W.W.Sm. and as the only other member of subsection *Taliensia* that grew nearby was *R. aganniphum*, it was assumed that this was the other parent. The wax of one plant, which appeared to be pure *R. proteoides* (Plate 2), had substantial amounts of both $C_{27}H_{56}$ and

Plate 1 Specimens from a population of *Rhododendron beesianum × R. roxieanum* var. *oreonastes*.

Plate 2 Specimens from a population of *Rhododendron proteoides × R. phaeochrysum* var. *levistratum*.

DAVID RANKIN

$C_{31}H_{64}$, and is thus a hybrid between C_{27} and C_{31} taxa. Other specimens (e.g. Plate 2b) had all the characteristics of the plant that was named *R. bathyphyllum* Balf.f. & Forrest, but was subsequently believed to be a hybrid of *R. proteoides × R. aganniphum* (Chamberlain, 1982). Their waxes had maxima at carbon number 31. The third specimen illustrated (Plate 2c) also appears to be a hybrid, although it does not conform to the pattern of *R. bathyphyllum*. Its wax distribution had its maximum at $C_{27}H_{56}$, and the presence of significant amounts of this compound links it with the plant shown in Plate 2(a). However, neither of the varieties of *R. aganniphum* has substantial amounts of $C_{27}H_{56}$ in its wax. We are therefore forced to the conclusion that this population of hybrid plants, including typical *R. bathyphyllum*, does not derive from the two species that grow close by. One parent is certainly *R. proteoides*; the second must be a C_{27} taxon, and the most probable candidate is *R. phaeochrysum* var. *levistratum*. Perhaps we should not be too surprised. If the parents were the two species growing adjacent to one another, we would expect to find a swarm of hybrids, not just a few plants. It is the chance arrival of pollen from a distance that has given rise to the small population.

Conclusions

Two sets of conclusions can be reached. The first set concerns the application of the methodology.

1 The distribution of the constituent compounds in the leaf wax can be used to distinguish similar taxa.

2 Information about the status of taxa can be obtained. In the present case the work indicates that some varieties may need to be recognised at a higher level and that others are not supportable.

3 Some specimens can be shown to be hybrids.

4 Indications about the parentage of hybrids can be obtained.

The second set of conclusions concerns the taxa that have been studied.

1 The varieties of *R. phaeochrysum*: var. *phaeochrysum* and var. *levistratum* should be considered as possible candidates for raising to subspecific or specific rank.

2 *R. phaeochrysum* var. *agglutinatum* may not be a single taxon, and the property of agglutinated indumentum should not be used in this case as a reliable taxonomic indicator.

3 Plants described as *R. alutaceum* var. *russotinctum* do not constitute a single taxon.

4 *R. alutaceum* var. *alutaceum* and var. *iodes* should be considered as being distinct at a higher level than that of variety.

5 Many long-established clones of *R. roxieanum* may be hybrids. New material from populations of single varieties should be used as the sources for further study. Varieties *roxieanum* and *cucullatum* are at present insufficiently represented to provide clear evidence about their status.

6 New material of *R. aganniphum* var. *aganniphum* and var. *flavorufum* is required to establish whether existing plants in cultivation represent the true taxa, or are hybrids.

7 The hybrid once known as *R. bathyphyllum* is not *R. proteoides × R. aganniphum*, but has a C_{27} taxon, probably *R. phaeochrysum* var. *levistratum*, as the second parent.

Acknowledgements

I thank Dr Anthony J. McAleese, Dr Brian A. Knights, Madeleine Chadwick, Stephen Peters and Fiona Sanderson for the experimental work. Plant material for study was provided by the Royal Botanic Garden Edinburgh, the Rhododendron Species Foundation, Peter A. Cox, Warren Berg and June Sinclair. Dr David F. Chamberlain initiated the project, identified specimens, and provided the specimens of hybrids, as well as much helpful information. Funding was given by the Royal Society, the Royal Society of Edinburgh and the American Rhododendron Society. Assistance was provided by the Academy of Sciences of the People's Republic of China.

References

CHADWICK, M. D., CHAMBERLAIN, D. F., KNIGHTS, B. A., MCALEESE, A. J., PETERS, S., RANKIN, D. W. H. & SANDERSON, F. (2000). Analysis of leaf waxes as a taxonomic guide to *Rhododendron* subsection *Taliensia*. *Ann. Bot.* 86: 371–384.

CHAMBERLAIN, D. F. (1982). A revision of *Rhododendron* II. Subgenus *Hymenanthes*. *Notes Roy. Bot. Gard. Edinburgh* 39: 209–486.

COX, P. A. & COX, K. N. E. (1997). *The Encyclopaedia of Rhododendron Species*. Glencarse, Perth: Glendoick Publishing.

EVANS, D., KANE, K. H., KNIGHTS, B. A. & MATH, V. B. (1980). Chemical taxonomy of the genus *Rhododendron*. In: LUTEYN, J. L. & O'BRIEN, M. E. (eds) *Contributions Toward a Classification of Rhododendron*, pp. 187–245. New York: The New York Botanical Garden.

The biology of cold tolerance in *Rhododendron*

Geoffrey R. Dixon[1], R. W. F. Cameron[2] & M. P. Biggs[3]

[1]*Department of Bioscience, Royal College, 204 George Street, Glasgow G1 1XQ (e-mail: geoffrey.dixon@strath.ac.uk),* [2]*Department of Horticulture & Landscape, Plant Science Laboratories, Whiteknights, PO Box 221, The University of Reading, Reading RG6 6AS, and* [3]*Genetic Research Instrumentation Ltd., Gene House, Dunmow Road, Felsted, Dunmow, Essex CM6 3LD, UK*

Acclimation and dormancy are essential processes by which woody perennials tolerate low and freezing temperatures. Global climate change may have significantly adverse effects on these processes. The genus *Rhododendron*, because of its size and the extensive environmental range that it colonises, offers a valuable model system whereby these processes may be investigated. Methods used in the investigation of acclimation and cold tolerance in *Rhododendron* are described together with some results illustrating the physiological and biochemical basis of acclimation. Acclimation may be quantified by measurements of cellular ion efflux, chlorophyll fluorescence and lipid content. Long-term field studies indicated the balanced cyclic nature of acclimation resulting from changes to the composition and architecture of plant lipids. The process of acclimation in the field could be disrupted by short periods of increasing autumnal temperatures. This is indicative of the potentially greater disruption that could be associated with gross climate change.

Introduction

Woody perennial plants evolved strategies to meet with the challenges of increasing cold in the autumn and winter and rising temperatures in spring. These contrasting thermal environments are dangerous to plants since cold stresses will cause substantial internal cellular damage in winter while in spring the resurgent growth may be killed completely. Plants respond to declining autumn temperatures through acclimation processes that develop tolerance to depths of cold that would even a few days previously have proved damaging. Acclimation increases through the autumn and early winter and dormancy begins with the onset of winter. During dormancy plants are well protected since short periods of conditions that might otherwise stimulate growth have no effect. Dormancy disappears towards the end of winter but plants may remain in an acclimated state and thereby protected from untoward short spells of increased temperature.

The generalised theory of the relationship between acclimation and dormancy hides a great deal of ignorance of the manner by which these processes commence, their mutual inter-relationships, and disappearance as growth re-commences. The need to understand these processes becomes crucially more important because woody plants are being exposed to gross and rapid shifts in seasonal conditions through the impact of world climate change, driven by man's impact on the environment.

Although climate change is generally described as 'global warming' there will be a critical impact on woody perennials resulting from effects on their abilities to acclimate, enter and shed dormancy and safely resume growth. Global climate change is likely to disrupt the carefully evolved physiological patterns of acclimation and dormancy that shield woody plants from low temperatures and crucially inhibit re-growth until the spring environment is safely warm enough for it to commence. Where acclimation and dormancy are disrupted and fail to provide protection, tissues could be killed before and during the re-commencement of bud growth. The repair of damage to early spring growth takes many months, and injured woody perennials may fail to reproduce as a consequence. If this damage is repeated over several seasons woody perennials will be severely weakened and eventually killed. It is crucial therefore to gain a detailed picture of the genetics, biochemistry and physiology of the processes by which tolerance to low temperatures has evolved in woody perennials. This knowledge may permit the prediction of the impact of global climate change as a cause of stress damage that could debilitate and possibly annihilate many of the ecologically important woody perennials.

The genus *Rhododendron* provides one of the largest groupings of woody plants, containing at least 850 species that have evolved to exploit an enormous geographical range. The alleged centre of diversity stretches from NW Yunnan through SE Tibet to western Szechwan; further east large numbers of species are found in Malaysia, Indonesia and New Guinea. The genus spreads north through eastern China into the Korean Peninsula, eastern Siberia and Japan, across the Bering Strait to western Alaska and into North America. Other species are found spreading westwards through Asia Minor into Europe and northwards into Scandinavia.

The contrasts of cold hardiness evolved in this genus are enormous. On the one hand is *R. lapponicum* (L.) Wahlenb., which grows in the circumpolar regions possibly having migrated at one stage as far as Canada and the northeastern USA, while on the other are the totally cold-tender East Asian tropical *Vireya* types. Within *Rhododendron* there is an unparalleled diversity of tolerance and susceptibility to cold and the possession of abilities to acclimate, acquire and shed dormancy. Consequently, *Rhododendron* provides an excellent model genus for scientific investigation of these processes.

Quantifying cold acclimation

The formation of ice within plant tissues and subsequent thawing result in massive symptoms of tissue disruption which can be initially assessed visually. Comparisons of the symptoms of cold stress between phenotypes have provided many of the vernacular hardiness classifications used for both herbaceous and woody types. Development of more subtle techniques permits the comparative quantification of states of acclimation and the cellular, subcellular and molecular processes which contribute towards frost tolerance (Dixon & Biggs, 1996a,b; Biggs & Dixon, 1998).

Plate 1 Frost injury to *Rhododendron* flowers.

Plate 2 Frost damage to *Rhododendron* foliage.

Measurement of conductivity

Relatively early in the study of acclimation it was recognised that during freezing the fluidity of cell membranes alters, particularly within parenchyma tissues of leaves and flowers, resulting in increased leakage of solutes. The excretion of solutes may be measured by determining changes to the conductivity of water into which frost damaged leaves are immersed (Cameron & Dixon, 1997, 2000). The efficacy of this method for tracing changes to the acclimation status of *Rhododendron* foliage is illustrated in Figure 1. Mature plants of the dwarf *R.* 'Hatsugiri' were grown in either controlled environments at either 5°C and short daylength (SD = 8 hours light) or 20°C and long daylength (LD = 16 hours light), or in unprotected field conditions. Foliage was removed at weekly intervals from each of the treatments and subjected to standardised freezing regimes, and the resultant efflux of

ions from the disrupted cells was compared. The application of 20°C and LD significantly accelerated the processes of de-acclimation, while those plants grown under natural field conditions only slowly became more susceptible to frost damage. Plants from the 5°C and SD treatment retained a state of deep acclimation for a month longer compared with either of the other treatments. The sudden diminution of acclimation in this treatment during April has been attributed, following further research, to innate circadian rhythms. These are present in certain genotypes and lead to reductions in acclimation irrespective of the environment applied to them.

Measurement of fluorescence emissions from chlorophyll b

The impact of low and freezing temperatures can be detected as changes to the integrity of the membranes that surround the thylakoids within chloroplasts. This damage

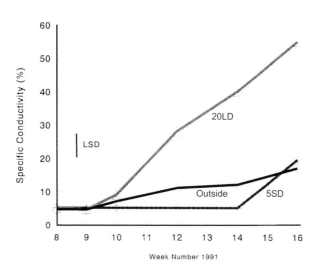

Figure 1 Freezing injury in acclimated and non-acclimated *Rhododendron* 'Hatsugiri' leaves. LSD = least significant difference, $P = 0.01$; 5SD = plants grown at 5°C, short days; 20LD = plants grown at 20°C, long days; outside = plants grown in the field.

may be quantified through alterations to the quanta of fluorescence emitted by Photosystem II (Percival & Dixon, 1997; Percival *et al.*, 1998). Quantification of such changes is achieved by exposing foliage to a period of darkness followed by spectrometric comparisons of fluorescence emissions. For simplicity damage is expressed as the ratio of variable fluorescence (F_v) to maximum fluorescence (F_m). Ratios of F_v/F_m below 0.75 indicate progressive damage to the thylakoid membranes and correlate with increasing stress within the chloroplast.

This technique has been applied to studies of environments that influence acclimation as illustrated in Figure 2. In the main experimental treatment plants of the dwarf *R.* 'Hatsugiri' were grown at either 20°C or 5°C with either SD or LD subtreatments. Samples of foliage were removed from each treatment and subtreatment. These samples

were exposed to controlled freezing at 0, –2, –4, –5, –6 or –8°C. Foliage grown at 5°C and SD possessed significantly greater tolerance to freezing conditions down to –60°C, whereas the chloroplasts from the other environmental treatments were substantially damaged below –40°C. Application of either SD or lower temperatures for plant growth (20°C and SD or 5°C and LD) permitted the retention of limited acclimation to –60°C as compared with the 20°C and LD treatment.

Measurement of fluorescence emissions from chlorophyll b permitted a detailed study of acclimation and de-acclimation in a range of *Rhododendron* species grown in the Younger Botanic Garden (a Specialist Garden of the Royal Botanic Garden Edinburgh), Benmore, Argyll, which contains the National Collection for the genus. Leaf samples were collected at monthly intervals from August 1993 to May 1995, subjected to stand-

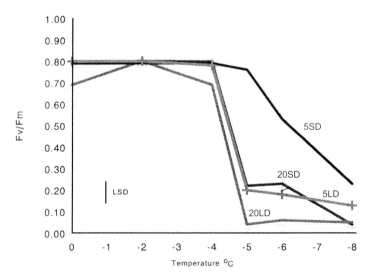

Figure 2 The influence of photoperiod and temperature on tolerance to freezing in *Rhododendron* 'Hatsugiri' leaves measured by fluorescence emissions. LSD = least significant difference, $P = 0.01$; F_v/F_m = ratio of variable fluorescence to maximum fluorescence; 20SD = plants grown at 20°C, short days; 20LD = plants grown at 20°C, long days; 5SD = plants grown at 5°C, short days; 5LD = plants grown at 5°C, long days.

ardised freezing conditions, the emission of fluorescence determined and the F_v/F_m values calculated.

Figure 3 illustrates the onset and decline of acclimation in *R. ponticum* L.: peaks of hardiness were attained in the December–January periods of 1993 and 1994, with an almost complete loss of hardiness in summer 1994. The symmetrical nature of the curves obtained highlighted the robust fitness of *R. ponticum*, and this undoubtedly contributes to its invasive capacities over large geographical zones.

Measurement of lipid composition and architecture

Alterations to the integrity of cell and thylakoid membranes during acclimation and de-acclimation as measured by conductivity and fluorescence suggest that concomitantly lipid composition may change. Indeed, such effects have been identified in limited analy-

ses of herbaceous plants (Kader & Mazliak, 1995). Extracting and quantifying lipids, especially those that may compose membranes in woody plants, is rendered more difficult because of the presence of complex phenolic substances resulting from secondary product metabolism.

Development of an analytical sequence involving the inhibition of lipases with *iso*-propanol and subsequent thin-layer and gas chromatography eventually provided repeatable and reliable results demonstrating changes to foliar lipids (Biggs & Dixon, 1995). Results obtained are illustrated in Figure 4, which for simplicity is restricted to one specific group of lipids, namely: 18:0 (stearic acid); 18:1 (oleic acid); 18:2 (linoleic acid); and 18:3 (linolenic acid) with 16:0 (palmitic) (the left hand part of the ratio refers to the number of carbon atoms in the molecule while the right hand value indicates the level of valency saturation or double and triple bonds).

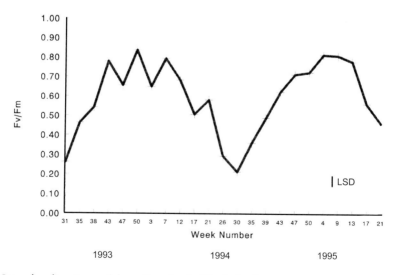

Figure 3 Natural acclimation and de-acclimation in *Rhododendron ponticum* over two seasons measured by fluorescence emissions. LSD = least significant difference, $P = 0.01$; F_v/F_m = ratio of variable fluorescence to maximum fluorescence. Plants grown at Younger Botanic Garden, Benmore, Argyll.

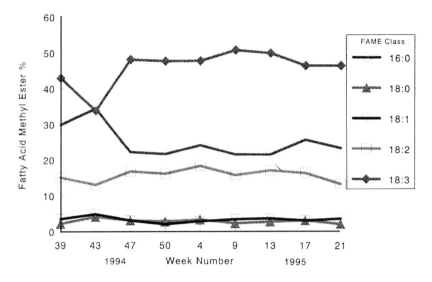

Figure 4 Fatty acid contents of *Rhododendron bureavii* leaves over one natural season. Each value is the mean from 20 replicates; each fatty acid was converted to its methyl ester (FAME) during analysis; 16:0 = palmitic acid; 18:0 = stearic acid; 18:1 = oleic acid; 18:2 = linoleic acid; 18:3 = linolenic acid. Plants grown at Younger Botanic Garden, Benmore, Argyll.

The relative proportions of these lipids present in foliage of *R. bureavii* Franch., grown at Younger Botanic Garden, were identified from September 1994 to May 1995. The period September to October 1994 was associated with a short spell of warm weather ('Indian summer') which resulted in de-acclimation of *R. bureavii* as identified by a significant reduction in the proportion of 18:3 (linolenic acid) and 18:2 (linoleic acid) and increase in 16:0 (palmitic acid). During de-acclimation it is thought that the cell and organelle membranes lose fluidity as a result of a reduction in the chain lengths and increased saturation of bonds within their lipid components. Such suggestions were supported by the evidence gained from the analyses of lipid components in the foliage of *R. bureavii*.

Discussion and conclusions

A comprehensive example of changing cellular, subcellular and biochemical processes elicited during the acquisition and loss of acclimation in a woody perennial, *Rhododendron*, has been achieved for the first time. This illustrates how plants are able to perceive environmental changes and then react to them. Such reactions can take place speedily, as seen with the rapid loss of acclimation during a period of milder weather or exposure to rapidly increased temperature and lengthening daylight. In this manner successful plants have the fitness to cope with increasingly stressful environments.

Continuing research utilises the genetic similarities not only between woody and herbaceous plants but also between plants and animals. In the latter, field relationships have been identified for the production of

Plate 3 Cold acclimation in *Rhododendron*.

Heat Shock and Cold Shock Proteins (HSP and CSP) and translated to studies of the mechanisms of cold tolerance in *Arabidopsis thaliana* (L.) Heynh. and rice (*Oryza sativa* L.). During both heat and cold stress, cells are damaged by the presence of active oxygen radicals; a protective response leads to the increased formation of enzymes such as glutathione reductase and ascorbate peroxidase (Fadzillah *et al.*, 1996). Studies are now focused upon identifying genetic triggers which result in increased production of such enzymes and their part in the acquisition of enhanced acclimation through the enhancement of phospholipid content in the membranes of cell organelles.

Understanding the processes of stress and acclimatory defence mechanisms in a woody perennial such as *Rhododendron* substantially improves estimates of the flexibility of natural ecosystems and the manner by which climatic perturbations may affect them. Industrially the research offers medium-term improvements in the manner by which new cold-tolerant cultivars could be bred, while in the shorter term it may prove feasible to develop diagnostic kits which will quantify the hardiness status of crops. This, coupled with effective weather prediction, would allow producers to make efficient use of the cryoprotectant materials that are now beginning to emerge onto the market (Graham *et al.*, 1997).

Acknowledgements

The research described here was funded by the Scottish Executive Environment and Rural Affairs Department (SEERAD), Edinburgh, and the Department for the Environment, Food and Rural Affairs (DEFRA), London. Helpful collaboration was enjoyed with staff at Younger Botanic Garden, Benmore, Argyll, Dr W. Christie, Scottish Crop Research Institute, Dundee, and staff in the late Biochemistry Department, SAC–Auchincruive.

References

BIGGS, M. P. & DIXON, G. R. (1995). Seasonal variation in the lipid content of *Rhododendron* leaves. *J. Exp. Bot.* 46: 66.

BIGGS, M. P. & DIXON, G. R. (1998). Temperature and light controlling cold acclimation of Rapid Cycling *Brassica* species. *Cruciferae Newsletter* 20: 89–90.

CAMERON, R. W. F. & DIXON, G. R. (1997). Air temperature, humidity and rooting volume affecting freezing injury to *Rhododendron* and other perennials. *J. Hort. Sci. Biotechnol.* 72: 553–562.

CAMERON, R. W. F. & DIXON, G. R. (2000). The influence of temperature, daylength and calendar date on cold tolerance in *Rhododendron*. *J. Hort. Sci. Biotechnol.* 75(4): 481–487.

DIXON, G. R. & BIGGS, M. P. (1996a). The impact of environmental change on acclimation and dormancy in *Rhododendron*. *Aspects Appl. Biol.* 45: 93–100.

DIXON, G. R. & BIGGS, M. P. (1996b). Measuring cold acclimation and reaction to climate change in perennial plants. In: *Analytical Techniques in Low Temperature Biology: Proceedings of the Annual Symposium of the Society for Low Temperature Biology at the University of Abertay Dundee*, p. 31.

FADZILLAH, N. M., GILL, V., FINCH, R. P. & BURDON, R. H. (1996). Chilling, oxidative stress and anti-oxidant responses in shoot cultures of rice. *Planta* 199: 552–556.

GRAHAM, L. A., LIOU, Y.-C., WALKER, V. K. & DAVIES, P. L. (1997). Hyperactive antifreeze protein from beetles. *Nature (London)* 388: 727–728.

KADER, J.-C. & MAZLIAK, P. (eds) (1995). *Plant Lipid Metabolism*. Dordrecht/Boston/London: Kluwer Academic Publishers, 588pp. ISBN 0-7923-3250-4.

PERCIVAL, G. C. & DIXON, G. R. (1997). Detection of salt and root deoxygenation stresses in *Alnus cordata* by measurement of leaf chlorophyll fluorescence. *J. Arboricult.* 23: 181–190.

PERCIVAL, G. C., BIGGS, M. P. & DIXON, G. R. (1998). The influence of sodium chloride and waterlogging stress on *Alnus cordata*. *J. Arboricult.* 24(1): 19–27.

The trials (and tribulations) of growing rhododendrons

Jim Gardiner

Curator, Royal Horticultural Society's Garden, Wisley, Woking, Surrey GY23 6QB, UK

During the early part of the 20th century the Rhododendron Horticultural Society, as it was affectionately known, was the meeting place where all the great growers and hybridisers of the day met. Floral Committee 'B' was a veritable *Who's Who* of the rhododendron world. During the 1930s many of the hybrids would have been sent to Wisley for trials from places like Exbury/Slocock/Waterers and Knaphill to be assessed for the Society's awards. With the arrival of Francis Hanger from Exbury, an active hybridisation programme took place from 1947 onwards. This paper, as well as profiling the value of the living collections within the garden context, highlights the value of the cultivated herbarium and botany department. The Rhododendron, Camellia and Magnolia Group of the RHS flourishes. With the publication of its Year Book a wide range of its activities are promoted, and local branches meet at regular intervals, bringing together a range of enthusiasts bonded together over their common interest.

In 1821, some 17 years after the Royal Horticultural Society (RHS) was founded, we leased 33 acres from the Duke of Devonshire on ground adjoining Chiswick House. The first RHS Garden was born and became a repository for plants introduced by the Society's collectors: David Douglas, John Reeves, Robert Fortune and Theodore Hartweg gathered from China and the Americas. (Many plants were also sent to our neighbours at Kew.)

During the middle of the 19th century the Society had a 'flirtation' with a Show Garden at Kensington, leaving the role of the experimental work, including plant trials, to be conducted at Chiswick. When the Society moved to Wisley in 1904, it immediately resumed this tradition of plant trials. Today we conduct over 60 plant trials each year, primarily on Wisley's Portsmouth Field.

Trials of seed-raised annuals, vegetables, perennials, glasshouse plants, bulbs, alpines and woody plants are assessed by 13 Standing and Joint Committees. These committees are composed of people who the Society recognises as having a huge practical and botanical knowledge of the plants they are being asked to judge. They may be from the horticultural industry at large or purely growing plants in their own gardens for a hobby. This 'intellectual ability' is of enormous value to us. As well as these committees being asked to judge plants during the period of interest, they also make recommendations on which plant trials we should be looking at in the future.

For instance, in the 1990s *Penstemon*, *Osteospermum* and *Diascia* were trialled and with the appropriate publicity were widely grown by the gardening public.

It is also important to draw together entries from as far afield as possible. Linda Jones and her colleagues in the Trials Office are responsible for scouring the plant and seed catalogues of the world for entries so that when planted they are seen as comprehensive and not just reflecting what is currently available in the UK. Forthcoming trials are advertised in our journal *The Garden* and on our website three years in advance of each trial to enable entries to be gathered together.

In the past there have been a number of awards to plants which, unless you were 'in the know', could be bewildering. Preliminary Commendation, Award of Merit and a First Class Certificate were available. However, in 1992 the Award of Garden Merit (AGM) was relaunched, producing at a stroke one award which the gardening public could readily identify with. The AGM is the plant's kitemark of quality, being awarded to plants that have the following attributes:

- Outstanding excellence for garden decoration or use.
- Availability in the trade.
- Good constitution requiring neither highly specialist growing conditions nor specialist care.

With regard to rhododendrons: these plants receive AGMs either after trial at Wisley or by committees deliberating at round table meetings. How did trials of rhododendrons arrive at Wisley? To understand this, it is worthwhile stepping back in time to 1915. When Charles Eley of East Bergholt in Suffolk visited his friend P.D. Williams of Lanarth in Cornwall in that year, it was suggested that they formed the Rhododendron Society, with Williams proposing Eley as the Honorary Secretary. It was formed with minimum organisation, with J.C. Williams (of Caerhays) as Chairman. The founding members were Williams, Eley and J.G. Millais of Compton Brow, Horsham. As well as notable garden owners from various corners of the British Isles being invited to join, Professor Isaac Bayley Balfour, W.J. Bean, Ernest Wilson and George Forrest were elected as honorary members.

The *Rhododendron Society Notes* were published from 1916. As it was mandatory for members to write, the notes were highly informative. The Society also organised the first Rhododendron Show in 1926, filling the RHS Hall at Vincent Square. The show was considered a huge success and was repeated the following year, but it did not attain the high standards of the previous year, due in part to a severe frost that taxed many of the exhibitors!

In 1927, under the Chairmanship of Lionel de Rothschild, the Rhododendron Association was formed, which enabled a larger number of interested people to take part in promoting rhododendrons. The Association published Year Books between 1929 and 1939 containing lists of species, expedition numbers, the stud book on hybrids as well as details of the establishment of plant trials of hardy hybrids introduced since 1918, to be conducted at Exbury.

Requests were sent to nurserymen to enter one plant of each for trial. A Joint Committee, composed of RHS and Rhododendron Association members, was empowered to inspect plants and make recommendations for awards.

These were planted in open ground adjacent to woodland to the north of Exbury House. During the first and second years, the Committee made several visits, with the first awards being recommended after their visit during the first week of June in 1933 –

Plate 1 *Rhododendron yakushimanum* (= *R. degronianum* ssp. *yakushimanum*) hybrid, Younger Botanic Garden, Benmore, Argyll.

1 First Class Certificate (FCC) and 14 Awards of Merit (AM) were proposed. The following year invitations were extended to Dutch nurseries.

In 1938 Mr de Rothschild proposed that the trial be relocated to Wisley, on the north-facing slope of Battleston Hill. Never having been cultivated in the strict sense before, this was no light task, with Scots pine and chestnut to be felled, old stumps excavated and the dense undergrowth of bracken and bramble cleared. Yet by 1940, the Committee met there for the first time, with six varieties being recommended for 'AM'. No further meetings took place until 1945, when deciduous and evergreen azaleas were subsequently trialled.

With the arrival of Francis Hanger as Curator at Wisley from Exbury, a new phase in rhododendron cultivation and introduc-

tion took place. Much of this attention centred around one plant, *Rhododendron yakushimanum* Nakai (= *R. degronianum* Carrière ssp. *yakushimanum* (Nakai) Kitam.) (Plate 1). This was first introduced by Lionel de Rothschild to Exbury in 1934; Hanger brought one of the two plants with him to Wisley. This was subsequently awarded an FCC when exhibited at Chelsea in 1947. Hanger described how it was planted on Battleston in 1945 and had outgrown the Exbury plant due to a 'liberal supply of spent hops and water'. When it was seen at the Chelsea Flower Show, Hanger reported that 'its white flowers commanded appreciation; however, if it had been seen a week earlier, it would have been more beautiful as the buds of the upstanding compact trusses were rich pink, fading to a pale pink as they developed to be finally pure white when opened'. The FCC clone is known as 'Koichiro Wada' after the source of this and many other plants from Wada's Hakoneya Nurseries in Japan.

From 1947 to 1968 Hanger hybridised both rhododendron and azaleas. The azaleas he introduced are generally known as the Wisley Rivers – 'Deveron', 'Derwent', 'Tay' and 'Dart', etc., 32 in all – and amongst the rhododendrons 'Tequila Sunrise' and 'Renoir' have been introduced.

Hanger was one of the first to recognise the value of *R. yakushimanum* as a parent. Its offspring retain a good compact habit, often maintaining the good foliage qualities of the parent. With hybridisation the range of flower colour is widened as well as improving tolerance to both cold hardiness and sunshine levels.

Breeding has continued apace. Windsor in the 1940s introduced 'Seven Stars', while Waterers began their programme, initiated by Gerald Pinkney and Percy Wiseman, in the late 1940s with the introduction of 'Golden Torch' and 'Pink Cherub', and latterly with their 'Seven Dwarf' series in the 1960s.

In the 1950s Arthur George of Hydon Nurseries introduced many, including 'Hydon Dawn' and 'Hydon Hunter'.

Harkwood Acres and Peter Cox in this country, Ken Janeck, David Leach in the USA and Hans Hachmann in northern Germany have all contributed significantly. Hachmann especially is producing excellent hybrids that are both interesting floristically and also for their attractive foliage. 'Edelweiss', 'Hachmann's Marlis' and 'Babette' are three examples.

Tired of wearing out Battleston Hill by endlessly trudging up and down like the Grand Old Duke of York, in the 1990s the late John Bond proposed the establishment of a new Trials Committee, the Woody

Plant Trials Sub-Committee, to take a fresh look at trialling. In 1996 a trial which contained about 120 cultivars of *Rhododendron yakushimanum* hybrids was planted in an attempt to kick start the interest among the gardening public. This was to:

- Assess and compare old and new cultivars and nominate outstanding performers for the AGM.
- Assess the performance between differently propagated plants of the same cultivar, whether micropropagated, grafted or from cuttings.
- Make a permanent record through herbarium specimens, photographs and written descriptions for the RHS Horticultural Herbarium at Wisley.

With the range of experts drawn together a thorough assessment was made and a 10-year trial was proposed.

Already results show distinctions in how micro-propped vs. grafted vs. cutting raised plants differ in their performance. For example, with 'Hachmann's Marlis' the grafted plants were flowering earlier than plants raised from cuttings, while with 'Babette' plants from cuttings were more floriferous.

The judging notes are valuable, not only for the vast reservoir of flowering data, growth responses, etc., but also for the judges' poetic prose – 'good doer, flowers, nice shape, dies prettily; good head of flower; not a pretty flower, sugar plum fairy in boots; flower colour a bit bright, would walk out of any good garden centre'.

All specimens are being lodged in the Horticultural Herbarium. The origins of this are traced to the early part of the 20th century, when the garden moved from Chiswick to Wisley. However, for the past 30 years, pressed specimens from awarded plants have been preserved. One of the aims is to create standard portfolios of all new cultivars. Currently it contains about 90,000 specimens, 25,000 colour images and 3500 paintings. Standard specimens are an invaluable part of the botanist's work and in this context some 500 rhododendrons are held here.

The living collection at Wisley today

The emphasis now is to display a balanced collection primarily of rhododendron hybrids. However, there are also small specialist collections of recent and reintroduced species. We can thank the Rhododendron, Camellia and Magnolia Group for sponsoring a collection of such species, which has plants collected by Cox and Hutchinson, Rushforth, McBeath, Millais and others. The Group also sponsored a collection of *Rustica pleno* rhododendrons which were planted in memory of Alan Hardy.

The entire plant collection at Wisley numbers more than 25,000 taxa, of which 1700 are *Rhododendron*. This enables us to display a limited collection of tender species and hybrids under glass, dwarfs on the Rock Garden, as well as mainstream and complementary and structure plantings in association with other woody, herbaceous and bulbous plants in the Wild Garden and on Battleston Hill.

Many changes have been seen over the past 10 years, following the severe storms of 1987 and 1990.

The Society will continue actively to promote the genus through its plant trials and its advisory work. This will include looking at the establishment of rhododendrons on Inkarho rootstocks to examine whether they are able to grow in pots and on alkaline soils; the establishment of demonstrations in its gardens, and the promotion of these in the literature; and, as part of its Bicentenary Plant Collection in 2004, *Rhododendron yakushimanum* 'Wisley Blush'. Despite the trials and tribulations of *Rhododendron* as a genus, the Society still recognises over 250 species and hybrids as having Awards of Garden Merit.

Commercial production of *Vireya* rhododendrons

Chris Fairweather

Beacon Gate, Beaulieu, Hants SO42 7YR, UK

The commercial collection of over 100 *Vireya* hybrids is described together with the origin of the collection. Routine procedures are described covering propagation, potting, compost, watering and feeding, together with observations on temperature and light requirements. The summer regime, which is aimed at producing flower buds and pruning to promote busy growth, is also described. The results of trials with the growth retardant 'Bonzi' are described. Problems with pests and diseases are mentioned together with the horticultural potential for the gardening public in the UK of this exciting group of plants.

My interest in *Vireya* rhododendrons started around 30 years ago when I was working for Exbury Gardens. On a summer visit to the Strybing Arboretum in San Francisco I was amazed to see at that time of year a bed of rhododendrons in full flower. This was my first encounter with the fascinating section *Vireya*. The curator kindly gave me a few cuttings, which duly rooted, and the collection started. Over the past 10 years many more rhododendrons have been added to the collection. Currently we have more than 100 hybrids together with a few species (see Table 1). About five years ago this became the official collection of *Vireya* hybrids under the direction of the National Council for the Conservation of Plants and Gardens (NCCPG). The nursery is run commercially and we have been interested to see if any of the many *Vireya* hybrids have a profitable future given the problems of the British climate.

Some of the growing techniques we employ at present will be demonstrated, with comments on our findings to date.

Propagation

All cuttings are taken from young plants in late summer and inserted into plug trays, which are placed on a heated propagation bed. Rooting percentages are generally high. This section of the *Rhododendron* family is easy to root. The apical bud is removed from cuttings to encourage branching.

Potting

Young plants are potted into 9 cm pots in early spring and placed in a frost-free environment. As temperatures rise and light levels improve, these grow away fast. With the apical bud having been removed, we hope to start with a bushy plant. We have been doing some trials with growth retardant to try to obtain this. Results to date have been encouraging.

Compost

Currently we use a mixture of coarse bark and coarse moss peat. The exact mixture is

Table 1 The hybrids and species in the collection.

Named hybrids			
Alexander	Golden Gate	Monte Christo	Sunset Fantasy
Alisa Nicole	Golden Girl	Moonwood	Taylori
Angelica	Gossamer White	Ne Plus Ultra	Terebintha
Aravir	Great Sensation	Nuigini Firebird	Thai Gold
Arthur's Choice	Groucho	Orange Queen	Toff
Big Softie	Gwenevere	Our Marcia	Triple
Blush Tumble	Halo	Penny Whistle	Triumphans
Bob's Crowning	Happy Wanderer	Pink Delight	Tropic Fanfare
Bold Janus	Hari's Choice	Pink Ray	Tropic Summer
Burgundy Surprise	Highland Arabes	Pink Thimble	US149
Calivar	Highland Jade White	Pleasant Company	Vladimir Bukovsky
Cameo Spice	Honey Star	Plum Pudding	V21
Captain Scarlet	Hooloo	Popcorn	V104
Caramia	Iced Primrose	Princess Alexander	V119
Carillon Bells	Inferno	Raspberry Cordial	V136 phaeopeplum ×
CC Lobster	Java Light	Razzmatazz	lochiae × leucogigas
Charming Valentino	Jean Baptiste	Red Mountain	Watermelon Dream
Chayya	Jimini Cricket	Red Rover	Wattlebird
Cherry Liquor	John Silver	Robert Bates	Will Silent
Christopher John	John West	Rob's Favourite	Yellow Ball
Clair Paravel	Just Peachy	Rosy Posy	110
Coral Flare	Just Pinkie	saxifragoides X	167
Coral Seas	Kiandra	Saxon Bonnie Bell	
Cordial Orange	Kisses	Saxon Glow/Blush	**Species**
Craig Farragher	Lady Di	Shantung Pink	acrophilum
Cream Serene	Lemon Light	Shantung Rose	carringtoniae
Cream Supreme	Lemon Lovely	Shepherd's Sunset	christi V103
Cyprian	Leonara Francis	Shogun	goodenoughii
Dr Sleumer	Lipstick	Show Stopper	jasminiflorum
Dresden Doll	Little Kisses	Silken Shimmer	konori
Fireplum	Little One	Silver Star	loranthiflorum
First Light	Little Pinkie	Simbu Sunset	macgregoriae
Flamenco Dancer	Littlest Angel	Solar Flare	praetervisum
Gardenia Odyssey	Lochiae Double	Souv de Mangels	rarilepidotum
George Budgen	Lochmin	St Valentine	saxifragoides V54 X
Gilded Sunrise	Magic Flute	stenophyllum × christi	tuba
Golden Charm	Magic Flute/JK	Strawberry Parfait	viriosum (lochiae)
	Miss Muffet	Sunny Splendour	zoelleri

Plate 1 *Vireya* rhododendron 'Java Light'.

Plate 2 *Vireya* rhododendron 'Popcorn'.

Plate 3 *Vireya* rhododendron 'Just Peachy'.

Plate 4 *Vireya* rhododendron 'Shantung Rose'.

50% coarse grade peat, 30% medium grade peat, 20% cambark potting, 3 kg per m³ Multicote fertiliser, 15 g iron per m³, plus Intercept to control vine weevil. The pH of this mix is around 4.5 to 5.0.

Feeding

With the addition of nine-month slow release fertiliser, no further feeding is required during the growing season. When the plants are checked in early spring they are given a very light dressing of sulphate of ammonia which improves leaf colour.

Temperature

During the winter months the temperature is kept between 8°C and 10°C as far as possible. The majority of vireyas tolerate this. A few develop red patches on the leaves, possibly indicating low temperature stress. When light and temperatures rise they all recover.

When the danger of frost is past the vireyas are put outside. These go onto drip irrigation, on a site that gets some shade and some sun during the day. Generally plants thrive in these conditions and appear to produce the all-important flower buds.

Light

The glasshouse containing all the young plants is given supplementary light during the winter months. The plants given this 12-hour day definitely benefit. A fan has been installed in the main house, which runs continually and ensures good air movement through the winter months, as like orchids vireyas do not like stagnant air conditions.

Watering

During winter all plants are kept quite dry and watered every 7 to 10 days as required. In late spring and summer watering increases to every two to three days, watering as the plants start to dry out. Waterlogged compost must be avoided at all costs as this causes fungal root rots.

Pruning

This is a vital operation in the production of bushy and well-budded plants. The apical bud is removed from the initial cutting. One-year liners are pruned back in spring. Larger plants potted into 2 or 3-litre pots are also pruned to encourage bushy growth. Last year we carried out some trials with the growth retardant 'Bonzi', and early results look encouraging.

Diseases

Greenfly are a minor problem that is easy to control. Mealy bug is the most troublesome insect pest. The retail aerosol spray 'Provado' gives the best control for this pest. Otherwise, apart from occasional mildews (*Phytophthora* sp.), we have few problems.

In selected hybrids we can produce attractive budded bushy plants. These should not be treated as houseplants. They are generally happy in a conservatory except for in the very high temperatures that can arise in the summer. They need to be kept frost-free over late autumn, winter and early spring, and then stood outside in an area where they get a mixture of some sun and shade during the day.

Currently the gardening public in the UK is mostly unaware of this exciting section of the *Rhododendron* family. While we are making every effort to promote them, it is an uphill task. We have a selection of vireyas planted at the Eden Project in Cornwall and have also had a successful display at the Hampton Court Flower Show in the NCCPG tent. The question is: will they become anything but plants for the true enthusiast? The answer to this is still not clear.

Species patterns in *Rhododendron* section *Vireya* from sea level to the snowline in New Guinea

George Argent

Royal Botanic Garden Edinburgh, 20A Inverleith Row, Edinburgh EH3 5LR, UK

Botanists and horticulturists have always argued about species: what they mean, where the limits lie and how one defines them. Vireya rhododendrons provide an interesting study in how the species have been historically defined and from rather limited studies how they are presently represented. The morphological species concept demands that there are discontinuities in correlated characters between species, but the degree of difference is very subjective and has led to the general terms 'splitters' and 'lumpers' to describe the extremes of those that describe species. The examples given here describe why one may be considered a splitter in one instance and a lumper in another and the rationale behind making apparently very different decisions in different cases.

In early post-Linnaean concepts where species were assumed to have been directly created and thus must be different, the morphology alone was examined and the basis of our classification was created on overall similarity. Great advances have been made since then in the understanding of the biological nature of species. This has meant that in many cases with additional information we can move from the morphological phase to a more advanced state. This may include genetic, often cytological, information, something of the general distribution and ecology, and the plasticity under artificial conditions. Study of the distribution and biology of vireyas provides an interesting exercise in theoretical species concepts but it is still in a very primitive state, with many species still known only from one or a very few collections and great areas remaining poorly or not explored.

Genetic material gets distributed mainly at two stages in the life of a plant, as pollen and as seed. In vireyas pollen is precisely transported by animals while the seed is imprecisely wind distributed. Both these methods of gene dispersal at the different stages of the life cycle have implications for the success and speciation of the group. Patterns of species variation that occur are: 1, widespread and relatively uniform; 2, widespread with considerable local variation; 3, widespread with reticulate variation (ochlospecies); 4, widespread and uniform; 5, restricted, showing evidence of recent evolution; 6, restricted but showing evidence of being in decline and relict. Examples of the different patterns are given, together with hypotheses on the reasons for the different situations.

Introduction

We are all taxonomists, we need to identify things around us and group them in ways which are useful, and it is probably largely due to this inherent universal trait that the way things are classified causes so much argument. Everybody has an opinion! Botanists and horticulturists seem inevitably at odds whenever a horticulturally important group is under study. The botanist is primarily interested in plants in the wild (or at least collected from the wild) while the horticulturist is more interested in how they perform in gardens. The modern botanist always has at the back of his mind an evolutionary entity, a successful biological breeding system manifested in morphological characters which have largely continuous variation within the species but some clear discontinuities between species. The horticulturist is often more finely tuned, recognising quite subtle differences in the performance of plants as seen growing in gardens. The range of cultivated wild plants is almost always much more limited than those in the wild. Often a species of *Rhododendron* may have been propagated from only a very few original collections (see Cullen, 2003; Lear & Martin, 2003) each from a different part of the range of the wild species and often then clearly distinct as garden plants. A common problem arises when these different entities, sometimes collected in different countries, are initially given different names by botanists, but then in the light of further intermediate collections (many of which do not get cultivated) later botanists want to amalgamate them. Horticulturists understandably cling to the well-known names for plants they still recognise as different, and object to the 'interference and inconvenience' of name changes.

The earliest *Vireya* species go back to the latter half of the 19th century when 'special creation' meant that species differences were thought to be absolute and depended solely on gross morphological attributes. Since that time several new disciplines have successively influenced species concepts. Cytology, genetics and the acceptance of evolution have had a profound effect on the way botanists regard species. The latest molecular studies are just beginning to impact on species concepts in vireyas (see Brown, 2003). They may give some apparently absolute measurements of difference (or similarity, depending on one's vantage point), although it has to be borne in mind that these studies look at only a very small number of genes compared with morphological characters which express the attributes of thousands. In fact botanists have not always been very ready to accept that the present-day products of evolution present a continuum of difference and that species do not fit neatly into equivalent boxes. Hilliard & Burtt (1971) made the perspicacious comment: ' … any group of adequate size will contain species of many different kinds …'. The vireyas illustrate this in exemplary fashion.

With international travel now so cheap and easy, many more botanists are seeing plants in the wild, so that species can be seen in their ecological context and field variation can be studied at first hand far more extensively than in the past. The cultivation of large collections of vireyas has also played a part in our understanding of species concepts and relationships. John Rouse showed clearly the fact that almost all vireyas can be crossed with each other with two main exceptions (Rouse & Williams, 1989). If style lengths are too different, cross-pollination fails. Pollen from short-styled species will not reach the ovules of long-styled species to fertilise them and conversely if pollen from long-styled species is used on very short-styled plants the pollen tubes grow on beyond the ovary and get 'lost'. The other exception is the small group of mostly mainland pseudovireyas such as *R. kawakamii* Hayata var. *flaviflorum* Liu & Chuang and *R. santapaui* Sastry et al., which form an isolated group that appears to be incompatible with all others of section *Vireya* (although apparently compatible amongst themselves), and probably they should not be included within this section. The fact that the vast majority of vireyas can hybridise with one another has important implications for the success of the group and the present distributional pattern. In some respects this vast bulk of *Vireya* species as named might be considered one enormous coenospecies in the concept of Turesson (Turesson, 1922). They have been selected as ecotypes, finely tuned to their present situations but in many cases weakly linked to and exchanging genes with other *Vireya* species. They have the potential through these hybridisations to exchange genes throughout much of their range, maintain high levels of heterogeneity and rapidly exploit changing circumstances. They reach reproductive maturity relatively quickly – as little as 15 months for *R. rarilepidotum* J.J.Sm. in cultivation (Rouse, personal communication) – and can produce prodigious quantities of seed over a considerable number of years. The frequent occurrence of hybrids and hybrid seed (Argent *et al.*, 1988; Binney, 2003) is the proof of this ability to rapidly exploit a changing situation.

Botanists have never followed the concept of potential breeding incompatibility as fully as zoologists do to define species. Vireyas provide many good examples of species which are potentially compatible but nevertheless show considerable fealty while growing close to or even intermixed with other *Vireya* species. Flower size differences have already been mentioned as one isolating factor; different pollinators and different flowering times and patterns are two others that are undoubtedly important. Pollen is one of the stages at which genetic material is distributed, and the degree of 'loyalty' a pollinator shows must be very important in maintaining *Vireya* species uniformity or the degree to which hybrid pollinations are made. Under conditions of habitat disturbance the balance can quickly shift to a predominance of hybrids whether this disturbance is natural or manmade. Hybrid swarms do occur and have been observed between *R. macgregoriae* F.Muell. and *R. zoelleri* Warb. in Papua New Guinea (personal observation) and between *R. buxifolium* Low ex Hook.f. and *R. rugosum* Low ex Hook.f. (*R. × coriifolium* (Sleumer) Argent) on Kinabalu in Sabah (Argent *et al.*, 1988). The first was due to wholesale destruction of the native forest habitat in the vicinity of the two parents. The second is a result of the natural open granitic slopes of a geologically young and actively growing mountain, giving rise to a vacant intermediate habitat between those of the parent species.

Seed is the other stage at which genetic material gets distributed, and the small, wind-borne seed might be thought to be excellent in distributing vireyas. However, very few of the species presently designated have wide distributions, and very few of these distributions show wide disjunctions. It might be expected that wide or disjunct distributions would be likely in plants growing mostly in the tropical belt of violent thermals which must quite frequently cast the tiny seed many thousands of feet into the air. *Vireya* seed will maintain dormancy if stored dry for months and often years (Moyles, personal communication and Galloway, unpublished experiments at the Royal Botanic Garden Edinburgh). It would appear that the moist conditions in clouds are not so favourable to the maintenance of viability. It also seems apparent that the long tails that give seed buoyancy are not designed for long-distance dispersal but to aid local dispersal in the relatively still forest conditions that many vireyas live in. Thus forest species such as *R. lowii* Hook.f. have the longest tails while those from exposed terrestrial situations have short tails as in *R. retusum* (Blume) Benn. or virtually no tails at all as in *R. abietifolium* Sleumer (Argent *et al.*, 1988).

The different kinds of species

1. Widespread and relatively uniform

The first *Vireya* to be described was *R. malayanum* Jack in 1822. Since Jack's discovery of this species in Sumatra it has been recorded from Thailand, Peninsular Malaysia, Borneo, Sulawesi and Seram (Sleumer, 1966). It shows considerable ecological variation from lowland heath forest habitats to exposed montane situations. These are correlated with rather 'weak' morphological variation such as flower colour, leaf size and position of the inflorescence, but it has not differentiated to the extent of 'breaking up' into clear species units to any great extent. Sleumer (1966) recognised five forms and segregated *R. micromalayanum* Sleumer which was based on small leaf size, although there is some suggestion that it differs in leaf anatomy from *R. malayanum* (unpublished student observations). An exception to this apparent lack of differentiation is on Kinabalu where it is

replaced by two related but clearly distinct species, *R. acuminatum* Hook.f. and *R. lamrialianum* Argent & Barkman. Records of *R. malayanum* itself from that mountain have not been substantiated (Argent *et al.*, 1988).

Another *Vireya* taxon which, arguably, has a wide distribution is the *R. brookeanum* Low ex Lindl./*javanicum* (Blume) Benn. complex in the western part of Malesia (Plate 1). Here we have *R. javanicum* from Java, well known in cultivation for more than a century, characterised by relatively short orange flowers, a glabrous, or at least not hairy, ovary, the leaves in weak pseudo-whorls and flower buds with the bract points (perulae) standing out away from the buds. Other forms included within this species by Sleumer (1966) were similar plants from Sumatra which, however, are much more variable in flower colour, and generally have spirally arranged leaves and more regularly appressed bracts. Philippine plants originally described as *R. schadenbergii* Warb. were also reduced to a variety of *R. javanicum* (Sleumer, 1966). These have quite distinctly differently coloured flowers but similar spirally arranged leaves and appressed bracts, and in the herbarium are hard to separate. *Rhododendron brookeanum* from Borneo was kept separate until examined during the fieldwork in Sabah. Here it was discovered that the main distinction used by Sleumer, the hairy rather than glabrous ovary, broke down and all variations between densely hairy ovaries, partially hairy ovaries and non-hairy ovaries could be observed. In fact this character had already been recorded as variable in Sumatra with the designation of *R. javanicum* var. *teysmannii* (Miq.) King & Gamble. The fact that these plants are relatively lowland rainforest epiphytes means that the only absolute isolation is between

Plate 1 Different forms of the *Rhododendron brookeanum/javanicum* complex all from Sabah, E Malaysia: (a) Nabawan form; (b) Kinabalu form; (c) Alab form.

the different landmasses. We cannot sample plants on a continuous basis but there is the potential for a continuous distribution throughout the rainforest areas. Thus although we have some very distinct forms in cultivation it is probably best to regard *R. javanicum* (the oldest name) to include *R. brookeanum* within one widespread and very variable species. The distribution might also support this, as the distribution of *R. javanicum* as defined in Sleumer (1966) very nearly encircles that of *R. brookeanum*. There is still a great deal of work to be done understanding this variation pattern and further work is needed to marry the demands for names for recognisably different entities in cultivation with an acceptable academic solution. The Philippine forms in particular may yet warrant specific recognition. Another solution might be to recognise *R. javanicum* as a distinct species but only including plants from Java, and allow *R. brookeanum* to have a wider distribution from Borneo to Sumatra at least (a not uncommon distribution in other plant species).

Other widespread species, which on the whole cause fewer problems, are *R. zoelleri* and *R. macgregoriae*. Both of these are relatively lowland, normally forest-growing, species which however grow well on the grassland created after forest clearance by man. *Rhododendron zoelleri*, which occurs throughout the length of the island of New Guinea, is remarkably uniform, although it does vary in flower size and anther length. It is one of only two species (out of 167) which occurs beyond the immediate confines of New Guinea as it is also recorded from the island of Maluku. *Rhododendron macgregoriae* is more variable, occurring in a wide range of colour forms from yellow through orange to pink and red. We do not know at present to what extent this variation is due to intermittent hybridisation but it seems very likely, although flower shape and the general species 'facies' remain remarkably constant. Several putative wild hybrids of *R. macgregoriae* have been collected (e.g. Ashburn White collected by Graham Snell) and *R. macgregoriae* has been the parent of many recent horticultural hybrids. Also of very widespread occurrence along the length of the main mountain chain of New Guinea is *R. culminicolum* F.Muell., although the variation in this 'species' has still to be more carefully studied. *Rhododendron culminicolum* var. *nubicola* (Wernham) Sleumer appears

to be a good species as originally described. It is quite distinct in both morphology and ecology (personal observation). Reports of *R. culminicolum* from New Ireland have proved unfounded (Argent, in press).

2. Widespread with considerable local variation

Rhododendron jasminiflorum Hook. sensu Sleumer (1966) occurs in Peninsular Malaysia as var. *punctatum* Ridl., in Sumatra as var. *heusseri* (J.J.Sm.) Sleumer, in Borneo as var. *oblongifolium* Sleumer and in the Philippines as var. *copelandii* (Merr.) Sleumer. This is a species in subsection *Solanovireya* H.F.Copel. with a long white hypocrateriform corolla which has differentiated significantly on each of the landmasses. *Rhododendron jasminiflorum* var. *heusseri* is probably the most distinct and should probably be recognised as a species in its own right. It has very distinctive indumentum on the stems, characteristic leaf shape and inflorescences which usually have only 2 or 3 flowers (although the type specimen in the Natural History Museum has 7). This 'variety' was first described as a form of *R. longiflorum* Lindl. and the fact that hybrids between *R. jasminiflorum* and *R. longiflorum* have been recorded from Sumatra has added to the confusion. Field observations in the north of Sumatra have shown this to be a uniform and easily recognisable taxon, growing normally as a high epiphyte in tall submontane forest. During large-scale disturbance caused by forest clearance it has hybridised with *R. longiflorum* on occasion, although two recent expeditions have failed to find any evidence of these hybrids persisting at the reported locality around Lake Toba. The remaining 'varieties' are still usually readily recognisable and because of the distinct geographical locations separated by sea they should at least be recognised at subspecific level.

3. Widespread with reticulate variation (ochlospecies)

The concept of the ochlospecies was developed by White (1962) as a result of his studies on the Ebenaceae, and a general definition was stated (White, 1998) as: 'A very variable (polymorphic) species, whose variation, though partly correlated with ecology and geography, is of such a complex pattern that it cannot be satisfactorily accommodated within a formal classification'. Cronk (1998) further elaborated 10 traits of which a species must exhibit at least six to qualify as an ochlospecies. *Rhododendron quadrasianum* Vidal would appear to qualify. It is certainly complex in its pattern of variation. Sleumer (1966) quoted Copeland as saying '… a special form of *R. quadrasianum* occurs on almost every mountain in the Philippines' and goes on to say 'I had to reduce his infraspecific taxa to a more reasonable number in order to avoid overlapping'. Sleumer clearly found the variation pattern difficult to deal with when writing the *Flora Malesiana* account. He included similar forms from Borneo and Sulawesi within this complex, and others of great similarity occur in New Guinea. The *R. quadrasianum* complex are plants with small leaves and small, red, orange or yellow flowers, mostly bird pollinated and with a tendency to flower continuously. In the detailed observations that were possible during the fieldwork undertaken for the *Rhododendrons of Sabah* (Argent *et al.*, 1988), it was observed that the members of this complex in Borneo as recognised by Sle-

umer behaved as at least three distinct species with additional local variation similar to that observed by Copeland for the Philippines (Copeland, 1929) (Plate 2). Since the object of this work was a local flora these were all recognised and can be separated with confidence at the individual island level. However, if one takes a broader geographical view it quickly becomes difficult to construct conventional taxonomic keys to identify all the different forms from different mountains. Of Cronk's (1998) 'strong traits' it can be argued that it conforms to all four. Of the 'weak traits' (Cronk, 1998), it partially fulfils the first, being geographically widespread although not particularly ecologically widespread. It certainly fulfils the

Plate 2 Forms of the *Rhododendron quadrasianum* complex from different mountains in the Philippines.

second as instanced by the detailed studies in Sabah (Argent *et al.*, 1988). Regarding the third weak trait (no. 7, Cronk, 1998) of having distinct and monotypic satellite species, there are good examples of these in *R. meliphragidum* J.J.Sm. in Seram (a geographical outlier) and *R. ericoides* Low ex Hook.f. (an ecological outlier). Of the fourth weak trait (no. 8) a good example would be that of *R. ericoides* on Kinabalu with *R. quadrasianum* var. *angustissimum* Sleumer on Mt. Mulu appearing very similar, and the two having been confused in the past. *Rhododendron* is certainly a large genus and *R. quadrasianum*, while not having a long synonymy compared with many other species, is complex by *Vireya* standards. Thus *R. quadrasianum* qualifies as an ochlospecies when subjected to the tests as set out by Cronk (1998).

4. Widespread and uniform

In complete contrast to *R. quadrasianum* is *R. bagobonum* H.F.Copel., unusual in that it has a wide distribution from the southern Philippines, from where it was first described, to Borneo (Argent *et al.*, 1988), Sulawesi (Keith Adams collections at Pukeiti), and Seram (personal observation). It shows little variation apart from small differences in leaf shape and corolla length. It has often been confused in the past with *R. quadrasianum sensu lato* but it is very distinct, having totally different scales on the leaves and a different style/ovary ratio. More important from a biological point of view is the fact that the stamen arrangement is also totally different. In *R. quadrasianum* the stamens are grouped on the lower side of the mouth of the flower which with the protandrous condition (of all vireyas) means cross-pollination regularly occurs. In the case of *R. bagobonum* the stamens form a radial pattern obstructing

the whole corolla mouth. In the event of a non-visit by a pollinator or even one that is only partially successful in removing pollen, self-pollination occurs, as the stigma – when it becomes receptive – grows up into the close-knit pattern of anthers between which the pollen mass is strung. There is good evidence that *R. bagobonum* donates genes as the male parent in such natural hybrids as *R.* × *planecostatum* (Sleumer) Argent but routinely self-pollinates and behaves as an apomict. Proof of this is the setting of 'good' fruit under all but the most unfavourable of growing conditions in cultivation, when pollinators are often totally lacking. This very unusual behaviour of the flowers appears to have allowed this species to adopt the sort of distribution that might be expected from tiny seeds carried in thermals whilst retaining its morphological identity in a very conservative fashion.

5. Restricted, showing evidence of recent evolution

Rhododendron buxifolium Low ex Hook.f. is one of Kinabalu's true endemics. It was placed by Sleumer (1966) in subsect. *Euvireya* H.F.Copel. but transferred to subsect. *Pseudovireya* (Clarke) Sleumer (Argent *et al.*, 1988) on the grounds that the scales were more typical of *Pseudovireya* scales and the fact that, although without a true fringe of hairs on the bracts, this species does have a small area that is fringed near the apices of these organs. It is an abundant plant on the granite dome of Mt. Kinabalu and is unlikely to be found elsewhere due to the fact that it does not grow below 3100 m. It is certainly an anomalous plant, having most unusual flower shape for species in section *Pseudovireya*, especially from Borneo. It appears to have arisen relatively recently (since Kinabalu is a very young mountain), probably by hybridisation between a *Pseudovireya* and a *Euvireya*.

Rhododendron lamrialianum Argent & Barkman (2000) is a new name for a well-collected plant from Kinabalu. It is similar to *R. buxifolium* in not having a complete fringe of hairs around the bracts but is otherwise fairly typical of subsect. *Malayovireya* Sleumer although with less dense scales on the leaves than is usual in that subsection. This lack of a complete fringe of hairs and the somewhat intermediate indumentum suggest it may be of hybrid origin, but it now appears to be a good species with different subspecies on Mt. Kinabalu and Mt. Trus Madi, the only two mountains from which it is known.

6. Restricted but showing evidence of being in decline and relict

The first of these species, *R. chamaepitys* Sleumer, is restricted to a very small area at the summit of Mt. Lambir in Sarawak, one of the old and declining mountains characteristic of the main range in Borneo. It is typical of subsection *Solenovireya* with its hypocrateriform corollas, and is apparently closely related to *R. jasminiflorum* – it could easily be taken as yet another subspecific taxon in that group. It does however have very distinctive foliage. It appears 'trapped' on this island of submontane shrubbery, and in decline as it is isolated from both other species and other suitable habitats. It might be thought to be an example of an exception to the generalisation made in the early part of this paper that *Vireya* species have an aggressive potentiality for change through low-level gene exchange with other species. However, it could be argued that this isolation from hybridisation

in a very restricted site is leading to inbreeding depression and is one of the threats to its survival. *Rhododendron ultimum* Wernham may be similarly isolated, growing at very high altitude on Mt. Carstensz, the highest mountain in New Guinea, but this isolation at high altitude allows easy discrimination from other related taxa. Despite doubts about its distinctiveness from *R. brassii* Sleumer (in the herbarium), field observations show these species to be quite different in flower colour, habit and habitat (Plates 3 and 4): *R. ultimum* is low growing, rarely above 15 cm high, whilst *R. brassii* is a lanky shrub often reaching more than 1 m. The habitats are also totally different, *R. ultimum* growing at higher altitude in open alpine meadow-like areas whilst *R. brassii* grows in subalpine shrubberies.

Concluding remarks

Vireyas exploit many different habitats, from lowland rainforest to alpine meadows. Lowland species, because of their potential for continuously linked habitats, are more difficult to treat as separate species, despite considerable variation. Plants on isolated mountain tops are usually easy to discriminate but there are all sorts of differences in the biological nature of this variation which make taxonomy the intriguing discipline that it is. The naming of a species is just the start of the process of describing and understanding biological phenomena. Understanding the evolutionary process that results in present-day variation is a highly academic pursuit. Description of that variation must result in utilitarian classifications which end users can use in a practical fashion but which still reflect the underlying evolutionary processes.

Plate 3 *Rhododendron ultimum* from Mt. Jaya, West Papua, Indonesia.

Plate 4 *Rhododendron brassii* from near Wamena, West Papua, Indonesia.

References

ARGENT, G. (in press). *Folia Malaysiana* 4(1).

ARGENT, G. & BARKMAN, T. (2000). Two exciting new species of *Rhododendron* section *Vireya* (Ericaceae) from Mount Kinabalu, Sabah. *The New Plantsman* 7(4): 209–217.

ARGENT, G., LAMB, A. L., PHILLIPPS, A. & COLLENETTE, S. (1988). *Rhododendrons of Sabah*. Sabah Parks Publication No. 8, 1–146.

BINNEY, D. (2003). Rhododendron collecting in Sulawesi. In: ARGENT, G. & MCFARLANE, M. (eds) *Rhododendrons in Horticulture and Science*, pp. 111–114. Royal Botanic Garden Edinburgh.

BROWN, G. K. (2003). Vireya rhododendrons: an insight into their relationships. In: ARGENT, G. & MCFARLANE, M. (eds) *Rhododendrons in Horticulture and Science*, pp. 95–110. Royal Botanic Garden Edinburgh.

COPELAND, H. F. (1929). Philippine Ericaceae I: The species of *Rhododendron*. *Philipp. J. Sci.* 40(2): 133–179, pls 1–16.

CRONK, Q. C. B. (1998). The ochlospecies concept. In: HUXLEY, C. R., LOCK, J. M. &

CUTLER, D. F. (eds) *Chorology, Taxonomy and Ecology of the Floras of Africa and Madagascar*, pp. 155–170. Royal Botanic Gardens, Kew.

CULLEN, J. (2003). The importance of living collections in the classification of rhododendrons. In: ARGENT, G. & MCFARLANE, M. (eds) *Rhododendrons in Horticulture and Science*, pp. 121–129. Royal Botanic Garden Edinburgh.

HILLIARD, O. & BURTT, B. L. B. (1971). *Streptocarpus: An African Plant Study*. University of Natal Press.

LEAR, M. & MARTIN, R. (2003). Putting a name to garden hybrids: deciphering and cataloguing rhododendrons at Exbury. In: ARGENT, G. & MCFARLANE, M. (eds) *Rhododendrons in Horticulture and Science*, pp. 178–185. Royal Botanic Garden Edinburgh.

ROUSE, J. L. & WILLIAMS, E. G. (1989). Style length and hybridization in *Rhododendron*.

In: SMITH, J. C. (ed.) *Proceedings of the Fourth International Rhododendron Conference, Wollongong, N.S.W.*, pp. 74–82.

SLEUMER, H. O. (1966). *An account of Rhododendron in Malesia*. Groningen, The Netherlands: P. Noordoff Limited.

TURESSON, G. (1922). The species and variety as ecological units. *Hereditas* 3: 100–113.

WHITE, F. (1962). Geographic variation and speciation in Africa with particular reference to *Diospyros*. In: *Taxonomy and Geography. Publ. Syst. Assoc.* 4: 71–103.

WHITE, F. (1998). The vegetative structure of African Ebenaceae and the evolution of rheophytes and ring species. In: FORTUNE-HOPKINS, H. C., HUXLEY, C. R., PANNELL, C. M., PRANCE, G. T. & WHITE, F., *The Biological Monograph: The importance of field studies and functional syndromes for taxonomy and evolution of tropical plants*. Royal Botanic Gardens, Kew.

Rhododendron conservation in China

Li De Zhu[1] & David Paterson[2]

[1]Director of Kunming Institute of Botany, and [2]Deputy Director of Horticulture, Royal Botanic Garden Edinburgh, 20A Inverleith Row, Edinburgh EH3 5LR, UK

A joint venture between the Royal Botanic Garden Edinburgh, Kunming Institute of Botany and Lijiang Alpine Plant Research Institute was launched in May 2002. The overriding objective of the joint venture is the conservation of plant biodiversity in the Yulong Xue Shan (Jade Dragon Snow Mountain) and, indeed, throughout the Hengduan Shan mountain range of SW China. The project is funded by both British and Chinese governments and by a number of British companies that have commercial interests in China. The commercial sponsors (BP, British Airways and BHP Billiton) all have clear environmental policies and are willing to support the establishment of a field station high on the flanks of the Yulong Mountain.

The biodiversity of NW Yunnan

The mountain range known as the Hengduan Shan forms a huge geophysical barrier separating the flat lands of eastern China from the Tibetan plateau and main peaks of the central Himalaya. The Tibetan plateau is drained by the great rivers, the Mekong, Salween, Irrawady and Ganges. These rivers generally flow from north to south and disgorge their contents into the Indian Ocean or South China Sea. The Yangtze River, which flows through NW Yunnan, also follows a course from north to south, but in the prefecture of Lijiang the river turns on itself and flows north below the western flanks of the Yulong Mountain. At this point the Yangtze cuts a great defile that separates the Yulong Mountain from the main Tibetan plateau. This creates an 'island' effect to the extent that the species found on the Yulong Mountain have evolved with a degree of botanical isolation that is not present in other parts of

the Hengduan Shan range. This has resulted in a particularly large degree of endemism.

The province of Yunnan is known throughout China as 'The kingdom of plants and animals'. It has about 16,000 species of vascular plants, representing approximately 50% of the botanical diversity of China. In the prefecture of Lijiang there are about 3700 species of flowering plant. Some 3200 species in 785 genera in 145 families have been recorded on the Yulong Xue Shan. Many of these species are now under threat, and the work of this new joint venture aims to prevent further species loss and secure a programme of stabilisation for threatened taxa.

Historical background and the new joint venture

The flora of the Yulong Xue Shan has long interested western botanists. George Forrest and Joseph Rock both spent a considerable amount of time studying the flora of the

Plate 1 Non-sustainable forest management.

Plate 2 Sustainable forest management.

172

Plate 3 Yulong Xue Shan.

Plate 4 Dawn of a new era for conservation in China.

Lijiang prefecture and made extensive collections, mainly from the eastern flanks of the Yulong Xue Shan. Joseph Rock set up home in the Snow Village, which is located on the western edge of the Lijiang plain immediately below the eastern flanks of the mountain.

In 1937 Professor K.M. Feng recommended the building of a field station at Lijiang to be used as a base for the study of the 'unique flora of the mountain' and as a facility for the cultivation of alpine and subalpine plants. From 1958 to 1970 the Lijiang Botanic Garden was developed but, for various reasons, the project was abandoned. In September 1995 the Lijiang prefectural government drew up proposals for a 'new' botanic garden in Lijiang. Unfortunately, again for a number of reasons, the proposal failed during the early stages. In March 1999, the Royal Botanic Garden Edinburgh and Kunming Institute of Botany, after several months of discussion, fact-finding and local negotiations, formally commenced a joint effort to raise funds for a three-phase programme to build a Field Station, Botanic Garden and Nature Reserve on the Yulong Mountain.

The British Foreign and Commonwealth Office provided 'seed' money and through the efforts of Sir Anthony Galsworthy, former British Ambassador to China, and his team of diplomats, sufficient commercial sponsorship was raised to fund the first phase of the development. The 'Foundation Stone Laying Ceremony' took place in May 2001, after which work commenced on the construction of buildings for the Field Station. The other major elements of the joint venture will be funded by the China Biodiversity Action Plan which arose as a result of China's participation in the Rio Summit.

The Field Station covers an area of 150 Chinese Mu (approxmately 10 ha) and is located at an altitude of 3200 m, with cultivation plots and growing facilities up to 3500 m. The main Botanic Garden occupies a site on the lower slopes of the mountain, at an elevation of 2600–3000 m, and covers an area of 400 Mu (c.26 ha). The Nature Reserve covers an area of 3700 Mu (246 ha) and contains a number of habitats in the elevation band 2600 to 4500 m.

Conservation initiatives

The Lijiang Botanic Garden and Field Station project will enable a number of conservation initiatives to be undertaken, and provide the Royal Botanic Garden Edinburgh and Kunming Institute of Botany with an ideal facility for the cultivation of alpine and subalpine plants.

Specific initiatives

- *Ex situ conservation of alpine plants from NW Yunnan*. The China Biodiversity Action Plan delegates the responsibility for the conservation of alpine plants native or endemic to NW Yunnan to the Lijiang Botanic Garden.
- *Ex situ conservation of plants in the Hengduan mountain range*. The Lijiang Botanic Garden and Field Station will provide a 'safe site' for the cultivation of rare and endangered species from many locations throughout the Hengduan Shan range.
- *Rehabilitation and repatriation of rhododendron species*. The *Rhododendron* collection within the Royal Botanic Garden Edinburgh contains many plants that were originally collected from the Hengduan Shan mountains. Some of these

Plate 5 Conservation through partnership. Professor Stephen Blackmore (Regius Keeper, RBGE) in attendance at the inauguration of the Lijiang Botanic Garden and Field Station Project.

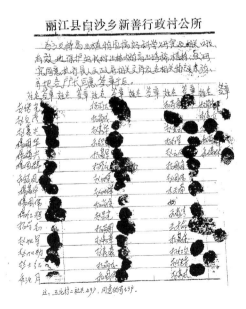

Plate 6 Securing local support. The head of each household from the neighbouring villages agrees to changes in land use.

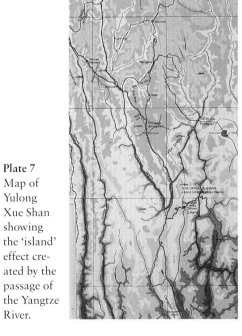

Plate 7 Map of Yulong Xue Shan showing the 'island' effect created by the passage of the Yangtze River.

Plate 8 Loss of forest habitat due to clear cut forestry.

Plate 9 Pristine mountain environment on the Yulong Xue Shan that is worthy of protection.

species have become rare in the wild and many others now occupy a smaller range. Species known to be in decline on the Yulong Mountain will be targeted for stabilisation and rehabilitation in an attempt to halt population decline. This will be aided by repatriation of germplasm from the Royal Botanic Garden Edinburgh collections. Similar repatriation programmes will be initiated to generate a species-rich cultivated *Rhododendron* collection within the exhibition area of the Lijiang Botanic Garden.

- *Joint research on environmental change and biodiversity conservation of the Tibetan plateau*. Using the Lijiang Botanic Garden and Field Station as a base, a number of expeditions will be mounted to known locations on the Tibetan plateau to set up long-term research programmes to monitor environmental changes that are taking place there. The results of this research will enable targeted biodiversity conservation work to take place, using the Lijiang Field Station as a base for such activity.

Putting a name to garden hybrids: deciphering and cataloguing rhododendrons at Exbury

Michael Lear[1] & Rachael Martin[2]

[1]Lear Associates, 8 East Street, Oxford OX2 0AU, and [2]Exbury Gardens, Exbury, Southampton SO23 1AZ, UK

The garden at Exbury in Hampshire and the collection of rhododendrons that it contains are the inter-war creation of Lionel de Rothschild, who was one of the most influential *Rhododendron* breeders and garden owners of the 20th century. Since his untimely death in 1942, the garden has been cherished by the Rothschild family and dedicated Head Gardeners. Despite over 500 hybrids being raised and named at Exbury, and the garden receiving a huge number of Royal Horticultural Society awards, by 1990 there was no overall inventory for the plants growing there. When members of the public asked how many rhododendrons there were growing in the 250 acres of garden only wild guesses could be offered.

In 1990 Lear Associates were invited to catalogue the collection, and a database in ARev (Advanced Revelation) was developed specifically for the task. This system has been designed to cope with all types of rhododendrons, whether species, grex swarms or clones, with the possibility of making a full description of an accession's inflorescence and foliage. The results of 11 seasons of fieldwork based on consistent field recording of *Rhododendron* morphology provide extensive data on corolla size, ratio, colours and phenology, etc. Alongside the accession file, a taxon profile of almost 5000 different types of rhododendrons has been built up to provide a definitive record. A stripped-down version of the database has also been developed for use on Psion Organisers™ in the garden.

The main purpose of the project has been to verify the identification of as many rhododendrons in the collection as possible and to improve knowledge of them. It has been a priority to locate and where possible re-apply 'LR' (Lionel de Rothschild) grex numbers. In a garden that was as much a nursery as a garden, it was important to distinguish sibling from clonal plants. It follows that it was desirable to record the specific plants which had been shown (and perhaps awarded) and to determine whether they were the same as the named varieties propagated and distributed to other gardens and the trade. A diagnostic tool, known as 'Matchmaker', has been developed which uses attributes such as flower colour, shape and size to present a list of possible names that might be applied. Attempts are being made to write keys to groups of closely related hybrids.

The methodology and system developed at Exbury has now been applied to other collections, including those of the even more poorly understood Victorian and Edwardian hybrids and cultivars. In the absence of precise cultivar names the systematic analysis of aesthetic features which predominate particular phases of *Rhododendron* breeding and landscaping makes this approach a useful tool in historic garden conservation and interpretation.

Introduction

The connection between the name Rothschild and rhododendrons has a legacy originating at the start of the 20th century, with Lionel de Rothschild. Through the influence of W.J. Bean, P.D. Williams and, most especially, J.C. Williams of Caerhays, Lionel was inspired not only to grow rhododendrons but also to create new ones, the possibilities being fuelled by the extended range of species recently introduced from China. Lionel grew species in pots at his garden at Gunnersby, West London, which enabled him to force items like *R. griersonianum* Balf.f. & Forrest into flower in March. However, it was the acquisition in 1919 of neglected coppice woodland and acid New Forest gravels of Exbury on the Hampshire coast which provided the rare opportunity to expand. Apart from some hardy hybrids and some Georgian trees, Lionel had a blank canvas. In his first year he bought from nurserymen – Gomer, Waterers, Gill, Veitch, Reuthe and Harry White of Sunningdale – mainly to build up a collection of species and selected foundation hybrids. He even drove to Orléans in France where he had purchased *R. calophytum* Franch. which had been raised from Wilson's seed.

For the collection of rhododendrons Exbury can be roughly divided into three areas. The first is the area running southwest below the house, appropriately called the 'Home Wood'. This area has the remnants of the Georgian landscape, and extends further to the Beaulieu River via the Winter Garden. The second, to the west of the house, beyond the Park, is Witcher's Wood, which was originally named 'Lowinsky Hill' after the acquisition in 1926 of the Thomas Lowinsky collection. The third and largest area, 'Yard Wood', is to the north, and also includes the Rock Garden and three old nursery areas. The whole area of Exbury covers 250 acres, more or less the same as Kew.

Lionel used his hybridising to extend his 'palette', and the pursuit of pure colours was a major preoccupation. For red he created 'Queen of Hearts', 'Gibraltar' and 'Fusilier'; for yellow, 'Hawk Crest', 'Idealist' and 'Fortune'; and for blue, *R. augustinii* Hemsl. 'Electra'. He was also very keen on pinks and sophisticated biscuit hues such as found in the various selections within the 'Naomi' and 'Carita' grexes. He also strove to increase reliability and abundance of flowering, and had deliberate breeding programmes for extending the flowering season into July and August.

His stud book enumerated 1210 crosses, and 462 have been named. He started naming with 'A', 'Abalone', 'Abott', 'Adelaide', 'Alix', 'Amalfii', and so on and then 'B', but this strict system was abandoned at some point. Zinc labels with acid etched writing were used to place the stud number physically on the plant, and many of these still exist today.

In one highly productive, legendary, day in May 1925 or 1926, Lionel made nine crosses, which included 'Lady Chamberlain' from LR178 obtained by crossing 'Royal Flush' × *R. cinnabarinum* Hook. 'Royelii'. 'Lady Rosebery' and 'Lady Berry' were also produced on that day, from LR179 and LR186, respectively. Another important date was 27 May 1924, for this was the day on which Lionel received his first award for showing a *Rhododendron*, a First Class Certificate (FCC) for *R. griersonianum*. This was all the more remarkable because George Forrest had discovered this species only seven years earlier, in June 1917.

It would be wrong to suggest that Lionel was obsessed with just hybrids: he sponsored plant collecting trips of George Forrest from 1920 to 1922, Farrer in 1919 and numerous Kingdon-Ward expeditions. There was a 100 ft × 50 ft teak house for tender maddeniis. He was responsible for the original introduction of *R. yakushimanum* Nakai (= *R. degronianum* Carrière ssp. *yakushimanum* (Nakai) Kitam.) to the West, when he imported two plants from Wada's Nursery to Exbury in 1934. One of the pair is still in Home Wood.

Lionel was the moving force behind and President of the Rhododendron Association (a successor to the Rhododendron Society) from its inception in 1928 to 1939. The constitution of the Association was a blueprint for the Royal Horticultural Society (RHS) Rhododendron and Camellia Group which exists today. The close friendship and rivalry between Lionel and the second Lord Aberconway at Bodnant, who became President of the RHS in 1931, 'made *Rhododendron* species and hybrids almost a British monopoly and a model for the gardening world overseas' (Postan, 1996).

Lionel died in 1942, aged only 60. The making of Exbury and its plant collection in only a two-decade period from 1919 must rank as one of the outstanding horticultural achievements of the 20th century.

Exbury passed to Lionel's eldest son, Edmund (Eddy), after the war and he still lives there today in his 85th year. Breeding and selecting was continued by Eddy, and his stud book contains 182 crosses. Many of these used his father's best crosses or repeated them. Much of Eddy's work was involved in flowering, selecting and naming his father's crosses that had not revealed their quality in his lifetime. Notable achievements have been 'Fred Wynniatt', 'Lionel's Triumph', 'Quaker Girl', 'Nimrod', 'Duchess of Rothesay', 'Bach Choir', 'Galactic' and 'Charlotte de Rothschild'.

Eddy's sons, Nicholas and Lionel, are both highly involved in the garden and continue to make new combinations. Exbury is very much a family affair; Eddy's younger brother, Leo, is Chairman of Exbury Gardens Ltd. What we see today is very much Lionel's legacy. Just as at Bodnant, the garden at Exbury has acted also as a nursery, in effect, a huge trial ground.

An incredible 238 *Rhododendron* species, hybrids and varieties from Exbury have received awards. To achieve this, showing rhododendrons at RHS shows was a major preoccupation and this continues to the present day. On the Sunday before the Chelsea Flower Show Doug Betteridge, who retired as Head Gardener in 1997, will be continuing that tradition as he still works two days a week in the garden. Other previous Head Gardeners at Exbury, including Fred Kneller, Arthur Bedford, Francis Hanger, Harold Comber and Fred Wynniatt, have been crucial to this successful partnership with the de Rothschild family.

The cataloguing project

In 1989 Alan Hardy rang to see if Lear Associates would be interested in doing some work at Exbury. Alan was a member of the Exbury Garden Board, and the Board wanted to know what they had in the garden. At the outset it was decided that the Exbury catalogue would include all woody plants except azaleas, to make the project manageable in its first phase. Although I had run the cataloguing system for the National Trust, encompassing 27 gardens, none matched Exbury in terms of size and complexity.

The garden was divided into 344 numerical areas, and we worked systematically through each area, tagging everything and finishing an area before moving on to the next. We tried to time the work so a maximum number of rhododendrons would be in flower to increase data capture. Stainless steel tags and wire were used to provide a cost-effective long-term recording technique.

Corolla colour was dealt with at two levels. There is of course the RHS Colour Chart, but this was found to be impracticable to take into the field. A customised colour chart was therefore created from swatches from the RHS chart to ensure consistent colour recognition. It is always important also to have free text to describe flare, spots, basal blotches and nectary colours.

Field sheets were designed to record plants in a systematic way. Let us go through the headings (which equate to the fields in the database). First we have flowering stage, for which we have six options: F0 No buds set; F1 Tight bud; F2 Showing colour (bud only); F3 Some colour; F4 Full out (with no flowers on the ground); F5 Flowers fading (some flowers on ground); F6 finished (almost all flowers on ground). When a flowering stage is set against a date, the phenology of the rhododendrons at Exbury can be recorded.

The number in truss is straightforward, but it is as variable on hybrids as it is on species so you have to hunt for a 'typical truss' or do some averages. There are 10 options for truss shape: round, dome, compact, etc.

Measuring individual flowers provided three figures: the width of a fully open (but not pushed flat) corolla, the overall length, and the distance from the sinus of the lobe to the base of the corolla (the tube measurement). Flower shape is very diagnostic, and we have seven main types, with combinations allowed. The number of lobes is normally 5 or 7 (this is important). The individual corolla lobe is a key feature and here we also have seven options – wavy, frill, pointed, round, notched, reflexed and overlapped. An additional character that I have recently discovered is the degree to which the lobes are narrow and do not overlap, creating a stellate appearance. The calyx has nine options and is very important, as is the condition of the ovary. Some ovaries are glabrous, some tomentose, some glandular and some a mixture of glands and hairs. The colour of the ovary is also something which is recorded more systematically now. The same situation exists with the style, which can be very diagnostic. The number of stamens should be recorded, and the projection of the stamens or range from shortest to longest is relevant in certain taxa. The pedicel has seven options and appears to be more significant than the rachis. Scent is recorded to three levels.

Leaf data are also recorded but I will not go into any detail here apart from to say that for the older hybrids, i.e. the hardy hybrids, the texture and whether the leaf is curled laterally or longitudinally are all very significant. Hair information is as important for hybrids as it is for species.

Other data are also recorded, such as what label is on the plant: of particular importance here is the original zinc stud number. Source, size and age are also recorded. Verification data about who decided on the name of the plant, and when, is obviously also essential.

The data gathering commenced in 1990 and has continued until now, work being carried out in the spring and early summer.

The database was custom built in Advanced Revelation (ARev, the software used in BG-BASE™) and although designed to accommodate all woody plants, it is specially structured to cope with *Rhododendron* species and hybrids.

In the database there is a taxon file which has 7790 entries, including around 5000 *Rhododendron* names. This is comprised of 447 *Rhododendron* taxa at species level, not all of which are in the garden, all the Rothschild-bred and selected plants including the entire stud books of Lionel and Edmund, and selected hybrids from elsewhere.

Working with hybrids can make working with species look straightforward: the complexity of hybrids can be mind numbing. The most complicated Exbury hybrid is 'Lady Romsey', which has been bred for five generations with five wild species in it, half being *R. yakushimanum*. This all has implications for the design of the database, as we wanted family trees to be constructed internally. Plants that are related through breeding are linked through their unique taxon numbers. Take *R. williamsianum* Rehder & E.H.Wilson, for instance. The taxon file can tell you what progeny this species has produced: when it has been the male or pollen parent there are 37 taxa, and when it has been the female or seed parent there are 16. Similarly, if I want to know if a particular combination of taxa has already been made, the database should be able to indicate this. For instance, if one crosses *R. fortunei* ssp. *discolor* (Franch.) D.F.Chamb. and *R. facetum* Balf.f. & Kingdon-Ward the result is the 'Ayah' grex.

Grex names have been used in the cataloguing at Exbury. A grex name covers all the range of progeny of any two plants, whether species × species, species × hybrid or hybrid × hybrid, and for either direction of the cross when pollen and seed parents are reversed. The grex or 'cultivar group' nomenclatural rank is perfect for understanding the sibling variation.

What has been described so far is encyclopaedic information. On the other side of the database is the accession file. This contains the plant by plant data collected in the garden and logged according to tag number. Through its name every plant in the accession file is linked to the taxon number. The accession file has three screens. The first is for general observations, planting date location, verifications etc. The second screen gives full observational information for that plant. The third screen is completed only if that plant is used for breeding.

The accession file has a maximum of 56 fields but it is rare that more than 30 are filled out. ARev has variable length fields, which is very useful when recording free text. By recording in detail, differences in apparently the same plants are made more clear, enabling clones to be distinguished from seedling variation in grex swarms. We have found that many hybrids have very poor published descriptions so the observations we have made in the field, constitute an improved and definitive description for many of the Exbury hybrids.

In a hypothetical situation, a cross is made producing a tray of 200 seedlings. Fifty are eventually planted out in one part of the garden, each with a number between 1 and 50. Five years or so later, they start coming into flower. Of the first to do so plants number 6 and 23 are best; they are earmarked for naming and showing, and possible names are considered. A year later, more of the grex swarm flower, with even better results: plant 33 is even better than 6 and 23. Attention shifts to plant 33 and it is therefore 'christened'. Plant 33 is shown under that name and it wins an award. It is propagated. It is hung with an

FCC label. A year later, plant 38 flowers: it is even better than 33. Here is the first dilemma: is plant 38 shown with a new name or is the label swapped from plant 33 to plant 38? If the label is swapped, what becomes of the plants that have been propagated from plant 33? Two years on perhaps an artist wants to illustrate the FCC form, but material is gathered from plant 6 by mistake. After a few decades, plant 38 may have been robbed of cuttings, and there is such demand that the propagator goes to plant 23 and they are sold as the FCC form. Other problems arise if the labels fall or are pruned off, or a gardener with the only first-hand knowledge leaves or retires; the cross is made again, or the plants are in different parts of the garden, making direct comparison harder. I am not saying this happened routinely at Exbury but the stabilisation of names can be fraught with difficulties in the originating garden.

Ultimately names of the hybrids must be checked, the range of grex material understood, the clones located within the grex. These are then like the 'type' plants. The perfect situation is when your 'type' plant is the oldest example in the garden, it is also the one that has always been cut from for RHS shows, it was the plant that was illustrated and that you have always propagated from.

As the database became fuller it became optimal to work within selective grexes. We made a 'pick list' from the database from the existing information and collected specimens of everything, even plants that were unnamed but looked similar. The plants were laid out in troughs in oasis (specimens will keep turgid for up to two weeks by this method). Specimens were grouped into affinities and the clones segregated, designating them clone A, B, C, etc., and putting the remainder in a miscellaneous section which represented the general grex swarm. When the names on the database are examined for the five plants of clone A, it is rare that they are all the same, but you decide on a name after comparison with descriptions on the Register, and in Phillips Lucas & Barber (1979) and Salley & Greer (1986). This was done for grex 'Naomi' and it was found that the clone 'Naomi AM' had been confused with the clone 'Naomi Exbury'. Of the 14 named clones of grex 'Naomi', we have located 13; 'Pixie' appears to have been lost from the garden. But two that were not previously thought to be in the garden, 'Carissima' and 'Nereid', were rediscovered. A key is now being written to the 'Naomi' grex.

Doug Betteridge retired in 1997, having started at Exbury in 1955, and has a wealth of knowledge in his head. This is a predicament of many gardens, and we were asked to take Doug to each of the tagged plants so that we could record his thoughts on the database, accession by accession. These sessions have proved invaluable in refining our understanding of the garden and in the accuracy of the naming of the plants. I would like to acknowledge his role with us over the last 12 years.

We now have 17,500 accessions, and to make looking them up in the field easier we have Psion Organisers™ with a stripped-down version of the database. The vast majority, 85%, of the *Rhododendron* collection at Exbury is of cultivated origin. Although 15% for wild taxa may not sound much, that is 2270 accessions, composed of around 130 species. One of the more notable ones is the purple-flowered *R. cinnabarinum* Hook.f. ssp. *tamaense* (Davidian) Cullen from KW21021.

A recent innovation is the 'Matchmaker' program: a specimen is taken, the characters are keyed in, and the database searches for the closest matches. The database thus becomes its own diagnostic tool. If however you key in the description of, say, a new French hybrid which has not been programmed into the system you will be disappointed. It is not yet perfect, and indeed it is still being developed, but we have the foundation for something really useful.

Concluding remarks

The Garden Board wants Exbury to become a centre of excellence in the understanding of rhododendrons, naturally centering on the Rothschild hybrids. There are 140,000 visitors to the garden annually and Exbury is in an excellent position to promote education of these visitors through its plant collection. Beverley Lear produced a *Visitor's Guide to Rhododendrons*, which is an attempt to show that there is much more to a rhododendron than a pretty flower and a name.

Looking beyond Exbury, the real issue with hybrid rhododendrons is that it is difficult to think of another plant group which has had more impact on gardens in the British Isles and at the same time is so anonymous. In many beautiful gardens one can be surrounded by Victorian or Edwardian rhododendrons which have no names. Rhododendrons are admittedly difficult plants to label but that is not the real problem. Identification, lack of adequate descriptions, and excessive naming by competing nurseries are at the crux of the issue. Some of the hybrids of course have not been registered. No doubt advances with DNA fingerprinting could revolutionise hybrid identification in the future. However, I hope that, as I have shown, obser-

vation beyond registration standards and systematic recording can make significant progress on hybrid identification and understanding.

The methodology we have developed at Exbury has been extended to earlier hybrids and gardens such as Langley Park near Slough, Castlewellan in Co. Down, Warley Place in Essex and Deepdene in Surrey. We have started to write identification keys, and have some drafts for white hardy hybrids, and work is currently underway on old *R. arboreum* hybrids.

The *Rhododendron* craze between 1830 and 1920 in northern Europe is equivalent to the tulipmania of two centuries earlier. Hybrids are barometers of period aesthetics. In the context of historic parks and gardens, the fact that old rhododendron hybrids are the horticultural equivalent of period wallpaper is now being increasingly appreciated by the Heritage Lottery Fund, English Heritage and the National Trust. Just calling a shrubbery 'hardy hybrid rhododendrons' is not good enough, and even if names are not achievable it is often possible to call it a '*caucasicum* Pallas hybrid'. To know the decade of planting or breeding, or which nursery produced them, is a better hook on which to base conservation grants and management. Clearly it also opens up more informed interpretation to the public. Hybrid and cultivar names are precious in historic collections but it is important to remember that unnamed plants are still worthy of conservation, especially if distinct.

There is a bed in the Winter Garden at Exbury which has been unchanged since it was planted in the early 1930s. It contains three different LR grexes. The subtle variation between and within the grexes creates an enchanting scene, in fact not dissimilar to natural variation in species in the wild. As such, this bed is not just a beautiful piece of planting but also a living example *par excellence* of the techniques of selection from the garden which was also the Rothschilds' production floor.

Finally, a comment on *R. rothschildii* Davidian which Davidian used to commemorate Lionel de Rothschild. This species is now thought to be a naturally occurring hybrid between *R. praestans* Balf.f. & W.W.Sm. and *R. arizleum* Balf.f. & Forrest. Hearing of the extent of hybridity in the wild, I think Lionel would not be turning in his grave at this revelation about his *Rhododendron*. He gave his life to the genus, in all its wonderful variation, both wild and cultivated. Perhaps understanding natural biodiversity and horticultural variation is not as different as first thought!

References

PHILLIPS LUCAS, C. E. & BARBER, P. N. (1979). *The Rothschild Rhododendrons*, 2nd edition (1st edition 1967). London: Cassel.

POSTAN, C. (ed.) (1996). *The Rhododendron Story: 200 Years of Plant Hunting and Garden Cultivation*. London: The Royal Horticultural Society.

SALLEY, H. E. & GREER, H. E. (1986). *Rhododendron Hybrids: A Guide to their Origins*. London: Batsford.

Ten years on from Rio: The impact of the Convention on Biological Diversity

David Rae

Royal Botanic Garden Edinburgh, 20A Inverleith Row, Edinburgh EH3 5LR, UK

The Convention on Biological Diversity (CBD) came into being in 1992. While many in the botanical community came to regard it as a legislative instrument designed to restrict the free flow of plant material it was, in fact, designed to be a convention to promote conservation. Central to the CBD are three objectives: the conservation of biodiversity, the sustainable use of biodiversity, and the fair and equitable sharing of the benefits derived from biodiversity. While all three objectives were designed to have equal status, it was the third that both caused most concern and received the greatest attention. Access to, and transfer of, plant material, along with issues of exploitation and benefit-sharing, became important issues that were discussed in numerous conference workshops. However, it has only been in the last six months that some sort of consensus concerning these issues seems to have been emerging, with the publication of *Results of the pilot project for botanic gardens: Principles on access to genetic resources and benefit-sharing, common policy guidelines to assist with their implementation and explanatory text* by Fernando Latorre García *et al.* A simpler, shorter, but in many respects similar, scheme has also been developed by the University of Bonn Botanic Garden, called the Access and Benefit-Sharing Scheme (ABS).

Two processes in fulfilment of the first objective (the conservation of biodiversity) are worthy of description. The first, under article 6 of the Convention, is the requirement for each country to produce a Biodiversity Action Plan. The UK government was the first to produce such a plan and there now exist action plans for rare and threatened species and habitats along with named individuals in agencies responsible for overseeing the actions required.

The second is the Global Strategy for Plant Conservation that originally arose from a resolution at the last International Botanical Congress, held in St Louis, Missouri. This was further developed as the Gran Canaria Declaration and it has now been discussed within the CBD framework and has emerged as 16 'outcome-orientated targets' aimed at (a) understanding and documenting plant diversity, (b) conserving plant diversity, (c) using plant diversity sustainably, (d) promoting education and awareness about plant diversity, and (e) building capacity for the conservation of plant diversity.

These two examples of positive actions for plant conservation, along with the measures for responsible acquisition, transfer and benefit-sharing, are starting to demonstrate the value of the Convention.

Introduction

In June 1992 more than 150 heads of state or governments attended the United Nations Conference on Environment and Development (UNCED), the so-called 'Earth Summit', in Rio de Janeiro. At this conference, the largest of its kind ever held, a number of conservation-related conventions, agreements and statements were proposed, debated and agreed. Agenda 21, the Convention on Climate Change and A Statement of Principles for the sustainable management of forests are examples of just a few of these. Another is the Convention on Biological Diversity (CBD). This paper describes the objectives and mechanism of the CBD, how the UK government has responded to one aspect of it, by producing the UK Biodiversity Action Plan, explains the issues surrounding access and benefit-sharing, and describes an international strategy derived directly from the CBD.

The Convention

At the UNCED 153 countries signed the new agreement, and it has now been ratified by more than 160 countries. It eventually came into force on 29 December 1993. In many ways the Convention must be regarded as a landmark in the environment and development field as, for the first time, it takes a comprehensive rather than sectoral approach to the conservation of the Earth's biodiversity and sustainable use of biological resources. It also recognises the fundamental issues raised in three key publications – *The World Conservation Strategy* (IUCN, UNEP & WWF, 1980), *Caring for the Earth* (IUCN, UNEP & WWF, 1991) and the *Global Biodiversity Strategy* (WRI, IUCN & UNEP, 1992) – that both biodiversity and biological resources should be conserved for reasons of ethics, economic benefit and indeed human survival. However, it goes beyond simply the conservation of biodiversity and the sustainable use of biological resources, to encompass such issues as access to genetic resources, the sharing of benefits derived from the use of genetic material, and access to technology, including biotechnology.

The Convention is based on 42 articles and at first sight appears to be long and complex. However, it is worth remembering that it is founded on just three fundamental objectives. These are:

- the conservation of biodiversity,
- the sustainable use of the components of biological diversity, and
- the equitable sharing of the benefits derived from the use of genetic resources, including ensuring relevant access to genetic resources, the transfer of relevant technologies and that appropriate funding is available.

A Guide to the Convention on Biological Diversity (Glowka *et al.*, 1994) is perhaps the single best publication explaining the Convention as it takes each article in turn and then explains it in logical and understandable terms.

Of the 42 articles 27 are to do with administrative procedures such as financial mechanisms, the settlement of disputes, voting rights and the adoption of protocols. The rest are highly pertinent to the three objectives listed above and include articles on, for instance, identification and monitoring, *in situ* and *ex situ* conservation, research and training, technology transfer, public education and awareness and technical and scientific co-operation.

COP and SBSTTA

Unlike many laws, treaties or conventions that once established are fixed forever, the CBD can, and does, get altered through time. The mechanism for these changes takes place through regular meetings of those who have ratified the Convention – the Conference of the Parties (COP). To date eight meetings have taken place (1994 Nassau, 1995 Jakarta, 1996 Buenos Aires, 1998 Bratislava, 1999 Cartagena, 2000 Nairobi & Montreal and 2002 The Hague) and at each a review of the implementation, financing and administrative arrangements have taken place. In addition to reviewing of the administration of the Convention major themes are debated in detail. Examples in recent years have included marine biodiversity, access to genetic resources, biosafety, intellectual property rights, alien species and identification and monitoring. As well as the presence of government representatives, COP meetings act as a forum for many special interest groups, NGOs and indigenous peoples, each keen that the Convention should act as a vehicle for their own particular interest in biodiversity conservation.

Another important official body, established under article 25 of the Convention, is the Subsidiary Body on Scientific, Technical and Technological Advice (SBSTTA) which meets regularly to provide the COP with expert advice on the implementation of the Convention and the conservation of biodiversity. To date seven meetings have been held, usually in Montreal, and 71 recommendations have been issued.

The UK Biodiversity Action Plan

While some have been quick to criticise the CBD for its bureaucracy there is no doubt that it has had a major impact and that this impact is increasing. It is bureaucratic but this should not prevent practical outcomes from taking place. Article 6 of the Convention, for instance, calls for Parties to 'Develop national action plans to conserve biodiversity'. In the UK the Department of the Environment was charged with the task of bringing together staff from central and local governments, agencies, collections, academic bodies and non-government organisations, to develop such an action plan.

In January 1994, less than two years after Rio, *Biodiversity, the UK Action Plan*, was published (Anon., 1994). This report gave an overview of biodiversity in the UK – it described the range of habitats and species in Britain, our science base and the extent to which biodiversity was being monitored and researched, the reasons for, and quantities of, habitat loss, and numerous examples of conservation projects, aspects of sustainable use and the importance of partnership and education. It concluded, in Chapter 10, with 59 aims for progress towards achieving objectives laid down elsewhere in the text. Significant amongst what came to be called 'the 59 steps' was item 33 – 'Prepare action plans for threatened species in priority order …'. Other significant points addressed the need to conserve habitats, develop public awareness and understanding and contribute to the conservation of biodiversity on a European and global scale. To achieve action on all these points a number of committees were set up, each reporting to the main Biodiversity Action Plan Steering Group. They included the Targets, Data, Public Awareness and Involvement and Local Biodiversity Sub-Groups.

Each of the sub-groups consulted widely and developed programmes for action. However, it is the work of the Targets Sub-Group which is described here. In charging this sub-group to develop action plans for threatened species and habitats, *Biodiversity, the UK Action Plan* had clearly stated that this should be in priority order – in other words, the most critically endangered species and habitat should be tackled first.

After considerable consultation with representatives from interested parties the sub-group came up with three species lists and key habitat listings.

For species, there was a long list of some 1250 species, a middle list of just under 300 species and a short list of 116 species. Categories for inclusion in these lists included one or more of the following:

- Threatened endemic and other globally threatened species.
- Species where the UK has more than 25% of the world or appropriate biogeographical population.
- Species where numbers or range have declined by more than 25% in the last 25 years.
- In some instances where the species is found in fewer than fifteen 10 km squares in the UK.
- Species which are listed in the EU Birds and Habitats Directives, the Bern, Bonn or CITES conventions, or under the Wildlife and Countryside Act 1981 and the Nature Conservation and Amenity Lands (Northern Ireland) Order 1985.

It was agreed that species which qualified for one or more of the above categories should be considered for the long list. The middle and short lists contained species which were either globally threatened or were rapidly declining in the UK (i.e. by more than 50% in the last 25 years).

Key habitats were selected from one or more of the following criteria: habitats for which the UK has international obligations; habitat at risk, such as those with a high risk of decline especially over the last 20 years, or which are rare; areas, particularly marine areas, which may be functionally critical, and areas important for key species. This gave a list of 38 key habitats for which costed action plans were initially prepared for 14, with plans for the remaining 24 due within two years.

A format for action plans was agreed (see Box 1) and those on the first list were published in 1995 (Department of the Environment, 1995). Other volumes appeared in 1998 and 1999, making a total of seven volumes in all.

Keen that positive conservation work did not cease with the publication of the action plans, the UK Biodiversity Group set up a system of representatives to oversee and stimulate activity. The *Champion's* job is to provide publicity and, possibly, financial support. It could be a company keen to sponsor a species. The *Contact Point* is usually a government agency such as Scottish Natural Heritage which acts as the first port of call for anyone wanting to know about a species or habitat. However, it is the *Lead Partner's* job to bring together a Steering Committee of relevant people, develop a work programme and generally encourage those nominated in the action plans to deliver the actions agreed to.

While the UK Biodiversity Action Plan process has undoubtedly created a lot of bureaucracy, there is no doubt that real progress is now being made with a number of key species and habitats.

Box I Layout of Biodiversity Action Plans for individual species.

Each plan is laid out in the same concise format with the following headings:

1. Current status

2. Current factors causing loss or decline

3. Current action

4. Action plan objectives and targets

5. Proposed action with lead agencies

   5.1 Policy and legislation

   5.2 Site safeguard and management

   5.3 Species management and protection

   5.4 Advisory

   5.5 Future research and monitoring

   5.6 Communication and publicity

   5.7 Links with other action plans

Access and benefit-sharing

An issue at the heart of the CBD, but which is still not fully understood by many, is that the genetic resources found within a country now belong to that country, and access to them can be controlled by the country. Up until 1993 all living things were thought to exist 'for the common good' – in other words, were freely available for anyone to take and use. Article 15 subsection 1 puts it as follows: 'Recognising the sovereign rights of states over their natural resource, the authority to determine access to genetic resources rests with the national government and is subject to national legislation'. This means that, perhaps not unreasonably, permission must be granted by a country before genetic resources, such as plant material, can be gathered by a third party, be that for research, education,

conservation or commercialisation. Such permission must be granted from a government-appointed ministry or agency and cannot be granted from a university, museum or field station as might have been done in the past. In addition, local stakeholders, such as indigenous people, also have to give their consent.

The purpose of article 15 was clearly to protect the rights of developing countries where, in the past, developed countries have made huge profits from their biodiversity, be it from direct logging or from pharmaceuticals derived from plant material. The article does not seek to stop the use of biodiversity in this way (in fact it encourages countries to make their biodiversity available to others) but it does give those countries holding such material the right to control access and transfer. Also, where there might be a profit resulting from the exploitation of biodiversity, then it states that such profit should be shared equitably between the source country and the developer. Article 15 subsection 7 states that: 'Each Contracting Party ... [should] share in a fair and equitable way, the results of research and development and the benefits arising from the commercial and other utilisation of genetic resources'. 'The fair and equitable sharing of the profits arising from biodiversity' would certainly include money where commercial activity has taken place, but could also include non-monetary benefit such as collaboration in research or sharing the results of research, capacity building or technology transfer.

While access and benefit-sharing issues certainly feature as one of the three main objectives of the CBD, many have felt that they dominated discussion to the detriment of the other two objectives. Botanic gardens, for instance, have been used to collecting plant material from around the world to use for research, education and conservation. In the past, also, botanic gardens of empire states were responsible for exchanging and acclimatising many plantation crops – a clear case of what would now be known (and condemned) as 'biopiracy'. Botanic gardens, however, were slow to realise the potential impact of the Convention, but have been working hard to understand and to implement it since about 1997. All the botanic garden networks and governing bodies, such as PlantNet (the Plant Collections Network of Britain and Ireland), Botanic Gardens Conservation International (BGCI) and the European Botanic Garden Consortium have held numerous conferences covering the Convention and have been working vigorously to help botanic gardens implement its requirements.

Before about 1997 there was considerable confusion about access, transfer and benefit-sharing, with staff in institutions such as botanic gardens believing that this aspect of the CBD imposed huge restrictions and bureaucracy. One thought did prevail though, and that was that it would be even more confusing for each botanic garden to develop its own policy. As a result of the various meetings and discussions two approaches seem to have emerged, one developed by the University of Bonn Botanic Garden and the other developed by an international consortium of botanic gardens led by the Royal Botanic Gardens, Kew.

The plan developed by Bonn resulted from a request by the German government to work up a plan dealing with access, transfer and benefit-sharing. The resultant scheme was simple and short and has been accepted by many European botanic gardens. While in theory it could be applied to the access of all kinds of plant material it was most appropriate for the exchange of plant material between botanic gardens through the *Index Seminum* seed exchange scheme, rather than the collection of plant material directly from the wild.

The Kew-led scheme, on the other hand, was predominantly concerned with the collection of plant material from the wild and was therefore somewhat more comprehensive than the Bonn scheme. Published in March 2001 after considerable international consultation, *Results of the pilot project for botanic gardens: Principles on access to genetic resources and benefit-sharing, common policy guidelines to assist with their implementation and explanatory text* (García *et al.*, 2001) has two main parts, as the title suggests. The 'Principles' section has chapters on the CBD and other laws, acquisition, use, written agreements, benefit-sharing, curation and preparing a policy. The 'Common Policy Guidelines' include sections to assist in the preparation of policies based on the 'Principles'. A chapter of explanatory text along with boxed case studies and 10 annexes follow these two major sections. The annexes include a model written agreement, a supply agreement and an example of a draft statement to obtain prior informed consent. The idea is that each botanic garden can adapt these for their own use.

While neither of these two approaches have been adopted wholesale by the botanic garden community there is no doubt that they have both distilled out information from the international debate and have moved the development of an international protocol forward greatly.

The Global Strategy for Plant Conservation

The Global Strategy for Plant Conservation arose from a resolution from the International Botanical Congress (IBC), held in St Louis, Missouri in August 1999, which called for plant conservation to be recognised as a global priority in biodiversity conservation. After the IBC a number of interested parties met in Gran Canaria in April 2000 during a meeting organised by BGCI to give further consideration to this issue and consider what next steps were possible or necessary. The outcomes of their deliberations were published as the 'Gran Canaria Declaration' which called for the development of a Global Strategy for Plant Conservation within the framework of the CBD. Thereafter, the proposals were presented at the CBD's COP meeting in Nairobi, and endorsed as its Resolution V/10. This resolution requested SBSTTA to give further consideration to the matter and to make specific recommendations at the next COP meeting, in 2002.

The Nairobi COP also called on the CBD Secretariat to arrange meetings with interested bodies to discuss the development of this global initiative in time for the SBSTTA meeting, and as a result two informal consultations were held, both with the Gran Canaria Group (Box 2). The first meeting took place in Montreal and the second, in May 2001, in London, organized by BGCI. At these meetings it was agreed that specific targets needed to be developed and adopted for inclusion in the Strategy, if it was going to have a significant impact. This was also additionally significant because, to date, few outcome targets have been prepared to measure the work of the CBD. Through these meetings 14 'outcome-orientated targets' were developed, aimed at:

(a) understanding and documenting plant diversity,

(b) conserving plant diversity,

(c) using plant diversity sustainably,

(d) promoting education and awareness about plant diversity, and

(e) building capacity for the conservation of plant diversity.

Following the May meeting in London a workshop was held at the International Association of Botanic Gardens (IABG) Congress in Cordoba which further considered the targets, and other comments and refinements were added by a range of international organisations (mainly those involved in the Gran Canaria Group) as well as by the CBD authorities in several countries. In September 2001 a paper was finalised for consideration at the 7th SBSTTA meeting held in November 2001 in Montreal. At SBSTTA this draft was debated comprehensively both during the working sessions of the conference and in the corridors. The result was a Recommendation (titled VII/8 Global Strategy for Plant Conservation) and the targets were further revised and updated to include an additional two (see Box 3).

The next stage involved working towards the next COP meeting which was due to be held in the Netherlands in April 2002. The Recommendation from SBSTTA recognised that further work on the Strategy was necessary, involving both national CBD authorities and the Gran Canaria Group working to refine the targets and the basis on which they had been developed. Following a comprehensive campaign to brief national delegations about the importance and significance of the Strategy it was duly accepted in the Netherlands.

The 16 targets are summarised in Box 3, but without the other elements outlined in

Box 2 The Gran Canaria Group.

The Gran Canaria Group is an informal grouping of representatives of a range of international and regional organisations and other major institutions involved in plant conservation, such as IUCN, WWF, IPGRI, FAO, BGCI, WCMC, IABG, the Smithsonian Institution, the Commonwealth Secretariat, the Royal Botanic Gardens of Edinburgh and Kew, Planta Europa and network organisations in Canada, the USA, Australia, Africa, South East Asia and other regions. A secretariat for the Gran Canaria Group is provided by Botanic Gardens Conservation International.

Box 3 Global Biodiversity Action Plan – targets.

Global targets to be achieved within 10 years are as follows:

Understanding and documenting plant diversity

1. A widely accessible working list of known plant species, as a step towards a complete world flora.
2. A preliminary assessment of the conservation status of all known plant species, at national, regional and international levels.
3. Development of models with protocols for plant conservation and sustainable use, based on research and practical experience.

Conserving plant diversity

4. At least 10 per cent of each of the world's ecological regions effectively conserved.
5. Protection of 50 per cent of the world's most important areas for plant diversity assured.
6. At least 30 per cent of production lands managed consistent with the conservation of plant diversity.
7. 60 per cent of the world's threatened species effectively conserved *in situ*.
8. 60 per cent of threatened plant species in accessible *ex situ* collections, preferably in the country of origin, and 10 per cent of them included in recovery and restoration programmes.
9. 70 per cent of the genetic diversity of crops and other major socioeconomically valuable plant species conserved, and associated local and indigenous knowledge maintained.
10. Management plans in place for at least 100 major alien species that threaten plants, plant communities and associated habitats and ecosystems.

Using plant diversity sustainably

11. No species of wild flora subject to unsustainable exploitation because of international trade.
12. 30 per cent of plant-based products derived from sources that are sustainably managed.
13. The decline of plant resources, and associated local and indigenous knowledge innovations and practices, that support sustainable livelihoods, local food security and health care, halted.

Promoting education and awareness about plant diversity

14. The importance of plant diversity and the need for its conservation incorporated into communication, educational and public awareness programmes.

Building capacity for the conservation of plant diversity

15. The number of trained people working with appropriate facilities in plant conservation increased, according to national needs, to achieve the Targets of this strategy.
16. Networks for plant conservation activities established or strengthened at national, regional and international levels.

the draft Strategy, such as its 'Scope, Rationale and Notes' and without 'Relevant Existing Initiatives'.

Conclusion

The international community's approach to biodiversity seems to have changed somewhat since Rio, 10 years ago. Biological diversity tends now to be considered as an essential part of efforts to eradicate poverty and achieve sustainable development. Conservation and development are no longer seen as conflicting but rather as mutually interdependent. This change has been in large part due to the process of international consensus-building instituted under the Convention and its inherent strengths of near universal membership, a comprehensive and science-driven mandate, international financial support for national projects, world-class scientific and technological advice and the political involvement of governments.

As the second World Summit on Sustainable Development (WSSD as it is now called), due to be held in Johannesburg, approaches it is worth considering the vast amount of work that has been achieved. While one could argue that it is likely that not one single species has been saved as a result of the CBD, nor has habitat destruction been stopped or reversed, it is just possible that some of the mechanisms have been put in place to make these happen in the near future.

In conclusion, while the CBD has generated a vast bureaucracy it is also true to say that a lot has been achieved. As well as the evidence of progress highlighted in this paper it is worth remembering that over 100 countries now have Biodiversity Action Plans and over $1.2 billion has been spent on biodiversity conservation through the Global Biodiversity Facility alone. While action plans and money alone don't necessarily equal success they are both needed before success can be achieved. There is no doubt that, internationally, the CBD has done more good for conservation than any other policy or piece of legislation.

References

ANONYMOUS (1994). *Biodiversity, the UK Action Plan*. London: HMSO, 188pp.

DEPARTMENT OF THE ENVIRONMENT (1995). *Biodiversity: the UK Steering Group Report*, Vol. 2, *Action Plans*. London: HMSO.

GARCÍA, F. L., WILLIAMS, C., TEN KATE, K. & CHEYNE, P. (2001). *Results of the pilot project for botanic gardens: Principles on access to genetic resources and benefit-sharing, common policy guidelines to assist with their implementation and explanatory text*. The Board of Trustees, Royal Botanic Gardens, Kew. ISBN 1-84246-023-4.

GLOWKA, L., BURHENNE-GUILMIN, S., SYNGE, H. in collaboration with MCNEELY, J. A. & GUNDLING, L. (1994). *A Guide to the Convention on Biological Diversity*. Gland & Cambridge: IUCN, xii + 161pp.

IUCN, UNEP & WWF (1980). *The World Conservation Strategy*. Gland: IUCN.

IUCN, UNEP & WWF (1991). *Caring for the Earth*. Gland: IUCN.

WRI, IUCN & UNEP (1992). *Global Biodiversity Strategy*. London: Commonwealth Secretariat, 92pp.

Growing rhododendrons at Pukeiti, New Zealand

Graham Smith

*Director, Pukeiti Rhododendron Trust, 2290 Carrington Road, R.D. 4,
New Plymouth, New Zealand*

The Pukeiti Rhododendron Trust was founded in 1951 with the gift of 150 acres of logged-over rainforest adjacent to the northwest slopes of the volcanic Egmont National Park. The area gradually grew to 900 acres and by 1980 the Trust had more than 2500 members. Whilst cool by New Zealand standards, Pukeiti is milder than any British garden and this is reflected by the range of plants that can be grown well. Recent trends in the climatic pattern indicate even warmer and wetter conditions will prevail in the future, resulting in further changes in the plant collections. Pukeiti is also committed to re-afforesting the bush and is 50 years into a 500-year project to replant the podocarps removed 70 years ago.

Introduction

Pukeiti is situated near the west coast of North Island at latitude 39°S at an altitude between 390 m and 500 m, 12 km from the Tasman Sea. Westerly winds ensure an even spread of moisture all year round with what might be called 'extreme rainfall', 4000 mm (145 in.) per year. Frosts are extremely light and infrequent whilst summer temperatures do not rise above 24°C.

A now 80-year-old *Rhododendron* 'Elegans' was one of the reasons for the Pukeiti site being chosen. It was planted by a roadman beside his hut and it was seen to be 'doing well' by William Douglas Cook in the late 1940s and he mentally noted that the high rainfall would 'ensure the death of any thrips' (Cook, 2001). Cook had been looking for several years for a site where rhododendrons could be grown to perfection when he saw the small isolated hill called Pukeiti (488 m) off the Carrington Road. One walk

through the beautiful woodland full of tree and filmy ferns was enough, and he knew he had found the site he wanted (Cook, 2001). It was not without a little subterfuge that founder members were recruited at £50 a time (a lot of money in 1950) and the Pukeiti Rhododendron Trust was formed with 20 founder members. The garden opened on 1 November 1951.

The garden centres around the entrance buildings from which a broad lawn gives views to rather formal but extremely colourful plantings. A hedge of *Camellia* 'Donation' and *Picea albertiana* S.Br. 'Conica' was planted in brick turrets but after 45 years they had to be replaced. Beds of hybrid rhododendrons and evergreen azaleas provide the traditional woodland garden scene, with *Cryptomeria japonica* D.Don as a backdrop. Evergreen 'Indica' and Belgian azaleas also form a semi-formal planting, with *Acer* 'Brilliantissima' in the background. A series of

Plate 1 Pukeiti: lawn view from the gatehouse.

meandering paths lead visitors away into woodland gardens where *R*. 'Winsome' with *Primula heladoxa* Balf.f. can be seen. The areas become more and more informal, with a mixture of native and exotic plants. Here an essentially different rainforest setting can be found in which both small and very large rhododendrons are presented. The 'large leaf' collection of rhododendrons is world class and well represented by plants with collectors' numbers from as early as 1953. *Rhododendron grande* Wight seedlings grow under the natural canopy of *Weinmannia* and *Beilschmedia* spp. The high rainfall and the light conditions provided by the evergreen canopy mean that plants grow better in open situations, and for them to perform well the canopy has to be thinned. For adventurous visitors there are 'bush' paths into the truly native vegetation, which is in the process of

being re-afforested with the aim of gradually returning it to its pre-colonial state.

Rhododendrons and azaleas at the garden

Many evergreen azaleas have been informally planted in a wild setting for contrast with ferns, which provide natural companion associations. Early flowering rhododendrons such as 'Chrysomanicum' and 'Snow Lady' form a September feature with *Magnolia campbellii* Hook. & Thoms. in the background. A 'wild garden' is the domain of the 'large leafs' and here *R. macabeanum* Watt ex Balf.f. and some of its hybrids fill the September scene with great pools of colour that seem to belong to the environment. Pukeiti's most famous large leaf – *R. protistum* Balf.f. & Forrest 'Pukeiti' from KW 21498, collected on the Myanmar 'Triangle expedition' of 1953 – flowers in August to September. New growth from the same plant follows two months later and is as spectacular in itself as the flower display. *Rhododendron magnificum* Kingdon-Ward flowers slightly later than the *R. protistum* and has looser trusses of rosy purple flowers and darker, heavier leaves with a thin indumentum from an early age. The distinctive pointed red buds of *R. magnificum* come into new growth a month earlier than *R. protistum* and can be vulnerable to early frosts. The later-flowering *R. sinogrande* Balf.f. & W.W.Sm. with *R. burmanicum* Hutch., which are usually out in October, enjoy mild, moist spring days with few frosts to damage the display. Pukeiti has named a few of the excellent seedlings that have masqueraded under species names, such as 'Ina Hair', n *R. macabeanum* seedling. A new self-sown seedling from 'Ina Hair' shows a lot of promise as a good garden plant but we have to recognise that few of these natural hybrids are ever going to become commercial. Something very new from N Vietnam, under the number AC 448, has foliage looking very similar to *R. protistum*; at this stage it looks enticingly subtropical. *Rhododendron montroseanum* Davidian 'Benmore' (KW 6261 from the 1924 Tibet expedition) is a famous Scottish plant and we have a young, hand-pollinated seedling at Pukeiti which shows the distinct long, silver-backed foliage and bright pink flowers characteristic of this cultivar.

The subsection *Arborea* is hugely successful almost anywhere in New Zealand. These plants grow quickly and are rarely damaged by early frosts. Self-sown seedlings abound in the bush margins. One of the most successful species at Pukeiti is *R. arboreum* Sm. in all its many forms, and the early plantings have become, after 45 years, tree-like thickets with great character. They also freely self seed and provide a nectar source for the native birds.

New introductions of *R. arboreum* from various wild collections have been added to the collection such as the Nepalese *R. arboreum* ssp. *cinnamomeum* (Lindl.) Tagg. var. *roseum* Lindl. introduced in the 1980s.

Some Kingdon-Ward collections that came to Pukeiti have produced outstanding plants in our climate. *Rhododendron elliottii* Watt ex Brandis is unsurpassed and provides two months of flower. *Rhododendron* 'Charisma' is the name we gave to a superior seedling under KW 20280 from the 1950 Assam/Tibet collection. Beautifully fragrant, it was listed as 'Ciliicalyx Alliance' but is not that species and is close to *R. ciliipes* Hutch. One of the finest red-flowered species and a plant that loves Pukeiti conditions is *R. elliottii*, KW 19083 form, collected on the 1949 Assam expedition. This is a very compact plant growing to about 1.8 m and free flowering in full sun. A wonderful group of hand-pollinated seedlings from *R. elliottii* (KW 20303) from the 1950 expedition in Assam is much more vigorous than the previous clone. They are grown with an underplanting of *Cardiocrinum giganteum* (Wall.) Mak. and *Cornus controversa* Hemsl. overhead (at 10 m × 12 m, it is the largest in New Zealand).

Rhododendron subsect. *Maddenia* is another major feature of the garden. These rhododendrons grow particularly easily and are amongst the most successful plants in the garden. The essence of Pukeiti is the dramatic *R. lindleyi* T.Moore with its divine scent drifting through the *Cyathea medullaris* (Forst.f.) Sw. tree ferns which provide the backdrop. *Rhododendron lindleyi* in its 'Ludlow and Sherriff' form is a more compact plant and has striking purple new growth. Both forms require excellent drainage. The most majestic of the maddenias has to be *R. nuttallii* Booth, which is a 'class plant' and successfully regenerates in the garden, so far showing no signs of hybridising. A recent introduction was a white *R. ciliicalyx* Franch. which has proved to be an outstanding garden plant but is not correctly named and so has been given the name 'White Gift'. This plant flowers in September and is strongly scented, and is earlier than all but the *R. veitchianum* Hook.f. Cubittii Group.

Some more recent species collections, particularly of the warm temperate *Choniastrum* and *Azaleastrum* sections which are well suited to the climate in Pukeiti, were the John Patrick's *Rhododendron* Venture collections from Taiwan. From these were raised some very interesting plants including some natural hybrids of *R. morii* Hayata (RV 9809) from the 1970 expedition, which is a lovely form. Even more recently collected from Yunnan is a fine *R. irroratum* Franch. (DT 317), a New Zealand collection by Sashal Dyal and Jeremy Thomson. These collectors have made several visits to the south of China but more importantly to Nagaland, NE India, where they recollected *R. macabeanum*, which is now growing well in the garden here. *Rhododendron stamineum* Franch. is not often seen but in the mild climate of Pukeiti it is an unusual late-flowering (Christmas) scented member of section *Choniastrum*. Even rarer is *R. vialii* Delavay & Franch. from Yunnan, unusual in the section *Azaleastrum* with its tubular red flowers and distinctive pink new growth. It is very prone to shoot borer attack in New Zealand, which is unfortunate, but it propagates easily and grows fast. One of the dwarfs we grow,

even if not a high alpine, is *R. pemakoense* Kingdon-Ward. It finds a good home on the raised terraces and is cut back every 10 years to keep it tidy.

There are many examples of hybridising done at Pukeiti and by other New Zealand growers to provide better plants for warm climates. In particular the scented maddenias are first class plants. Hybrid New Zealand rhododendrons are now plentiful and often much more suited to our mild climate. 'Lemon Lodge' is a good example, raised at Pukeiti from open-pollinated 'Prelude' and almost certainly crossed with our white *R. decorum* Franch. It is very floriferous and a good performer everywhere it has been tried in the country. It has received a New Zealand 'Award of Distinction'. Another excellent Pukeiti plant, which was open-pollinated from *R. yakushimanum* Nakai (= *R. degronianum* Carrière ssp. *yakushimanum* (Nakai) Kitam.) and is thought to have *R. aberconwayi* Cowan as the other parent, is 'Coconut Ice'. It is a compact plant that smothers itself every year with pink and white flowers and is easy to propagate and grow. The best compact red is 'Rubicon', bred by Ron Gordon by crossing 'Noyo Chief' with 'Kilimanjaro'. All seedlings were tall, vigorous reds except one, a compact, wonderful foliage plant with fine red trusses of flowers that are out for two months or more. Local plant breeder Felix Jury raised a number of tender hybrids using the fine 'Bodnant' form of *R. maddenii* Hook.f. ssp. *maddenii* (*polyandrum*) crossed with *R. cinnabarinum* Hook.f. hybrids. 'Moon Orchid' (× Sirius) has scented yellow flowers, red stems and disease-free foliage. 'Felicity Fair' is a sister seedling with larger, apricot flowers, stunning in full bloom. 'Bernice' (× 'Royal Flush Townhill') is the reddest of the group and is very heavy flowering; it grows to 2 m × 2 m and always looks good in the garden.

Foliage plays a significant part in the garden landscape and some of the rhododendrons are really striking. *Rhododendron degronianum* ssp. *degronianum* (*R. metternichii* Siebold & Zucc. spp. *pentamerum* (Maxim.) Sugim.) from wild source seed is an outstanding example of attractive foliage. The Taiwanese *R. pachysanthum* Hayata (RV 72001) is worth growing for the felted brown foliage alone. It was cultivated from the Patrick 1972 collection but it is slow growing. The tender *R. moulmainense* Hook.f. Stenaulum Group has magnificent, translucent, wine-coloured new growth which follows the very early, scented, lilac-pink flowers.

Attempts are made to provide epiphytic habitats such as tree fern stumps, and many self-sown seedlings appear on vertical banks, moss-covered logs and growing tree ferns. Growing plants as epiphytes is a speciality of Pukeiti, and observations of plants in the wild has given ideas we have tried in the garden. The high rainfall, evenly spread, makes establishing some species with critical drainage requirements very difficult. Only by planting some species completely above the ground have we been successful. One of the features of Pukeiti with its natural abundance of tree ferns is the ability to use epiphytic sites such as tree fern stumps. *Rhododendron dalhousiae* Hook.f. var. *dalhousiae* from Nepal is one species which does well sited on these fibrous bases. *Rhododendron beanianum* Cowan, like most of the *Neriiflora* subsection, also requires extremely good drainage and will not grow in the ground at Pukeiti except when planted on a tree fern base. *Rhododendron recurvoides* Tagg &

Plate 2 View of pools at Pukeiti.

Kingdon-Ward is happy planted amongst a jumble of old tree trunks and tree fern logs. *Rhododendron griersonianum* Balf.f. & Forrest seedlings growing naturally out of tree ferns, and flowering, are perhaps the ultimate epiphytes but they can be mixed with other self-sown plants such as *Clethra arborea* Ait. and *R. simsii* Planch.

Hydrangeas are ideal companion plants to have with rhododendrons at Pukeiti, and their long flowering season overlaps with them to provide interesting colour contrasts. Late-flowering *R. facetum* Balf.f. & Kingdon-Ward (SBEC 0183, Yunnan 1981) does superbly in the New Zealand climate and is enhanced by the association with *Hydrangea* 'Blue Prince'. Hydrangeas also contrast with the 'hot coloured' vireyas (rhododendrons of section *Vireya*).

Vireyas

Vireyas do well at Pukeiti with just a glass roof over their heads to keep off the heavy winter rain and occasional light frosts. In these open-sided shelters they do superbly well, and we have a very extensive species and hybrid collection. Many of these are wild collected, and have collectors' numbers, whilst the hybrids are mostly New Zealand and Australian bred. They provide flower over the full 12 months of the year and are a great attraction to visitors. The very first *Vireya* to attract attention in our collection was the repeat flowering Australian *R. laetum* J.J.Sm. × *zoelleri* Warb. seedling, which we named 'Simbu Sunset' after its fiery yellow and orange flowers. It is still growing after more than 20 years, having been cut back hard several times. 'Java Light' is another Australian seedling that we named as an outstanding garden plant: it flowers all year

round with vivid deep orange trusses. Os Blumhardt is a well-known breeder of warm climate plants and has produced some excellent *Vireya* hybrids such as the white 'Silver Thimbles' (*R. anagalliflorum* Wernum × *R. macgregoriae* F.Muell.). It is low growing and perpetual flowering. The dramatic red 'Inferno' lights up the scene when in flower. It is an *R. christi* Foerste hybrid that is much easier to please than the known parent species, a Blumhardt breeding trait! Pukeiti has produced some good hybrids such as 'Peaches and Cream' and 'Gilded Sunrise'. A × *R. konori* Becc. cross called 'Beverley McConnell' is a fine, scented, excellently foliaged plant which is now 1.8 m tall.

Visiting plants in their own habitat can be full of surprises for the grower. Observing the red *R. commonae* Foerste growing in the running waters of the Kain Swamp in the highlands of Papua New Guinea (PNG) proved truly unusual, and the multi-coloured forms of this species that occur here are not found anywhere else. I am pleased that running water is not a prerequisite for cultivating *R. commonae*. It performs extremely well in cultivation, with evidence of some hardiness. The cream form of *R. commonae* flowers several times a year and is a more open-growing plant, indicating long-established hybrid involvement. One of my favourite introductions is *R. rubineiflorum* Craven, also from PNG. It can be seen at Pukeiti growing on a tree fern log just as it did in the wild. I also collected *R. solitarium* Sleumer from the type locality, the summit of Mt. Kaindi (PNG), and it has proved to be an ungainly plant in cultivation, but with great foliage and beautiful blush-white, scented flowers. An exciting plant from Mt. Miap (PNG) looks likely to be an *R. herzogii* Warb. hybrid, possibly with *R. culminicolum* F.Muell. It is beautifully scented and attractive in habit and it has been christened 'Starburst' because of the rich pink shooting-star flowers. *Rhododendron herzogii* itself is white and scented with a pink flushed tube; it flowers heavily several times a year and the scent is superb. Our collections of this species come from Nigluma in PNG. *Rhododendron saxifragoides* J.J.Sm. is remarkable for its taproot that anchors it to the bottom of glacial bogs, high up at 3900 m on Mt. Giluwe, PNG. The cushion-like habit and single red flower are distinct but it is slow and difficult in cultivation. *Rhododendron* 'Saxon Glow' is the result of a successful attempt to use *R. saxifragoides* as a parent to produce dwarf, free-flowering vireyas, and at Pukeiti these are outstanding mound-like plants up to 80 cm high that cover themselves in red flowers several times a year. Using the natural habitat as a guide, *R. rarum* Schltr. was planted in an orchid basket where it displays its pendulous red flowers in a similar way to its wild habit of hanging out of a tree. This same treatment given to *R. stenophyllum* Hook.f. ex Stapf ssp. *angustifolium* (J.J.Sm.) Argent, A.L.Lamb & Phillipps from Mt. Kinabalu (Sabah, Malaysia) provides a delightful method of displaying the plant whilst also providing the essential free drainage it requires. The orange bells and needle-like foliage are very distinct and often attract comment. One of the finest yellow vireyas is *R. retivenium* Sleumer, also from Mt. Kinabalu. In its better forms it is stunning and fragrant but some forms are not so flamboyant. It is a vigorous grower with rigid, lanceolate foliage. Of the more recent introductions from Sumatra, *R. rarilepidotum* J.J. Sm. is proving to be an easy and very striking red-flowered shrub and must have great hybrid potential. The variable group comprising the *R. javanicum* (Blume) Benn. complex

has many variations, and the Kinabalu form is one of the most spectacular, with large trusses of vivid orange flowers. It can however be a sparsely branched plant and needs company to hide the bare stems. *Rhododendron alborugosum* Argent & J.Dransf. has just flowered for the first time at Pukeiti and appears to have promise, with good foliage and well-displayed white flowers. These appear on much younger plants than in its namesake *R. rugosum* Low ex Hook.f. After just 22 years we have flowered the yellow-flowered *R. lowii* Hook.f. collected by Keith Adams from Paka Caves, Mt. Kinabalu. This is said to be the largest leaved and flowered of all vireyas but it is slow growing and direct sunlight scorches the foliage.

Concluding remarks

Pukeiti is truly a 'Garden for all Seasons', with interest and beauty throughout the year. Now in its 51st year, it has established itself with a worldwide reputation not only for having some of the best growing conditions for rhododendrons and some of the most outstanding walks and vistas but also for its well-curated collection, with good records of many of its plants. This makes it valuable internationally both educationally and scientifically. The commitment to replenish the original Rimu-Rata forest (*Dacrydium cupressinum* Sol. ex Forst. and *Metrosideros robustus* A.Cunn.) is now becoming evident, but given that maturity is about 500 years it has a long way to go. In the meantime several hundred young trees are planted every year. The icon of Pukeiti's collection in its Jubilee year was the enormous *R. protistum* 'Pukeiti' (KW 21498). This plant has become a celebrity and regularly features in newspaper and magazine articles when in flower in the depths of winter. Some years it will have more than 200 soccer ball-sized pink trusses. A younger plant is now in another part of the garden and a whole valley has been planted with young seedlings from a more recent wild collection and should make a unique and spectacular feature in years to come.

The blending of native with exotic plants is Pukeiti's legacy for the future.

Reference

COOK, W. D. (2001). The Dream. *Pukeiti: A Garden for all Seasons* 51(2): 10.

Rhododendrons in the alpine garden

John Main

Cairnbeck, Newtown, Irthington, Carlisle, Cumbria CA6 4PF, UK

The rock garden and associated areas allow the alpine gardener to grow rhododendrons and other dwarf ericaceous shrubs in a garden situation. The rhododendrons mostly originate in the Himalaya, from Sikkim and China and from Taiwan, and there are also two European species from the Alps. The dwarf species and hybrids can be grown in drifts or groups and also as plantings surrounding a water feature. Some of the many different cultivars that are commonly used will be illustrated. *Rhododendron yakushimanum* (*R. degronianum* ssp. *yakushimanum*) with its many hybrids will be shown as they often form an important part of the Japanese style of garden.

In the rock garden the evergreen 'Kurume' azaleas, such as the lovely pink 'Hinomayo', are most useful, along with many of the dwarfer rhododendrons such as *R. campylogynum* and *R. impeditum*.

For the alpine gardener the peat garden will be the area of the garden where most of the rhododendron collection will be grown. Examples will be shown starting with the taller growers towards the rear, where species such as *R. augustinii* and *R. rigidum* can be used to form the backdrop. *Rhododendron forrestii* ssp. *forrestii* and the hybrid 'Yaku Fairy' can be planted directly into the blocks of peat. In front of the taller rhododendrons associated herbaceous and bulbous plants can be planted, including *Meconopsis grandis* and *M.* × *sheldonii*, *Nomocharis*, *Trillium* and hardy orchids. Dwarf and low-growing rhododendrons, such as *R. impeditum*, can be added to the front of the planting. Other companion plants which can be grown here are gentians and primulas, and other dwarf shrubs such as *Cassiope* and *Phyllodoce*.

Rhododendrons have a multiplicity of uses in the garden, from the wild garden to the rock garden, and some enthusiasts grow them in pots and containers, particularly if required for shows or if the gardener lives in an alkaline area.

In comparison to many other forms of gardening the alpine garden or rock garden is fairly recent in its conception, and this may be the reason why many alpine gardeners do not use rhododendrons, including dwarf forms, in their rock gardens. The James McNab Rock Garden in the Royal Botanic Garden Edinburgh (RBGE), dating from about 1870, contained few rhododendrons at that time for it was not until such dwarf species such as *R. fastigiatum* Franch. came into cultivation from the Himalayas and China (this species in 1940) – from collectors such as Forrest, Rock, Wilson, Kingdon-Ward, and Ludlow & Sherriff – that rhododendrons made any impact on the alpine gardener. The Victorian

Plate 1 Dwarf rhododendrons.

Plate 2 *Rhododendron yakushimanum* (= *R. degronianum* ssp. *yakushimanum*) trial, Wisley.

rock gardeners preferred to build copies of mountains in their gardens, with very little room left for planting as they wished the rock to dominate the scene and plants were seen as secondary.

Here in Scotland *Calluna* and *Erica* dominate the higher mountains whereas in the Sikkim Himalaya *R. anthopogon* D.Don and *R. setosum* D.Don provide the ground cover. In Europe the 'Alpenrose' *R. ferrugineum* L. is one of only two species. If there had been more dwarf rhododendrons perhaps it would have made a greater impact on the alpine gardener. In the RBGE mass plantings of dwarf rhododendrons are grown very much in a 'heather garden' style (Plate 1).

The one rhododendron which has made an impact, particularly with European landscape architects, is *R. yakushimanum* Nakai (*R. degronianum* Carrière ssp. *yakushimanum* (Nakai) Kitam.) (Plate 2) and its hybrids. An example of their use can be seen in the Araki Japanese Garden in Hamburg. Trials of *R. yakushimanum* hybrids at the Royal Horticultural Society (RHS) Garden Wisley have always been popular with members and visitors; perhaps if the RHS had more trials of other dwarf rhododendrons it could help to increase their popularity in our gardens. It would appear that evergreen azaleas have had a much larger impact in rock gardens, particularly colourful cultivars such as 'Hinomayo'. It is very important when considering planting dwarf rhododendrons in the rock garden that they are not planted too deeply.

The peat garden as a feature is much younger in its concept and it is here that most British gardeners will grow their dwarf species, in humus-rich soils using blocks of peat to form the terraces on which the plants will be grown instead of rocks. The aspect is very important for this feature, and it should have a much more northerly position to avoid the hot midday sun of the more usually south-facing rock garden.

By all means have your larger trees and shrubs at the rear to begin forming the picture. In front of this some of the slightly larger rhododendrons can be grown, including *R. augustinii* Hemsl. and *R. rigidum* Franch. along with dwarf *Pieris*. Species such as *R. forrestii* Balf.f. ex Diels ssp. *forrestii* and the hybrid 'Yaku Fairy' can be planted directly in the blocks of peat along with other dwarf ericaceous shrubs, the roots of which help to bind the blocks together, thereby increasing their life. Amongst these taller growing shrubs, taller growing herbaceous and bulbous plants can be accommodated and here *Meconopsis* × *sheldonii* G.Tayl. (Plate 3), *Nomocharis farreri* (W.E.Evans) R.Harrow, the North American woodlander *Trillium grandiflorum* (Michx.) Salisb. 'Roseum' along with the hardy orchid *Dactylorhiza elata* (Poir.) Soó, *Paris polyphylla* J.J.Sm. var. *yunnanensis* (Franch.) Hand.-Mazz. and the 'candelabra' primulas will provide a foil for the smaller rhododendrons etc. Unfortunately, due to wetter, colder springs dactylorhizas have become much more susceptible to attack from the fungus *Cladosporium*.

In recent years there have been many reintroductions, particularly through the help of our Chinese colleagues in Kunming; *Lilium lophophorum* Franch. is one example. However, we should be looking at ways to increase such plants when they have been reintroduced without having to continually go back into the wild for more. With bulbous plants like *Lilium* it is a fairly simple task to propagate from home-produced seed or from scales, both of which will give

Plate 3 *Meconopsis × sheldonii.*

Plate 4 *Rhododendron* 'Princess Anne'.

flowering-sized bulbs within three years.

As we near the front of the bed, homes can be found for dwarfs such as *R. campylogynum* Franch. and *R. impeditum* Balf.f. & W.W.Sm. and the reliable *R.* 'Blue Tit', *R.* 'Princess Anne' (Plate 4) and the more recent *R.* 'Oban'. But selection should not purely be *Rhododendron*, for the many other dwarf ericaceous shrubs are very much at home, including *Cassiope*, *Phylliopsis* × *hilleri* Cullen & Lancaster 'Pinocchio' and *Phyllodoce caerulea* (L.) Bab. Between these dwarf shrubs room can be found for the many autumn-flowering gentians.

We could all make better use of dwarf rhododendrons in our gardens for they are ideally suited to the smaller home gardener or for more mass planting on a larger scale.

A genetic and physiological study of *Rhododendron* cold hardiness

Rajeev Arora[1], Chon-Chong Lim[2], Steven L. Krebs[3] & Calin O. Marian[4]

[1]*Department of Horticulture, Iowa State University, Ames, IA 50011, USA,* [2]*Soft Landscaping, CyPark Sdn Bnd, No. 15 Jalan 16/11, 46350 Petaling Jaya, Malaysia,* [3]*The David G. Leach Research Station of the Holden Arboretum, 1890 Hubbard Rd., Madison, OH 44057, USA, and* [4]*Department of Biological Sciences, Florida State University, Tallahassee, FL 32306, USA*

Few genetic studies have been conducted on the inheritance of cold hardiness (CH) in woody plants. We initiated this research to develop a reliable and non-destructive method of determining CH in progeny populations segregating for this trait, and to apply this methodology to such populations. The distributions of leaf freezing tolerances – in F_2 and backcross populations derived from a *R. catawbiense* × *R. fortunei* cross (super cold-hardy and moderately cold-hardy species, respectively) – suggested that as few as three genes with strong additive effects were controlling variation in hardiness, and that differences in midwinter freezing tolerance were primarily due to differences in acclimating ability among the segregants. Experiments are underway to 'tag' these genes by locating DNA markers that are physically linked to and co-segregate with the CH trait in these populations. Observations from genetic studies led us to hypothesise that in *Rhododendron*, a woody perennial with prolonged juvenile or pre-flowering phase, physiological age factors influence adaptive traits such as cold tolerance. In support of this notion, a multi-year study indicated that leaf freezing tolerance (LFT) increased with developmental phase-change (juvenile to mature plants) in *Rhododendron*. Data from backcross (*n* = 20) populations derived from *R. catawbiense* and *R. fortunei* parents indicated a significant increase in LFT of ~ 5–6°C per year as the seedlings aged from two to five years old. We believe that the maximum (or 'true') cold-acclimation potential of these populations could not be realised until they reached reproductive maturity. Physiological studies in our laboratory indicate that levels of a 25 kDa dehydrin (a specific 'stress-protein') were closely associated with differences in LFT among cold-hardened F_2 segregants and among a wide array of *Rhododendron* species. However, this protein was not found to accumulate in *Rhododendron* species unable to exhibit seasonal cold hardening. It is suggested that the presence vs. absence of the 25 kDa dehydrin could serve as a genetic marker to distinguish between super cold-hardy and less cold-hardy rhododendron genotypes. Moreover, quantitative accumulation of this dehydrin during cold acclimation appears to be associated with the cold-acclimation ability in *Rhododendron* spp. Tissue-specific differential accumulation of this protein appears to be correlated with the differential sensitivity of various parts of the *Rhododendron* leaf to cold injury.

Introduction

Woody landscape plants residing in temperate zones experience adverse winter conditions during their annual cycle and, therefore, must develop sufficient cold hardiness in the autumn of each year (cold acclimation) in order to prepare for overwintering and surviving low midwinter temperatures. Lack of adequate cold acclimation (CA), a genetically determined trait, often limits where many landscape plants can be grown successfully, besides affecting their geographic distribution. Low temperatures often cause winter injury to many landscape plantings, thereby causing substantial economic losses. To be able to expand the range of landscape plants such as rhododendrons that could be grown in areas with cold winters, one must first acquire a basic understanding of the cold-acclimation process. However, few attempts have been made to study the genetic, physiological or molecular mechanism(s) of CA in woody perennials.

Why is *Rhododendron* a useful system for cold-hardiness research?

The genus *Rhododendron* has a number of attributes that make it amenable to cold-hardiness (CH) research. There are over 800 species of *Rhododendron* distributed throughout the Northern Hemisphere, ranging from tropical to polar climates and varying widely in CH. Many species in subsection *Ponticum*, such as *R. maximum* L. and *R. brachycarpum* D.Don ex G.Don, are leaf-hardy to −60°C and bud-hardy to −30°C, whereas cold-tender species such as *R. griersonianum* Balf.f. & Forrest and *R. barbatum* Wall. ex G.Don show both leaf and bud damage at temperatures approaching −18°C. Hardiness is correlated with provenance (altitude and latitude), which suggests that the trait has evolved through natural selection on genetic variability. A wide range of CH among species and the relative ease of cross-hybridisation in this genus makes it possible for the generation of progenies segregating for CH to be used in genetic studies.

Some *Rhododendron* species, notably those in section *Ponticum*, are also broad-leaved evergreens, which allows the use of leaves, year-round, for CH studies. This makes rhododendron a preferred experimental system for cold-hardiness research for the following reasons. By using leaves to estimate CH (as we have in our research), physiological complexities arising from the simultaneous changes in dormancy and cold hardiness in buds (Arora *et al.*, 1992) are avoided. Also, in contrast to flower buds, freezing tolerance in *Rhododendron* leaves is conferred without supercooling, enabling the use of leaves for freeze–thaw experiments and ion-leakage assays for the determination of leaf freezing tolerance (LFT). We have previously demonstrated that this method of estimating CH provided LFT rankings in an array of evergreen *Rhododendron* cultivars that were consistent with USDA hardiness zone rankings (Lim *et al.*, 1998a). Moreover, in breeding programmes, progenies are typically grown to maturity (5+ years) in order to assess winter damage to floral buds, and therefore improving cold hardiness via traditional breeding has been a rather slow process. However, the use of leaves for CH measurements from relatively young progenies (1–2 years) may allow less hardy progenies to be culled early on (without having to wait till reproductive maturity is attained), saving time, labour and resources.

Can LFT be used to predict flower bud hardiness in rhododendron plants?

So far in our genetic and physiological analyses of CH of rhododendrons we have used LFT (as opposed to bud hardiness) to classify rhododendron genotypes or progenies as more or less hardy. So, the question arises: considering the nursery industry in colder climates desires new hybrids and cultivars primarily with improved bud hardiness, is the use of LFT as an indicator of rhododendron CH appropriate? Data on the relationship between leaf and bud hardiness of about 37 *Rhododendron* species indicate that estimates of CH based on leaf performance can be used as a general measure of floral bud hardiness (Figure 1), although cold-hardened leaf tissues are typically hardier than floral buds, maximum hardiness of the latter being limited by the supercooling of bud primordia (Wisniewski & Arora, 1993).

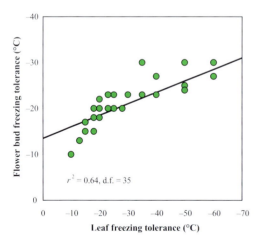

Figure 1 Regressions of floral bud freezing tolerance on leaf freezing tolerance (LFT) in *Rhododendron* species (37 plants in total); regression analysis of data presented by Sakai *et al.* (1986). (From Lim *et al.*, 1998b.)

Genetic study

Genetic populations and evaluation of LFT of progenies

A single F_1 plant, 'Ceylon' (*R. catawbiense* Michx. × *R. fortunei* Lindl.), was selfed to create an F_2 population. Non-acclimated (unhardened during summer) and cold-acclimated (midwinter-hardy) plants were evaluated for LFT by subjecting leaf-discs to a laboratory-based controlled freeze–thaw protocol and determining freeze injury using electrolyte-leakage assay (Lim *et al.*, 1998a). Much of the experimental variability commonly associated with field trials is significantly reduced by the controlled freeze method, and the somewhat subjective method of estimating CH by visual assessment of leaf injury is replaced with the more quantitative ion-leakage procedure. Ion leakage over a range of freezing temperatures typically displayed a sigmoidal response in *Rhododendron* leaves (Figure 2). We used the parameter T_{max}, defined as the temperature causing the maximum rate of injury, as an estimator of CH (Lim *et al.*, 1998a) (Figure 2).

Non-acclimated CH and cold-acclimation ability are independently controlled

Midwinter CH status is really the outcome of two properties in an individual plant: its freezing tolerance in the non-acclimated state and its acclimating ability. The cold-acclimation process is triggered by environmental cues such as shorter daylength and cooler temperatures beginning in late summer and extending into late autumn (Fuchigami *et al.*, 1971), and acclimating plants undergo physiological and structural changes that condition them to low midwinter temperatures.

The acclimating ability of a plant in our research was defined as the LFT in the cold-

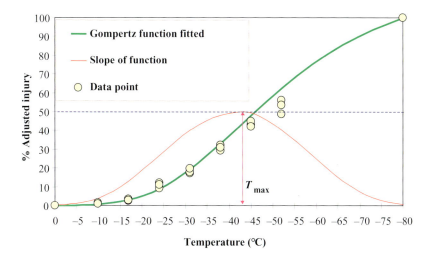

Figure 2 Per cent injury as a function of freezing temperatures in *Rhododendron* 'Ceylon' leaf tissues. Data are fitted to a Gompertz function where the slope of the curve (d Ad injury/d Temp) is the rate of injury. T_{max}, inflection point of the slope, is the temperature causing maximum rate of injury and is defined as leaf freezing tolerance (LFT). (From Lim *et al.*, 1998b.)

acclimated condition minus the LFT in the non-acclimated (NA) condition (Lim *et al.*, 1998b). As expected, NA-LFTs from summer-collected progenies in F_2 populations were much lower than CA-LFTs determined for the same individuals during the previous winter. There was, however, significant variation in NA-LFTs (–2 to –9°C in the F_2), and we were interested in determining whether these freezing-tolerance differences in the NA state had an impact on a plant's midwinter hardiness level. Scatter-plots of these CH components indicate that NA-LFT is not correlated with cold-acclimation ability (Figure 3A), whereas CA-LFT was highly correlated with acclimation ability (Figure 3B). These results suggest that the physiological processes involved in NA har-

diness and acclimation ability are independent of each other and that these two components of midwinter hardiness are under independent genetic control, a conclusion also drawn from research on potatoes (Stone *et al.*, 1993). The relationships also indicate that, of the two components, it is the variation in acclimating ability (–13.6 to –37.3°C in the F_2 progenies) rather than NA hardiness that accounts for the observed differences in T_{max} in rhododendrons. This is the meaningful variation from a CH breeding perspective, and from a more basic point of view it is quite possible that cold-hardiness genes confer freezing tolerance via their major effects on acclimation ability (induced CH) rather than on the non-acclimated freezing tolerance (constitutive CH).

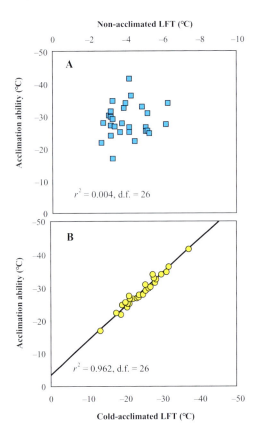

Figure 3 Relationships between the components of cold hardiness in F_2 progenies. (A) Non-acclimated LFT vs. cold-acclimation ability. (B) Cold-acclimated LFT vs. cold-acclimation ability. (From Lim *et al.*, 1998b.)

Physiological study

Dehydrins and cold acclimation

Dehydrins (also known as group 2 LEA family of proteins) are hydrophilic, heat-stable proteins that are induced in response to plant stresses possessing a dehydration component, such as salt, water or freezing stress (Close, 1996). Accumulation of dehydrin protein and transcripts during CA has been amply documented in a number of herbaceous species (Close, 1997), where freezing tolerance of cold-acclimated tissues typically does not exceed −15°C. Investigations of dehydrin expression and its association with CA in woody perennials, which exhibit significantly higher freezing tolerance in a cold-acclimated state (−50°C or lower) than herbaceous plants, are comparatively scarce. A functional role for dehydrins in freezing tolerance is suggested, in part, by their *in vitro* cryoprotectant properties (Lin & Thomashow, 1992; Wisniewski *et al.*, 1999). A direct relationship between dehydrins and freezing tolerance was demonstrated by the ability of constitutively regulated *cor* (*cold-regulated*) genes (some of which encode dehydrin proteins) in *Arabidopsis* to confer freezing tolerance without prior acclimation (Jaglo-Ottosen *et al.*, 1998). It has also been postulated that dehydrins may act as ion-sequesters (Palva & Heino, 1998) or as molecular chaperones (Campbell & Close, 1997; Close, 1997) under stressful conditions, stabilising proteins and membranes.

We conducted experiments to determine whether dehydrin profiles differ qualitatively and quantitatively among progeny from a population segregating for CH. This line of inquiry was derived from earlier work (see above), which documented that F_2 segregants from a cross between moderately cold-hardy and super cold-hardy *Rhododendron* species varied from −18 to −48°C in LFT (Lim *et al.*, 1998b). Remnant leaf tissue from selected progeny already evaluated for LFT was used for dehydrin analysis, in order to determine the relationship between biochemical phenotypes (dehydrin accumulation) and CH phenotypes in the F_2 population.

A 25 kDa dehydrin as genetic marker for freezing tolerance in *Rhododendron*

Comparisons of LFT and dehydrin profiles among F_2 segregants included a super cold-hardy parent (*R. catawbiense*), a moderately hardy parent (*R. fortunei*), the F_1 hybrid cultivar 'Ceylon' derived from the cross *R. catawbiense* × *R. fortunei*, and F_2 seedlings resulting from the self-pollination of 'Ceylon'. Initially, the progeny were classified by hardiness phenotype into 'low', 'medium' and 'high' LFT groups. Each group differed significantly from the adjacent group by a mean leaf T_{max} of ~ 10°C (Figure 4). Leaves from five random progeny in each group were pooled equally on a fresh weight basis, and the three phenotypic 'bulks' were extracted for total proteins in borate buffer (Arora *et al.*, 1992; Lim *et al.*, 1999) and analysed for differences in dehydrin profiles through immunoblot analysis (Western blotting) using anti-dehydrin antibody (Arora &

Wisniewski, 1994; Lim *et al.*, 1999). Once bulk differences in dehydrin expression were evident, leaves from three individuals in each F_2 bulk were extracted and evaluated separately.

Coomassie-stained SDS–PAGE profiles of total proteins and their anti-dehydrin immunoblots for the two parents, F_1 and F_2 'bulks' are presented in Figures 4(A) and (B). The highest optical density (O.D.) value was derived from a 25 kDa dehydrin (through densitometry scans of immunoblots; Lim *et al.*, 1999) that was present in the super-hardy *R. catawbiense* parent, absent in an equal loading of protein from the moderately hardy *R. fortunei* parent, and present at intermediate levels in the F_1 hybrid 'Ceylon', which also displayed an intermediate LFT. Levels of the 25 kDa dehydrin in cold-acclimated leaves from *R. catawbiense* were about five-fold higher than in non-acclimated leaves from the same source;

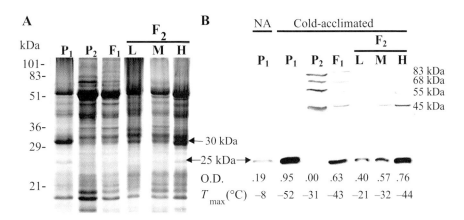

Figure 4 (A) SDS–PAGE profiles of total soluble proteins from cold-acclimated leaves (30 μg per lane). P_1 = *Rhododendron catawbiense*, P_2 = *R. fortunei*, F_1 = *R. catawbiense* × *R. fortunei* = 'Ceylon', F_2 = 'Ceylon' selfed. L, M and H correspond to the 'low', 'medium' and 'high' freeze-tolerant F_2 bulks. (B) Anti-dehydrin immunoblots of parents, F_1 and bulks of F_2 progenies. Protein (15 μg) was loaded in each lane. NA = non-acclimated, O.D. = optical densities, T_{max} = quantitative measure of leaf freezing tolerance. (From Lim *et al.*, 1999.)

however, the relatively abundant 25 kDa dehydrin on immunoblots appeared as a faint band on the corresponding protein gel (Figures 4A and B). The reason for apparent lower dehydrin intensities on SDS–PAGE is not clear. Among F_2 seedlings, the 25 kDa dehydrin was the only protein clearly associated with differences in CH. Comparisons of the F_2 tissue bulks grouped by phenotypic class – 'low', 'medium' and 'high' LFT – indicated a 50 to 100% increase in the 25 kDa dehydrin level as the CH status increased (Figure 4B). When individual progeny from these F_2 bulks were evaluated, most but not all of the offspring displayed increased dehydrin accumulation at higher levels of LFT (Figure 5A). Regression of T_{max} on dehydrin O.D. values for this population (Figure 5B) resulted in a significant positive relationship

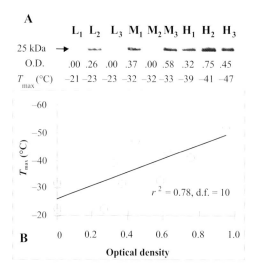

A

	L_1	L_2	L_3	M_1	M_2	M_3	H_1	H_2	H_3
25 kDa →									
O.D.	.00	.26	.00	.37	.00	.58	.32	.75	.45
T_{max} (°C)	−21	−23	−23	−32	−32	−33	−39	−41	−47

$r^2 = 0.78$, d.f. $= 10$

B

Figure 5 (A) Anti-dehydrin immunoblots of nine individual F_2 progenies. Protein (15 μg) was loaded in each lane. (B) Regression analysis of LFT on dehydrin O.D. in the population comprised of parents, F_1 and nine F_2 progenies. O.D. = optical densities, T_{max} = quantitative measure of leaf freezing tolerance. (From Lim *et al.*, 1999.)

$(r^2 = 0.78$, d.f. $= 10$, $P < 0.05)$, an indication that 25 kDa dehydrin levels were reasonably predictive of acclimated LFT status in this population.

Genetic interpretation of dehydrin profiles

Because the immunoblots are detecting temperature-induced proteins, the parental differences in dehydrin profiles (presence vs. absence of a specific molecular weight protein) could be attributed to regulatory genes rather than dehydrin-encoding structural genes. For example, *R. catawbiense* and *R. fortunei* could share identical dehydrin genes under the control of regulatory genes, which respond differently to low temperatures. However, studies with cold-responsive regulatory elements (transcriptional factors) suggest that they promote coordinate expression of a suite of cold-regulated genes, some of which also encode dehydrin or dehydrin-like proteins (Jaglo-Ottosen *et al.*, 1998). Moreover, there has been no evidence in the literature, to date, supporting the scenario whereby an identical dehydrin gene common to two cold-acclimating species is differentially induced by cold in only one of them. Therefore, it appeared likely that the dehydrin profiles observed in acclimated leaf tissue from each parent, in our study, represented the expression of the full set of cold-regulated dehydrins, and that the presence vs. absence (+/−) of the 25 kDa dehydrin was due to structural gene differences rather than regulatory genes. Moreover, accumulations of several dehydrins in the F_1 (based on O.D. or visual estimates) were intermediate to parental levels, suggesting a gene dosage effect, which could result from co-dominant expression of 'presence' and 'absence' alleles at the corresponding loci. Based on the above scenario, we had suggested that the presence or absence of 25 kDa dehydrin could serve as a genetic marker to distinguish between super-hardy and less hardy *Rhododendron* species (Lim *et al.*, 1999). However, with improved protein extraction protocols developed recently in our laboratory we were able to detect the 25 kDa dehydrin in the cold-acclimated leaves of *R. fortunei* also (see below; Table 1), although at a significantly lower level than in *R. catawbiense*. Therefore, it now appears that differential accumulation of 25 kDa dehydrin (and perhaps differential hardiness) in *R. catawbiense* and *R. fortunei* may be due to differential gene regulation of 25 kDa dehydrin rather than structural gene differences as previously thought.

25 kDa dehydrin is highly conserved across diverse *Rhododendron* species

Winter acclimated leaves from a total of 11 *Rhododendron* species were obtained from The Holden Arboretum's David G. Leach Research Station in Madison, Ohio (Table 1). The species surveyed represented evergreen taxa within two highly divergent (evolutionarily) *Rhododendron* subgenera – subgenus *Hymenanthes*, the non-scaly-leaved or elepidote rhododendrons, and subgenus *Rhododendron*, the scaly-leaved or lepidote species. Cold-acclimated leaves were collected during late December and January from plants growing either in the field or in containers held in cold storage. Non-acclimated leaves

Table 1 *Rhododendron* species used in this study, their corresponding cold-acclimated leaf hardiness, and various dehydrin species that accumulate in acclimated tissues.

Species	Leaf freezing tolerance (°C)[a]	Dehydrins (kDa)
Subgenus *Hymenanthes*		
R. brachycarpum	−60	25, 28, 46, 50, 64
R. catawbiense 'Catalgla'	−53	25
R. dichroanthum var. scyphocalyx	−23	25, 28, 30, 32
R. fortunei 'Gable'	−38	25, 28, 41, 46, 50, 64
R. maximum	−52	25, 30, 50, 73
R. metternichii	−48	25, 28, 50
R. vernicosum 'Gable's Vernicosum'	−25	25, 28, 37, 50
R. yakushimanum[b]	−40	25, 50
Subgenus *Rhododendron*		
R. brookeanum	−7	41[c]
R. mucronulatum	−50	25, 34, 50
R. russatum	−40	25, 32, 41

[a] Italics indicate LFT estimated in our laboratory as T_{max} (the temperature causing maximum rate of injury); unitalicised data are leaf LST values (lowest survival temperatures) from Sakai *et al.* (1986).
[b] = *R. degronianum* ssp. *yakushimanum*.
[c] Does not increase in cold-acclimated leaves compared with non-acclimated leaves.

were collected from the same individuals during summer (July–August).

A total of 11 dehydrins was observed in the survey of 11 *Rhododendron* species (Table 1). They ranged in molecular weight from 25 to 73 kDa, and were detected, for the most part, in both non- and cold-acclimated leaf tissues. With one exception, accumulation of the 11 dehydrins was higher in cold-acclimated than in non-acclimated leaves (Table 1). Our data are consistent with previous observations (Close *et al.*, 1993) that dehydrins belong to a multi-gene family that can vary within and among plant taxa. Although intergeneric and intraspecific variability of dehydrins in plants has been documented (Muthalif & Rowland, 1994; Wisniewski *et al.*, 1996; Sarhan *et al.*, 1997), data on interspecific variability in herbaceous and woody species are scarce. The presence of 11 dehydrins distributed among a diverse group of *Rhododendron* species that vary in their cold hardiness suggests that both quantitative and qualitative differences in dehydrins could affect winter survival.

Most notably, the 25 kDa dehydrin was found in all but one of the taxa (90%). The species that lacked this variant, *R. brookeanum* Low ex Lindl., is a tropical Indonesian plant that may not be capable of cold acclimation. The only other band that appeared in *R. brookeanum* immunoblots, a 41 kDa dehydrin, did not appear to be upregulated in cold-acclimated leaf tissue (Table 1). Although there is little knowledge about the cellular function of dehydrins during freezing events, it is nonetheless intriguing to speculate here that the 25 kDa protein may play some key role in cold hardiness, based on its evident conservation over evolutionary time. It appears that there may have been some selection pressure to maintain this particular dehydrin at high frequency among *Rhododendron* species.

Conclusion

As discussed earlier in this paper, the differential quantitative accumulation of 25 kDa dehydrin could serve as a genetic marker to distinguish between 'super-hardy' and 'less hardy' *Rhododendron* genotypes. This work was based on a segregating population from a controlled, interspecific cross, where F_2 progeny shared a similar genetic background. Taken together with the finding that the 25 kDa protein is the only highly conserved dehydrin in our species survey, these data are concordant with our hypothesis that this particular dehydrin plays a key, but unknown, role in *Rhododendron* cold hardiness. From a physiological perspective, overall plant cold hardiness (or winter survival) is a complex trait and is an outcome of the interaction between several biochemical, physiological and morphological factors. Research to date involving both mapping and transformation strategies has not identified single structural genes that 'confer' the cold-hardy phenotype. In all likelihood, the 25 kDa dehydrin alone is not sufficient to confer tolerance to freezing stress, but it may be a necessary component of biochemical interactions with other cryoprotectant metabolites.

Acknowledgements

This research was supported, in part, by grants from the American Rhododendron Society.

References

ARORA, R. & WISNIEWSKI, M. E. (1994). Cold acclimation in genetically related (sibling) deciduous and evergreen peach (*Prunus persica* [L.] Batsch). II. A 60-kilodalton bark protein in cold-acclimated tissues of peach is heat stable and related to the dehydrin family of proteins. *Plant Physiol.* 105: 95–101.

ARORA, R., WISNIEWSKI, M. E. & SCORZA, R. (1992). Cold acclimation in genetically related (sibling) deciduous and evergreen peach (*Prunus persica* [L.] Batsch). I. Seasonal changes in cold hardiness and polypeptides of bark and xylem tissues. *Plant Physiol.* 99: 1562–1568.

CAMPBELL, S. A. & CLOSE, T. J. (1997). Dehydrins: genes, proteins, and associations with phenotypic traits. *New Phytol.* 137: 61–74.

CLOSE, T. J. (1996). Dehydrins: emergence of a biochemical role of a family of plant dehydration proteins. *Physiol. Plant.* 97: 795–803.

CLOSE, T. J. (1997). Dehydrins: a commonality in the response of plants to dehydration and low temperature. *Physiol. Plant.* 100: 291–296.

CLOSE, T. J., FENTON, R. D. & MOONAN, F. (1993). A view of plant dehydrins using antibodies specific to the carboxy terminal peptide. *Plant. Mol. Biol.* 23: 279–286.

FUCHIGAMI, L. H., WEISER, C. J. & EVERT, D. R. (1971). Induction of cold acclimation in *Cornus stolonifera* Michx. *Plant Physiol.* 47: 98–103.

JAGLO-OTTOSEN, K. R., GILMOUR, S. J., ZARKA, D. G., SCHABENBERGER, O. & THOMASHOW, M. F. (1998). *Arabidopsis* CBF1 overexpression induces *cor* genes and enhances freezing tolerance. *Science* 280: 104–106.

LIM, C. C., ARORA, R. & TOWNSEND, E. D. (1998a). Comparing Gompertz and Richards functions to estimate freezing injury in *Rhododendron* using electrolyte leakage. *J. Amer. Soc. Hort. Sci.* 123: 246–252.

LIM, C. C., KREBS, S. L. & ARORA, R. (1998b). Genetic study of freezing tolerance in *Rhododendron* populations: Implications for cold hardiness breeding. *J. Amer. Rhododendron Soc.* 52: 143–148.

LIM, C. C., KREBS, S. L. & ARORA, R. (1999). A 25 kD dehydrin associated with genotype- and age-dependent leaf freezing tolerance in *Rhododendron*: a genetic marker for cold hardiness? *Theor. Appl. Genet.* 99: 912–920.

LIN, C. & THOMASHOW, M. F. (1992). A cold-regulated *Arabidopsis* gene encodes a polypeptide having potent cryoprotective activity. *Biochem. Biophy. Res. Commun.* 183: 1103–1108.

MUTHALIF, M. M. & ROWLAND, L. J. (1994). Identification of dehydrin-like proteins responsive to chilling in floral buds of blueberry (*Vaccinium*, section *Cyanococcus*). *Plant. Physiol.* 104: 1439–1447.

PALVA, E. T. & HEINO, P. (1998). Molecular mechanisms of plant cold acclimation and freezing tolerance. In: LI, P. H & CHEN, T. H. H. (eds) *Plant Cold Hardiness: Molecular Biology, Biochemistry, and Physiology*, pp. 1–14. New York: Plenum Press.

SAKAI, A., FUCHIGAMI, L. & WEISER, C. J. (1986). Cold hardiness in the genus *Rhododendron*. *J. Amer. Soc. Hort. Sci.* 111: 273–280.

SARHAN, F., OUELLET, F. & VAZQUEZ-TELLO, A. (1997). The wheat wcs120 gene family: a useful model to understand the molecular genetics of freezing tolerance in cereals. *Plant Physiol.* 101: 439–445.

STONE, J. M., PALTA, J. P., BAMBERG, J. B., WEISS, L. S. & HARBAGE, J. F. (1993). Inheritance of freezing resistance in tuber-bearing *Solanum* species: evidence for independent genetic control of nonacclimated freezing tolerance and cold acclimation capacity. *Proc. Natl. Acad. Sci.* 90: 7869–7873.

WISNIEWSKI, M. & ARORA, R. (1993). Adaptation and response of fruit trees to freezing temperatures. In: BIGGS, A. R. (ed.) *Cytology, Histology and Histochemistry of Fruit Tree Diseases*, pp. 299–320. Boca Raton, FL: CRC Press.

WISNIEWSKI, M., CLOSE, T. J. & ARORA, R. (1996). Seasonal patterns of dehydrins and 70 kDa heat-shock proteins in bark tissues of eight species of woody plants. *Physiol. Plant.* 96: 496–505.

WISNIEWSKI, M., WEBB, R., BALSAMO, R., CLOSE, T. J., YU, X-M. & GRIFFITH, M. (1999). Purification, immunolocalization, cryoprotective, and antifreeze activity of PCA60: a dehydrin from peach (*Prunus persica*). *Physiol. Plant.* 105: 600–608.

Rhododendron husbandry and hygiene: essential steps towards survival

Caledonian Mycological Enterprises, Edinburgh EH4 3HU, and Royal Botanic Garden Edinburgh, 20A Inverleith Row, Edinburgh EH3 5LR, UK

A general survey of the diseases and disorders of *Rhododendron* is presented, with possible treatments offered. A plan for careful husbandry is advocated.

Introduction

Rhododendrons are generally rather hardy plants and although it is true that a complete computer-generated book devoted to fungi on rhododendrons (Farr *et al.*, 1996) has been produced, these shrubby ericaceous plants are generally fairly free of injury and disease.

There have been some epidemics in recent years, for example of mildews as outlined by Helfer (2003) and of anthracnose by Vinnere *et al.* (2002) in Sweden, but such are comparatively rare in the UK. Equally, root rot caused by fungi such as *Armillaria* spp., although widespread, is generally a result of the history of the site. Rhododendrons, especially the elepidote forms, appear to be rather resilient, and some of the most conspicuous damage and injury to these plants is caused by physical factors such as chilling winds producing irregular leaf banding followed by necrosis, cold effects, such as low soil temperature with its associated bark splitting, stem girdling and leaf fall, the last often some time after the damage has been done, and pollution. Any damage predisposes a plant to fungal attack.

Rhododendrons show very distinctive symptoms in nutrient-deficient media, particularly those low in nitrogen and potash, which are accentuated in some series, e.g. *Barbatum*. It must be realised that rhododendrons require calcium for the formation of cell walls in the same way as other plants. Both iron and calcium can sometimes be lacking in the kinds of acidic soil mixes in which rhododendrons generally thrive. A chelating agent may therefore be necessary to alleviate chlorosis under these conditions. Making sure the root ball is open and not restricted – as can occur in some containerised plants – will often be sufficient to reduce such deficiencies. Growing rhododendrons in heavily charged calcareous soils or close to buildings from which calcareous material leaches can prove debilitating, possibly interfering with nutrient input from the mycorrhizal fungi found in the fine feeding roots.

Pollution from wind-blown biocides, acidic particulate matter and ozone, all from human industry and domestic activity, can all debilitate shrubs such that, as with other physical damage, the specimen plants then become susceptible to fungal attack. As rhododendrons make good cover for game and can act as effective windbreaks they have often been planted in extreme conditions. It is not infrequent to see rhododendrons planted by the sea becoming scorched by salt-spray; indeed in some years these salt-laden winds can penetrate several kilometres inland.

Bacteria and viruses

Crown gall, *Agrobacterium radiobacter* E.F.Sm. var. *tumaefaciens* Towns., is known in rhododendrons but is much rarer than in many other shrubs. It forms irregular, rather ugly-looking swellings on twigs and branches, especially at the junction of the stem and root system. It can be easily eliminated, as other fasciations of less well-documented origin, by cutting off infected branches and burning.

Like most plants, probably all species of *Rhododendron* support viral infections, some possibly host specific, but little systematic work has been carried out in this area. Necrotic ring spot is probably the most widespread virus infection, and can be detected at a very early stage by holding a leaf of the suspected host up to the light. Rings within the leaf will be seen. Plants infected will soon decline in vigour and become stunted. The best way of dealing with this problem is to uproot the affected plant and burn it. Former colleagues have often suspected that some forms of otherwise inexplicable stunt in *Rhododendron* are caused by a virus.

Whether virus infection is suspected or not, always clean pruning instruments before taking cuttings. As indicated in detail below, reducing the refugia for plant-sucking insects will also reduce the risk and subsequent spread of any insect-transmitted diseases.

Insect-associated problems

Insect activities should always be taken into account when considering fungal activity as frequently damage caused by insect activity weakens the host, or in many cases more directly assists entry into otherwise protected tissues. However, fungal attack may not be the prime cause of a field character. Thus in the case of capsid bug attack, probably from *Lygocoris*, puckering is typically exhibited as a result of the original feeding pattern, where damage occurs to several leaf lamina as they were stacked within the unopened bud. Chewed leaves are the result probably of the vine weevil (*Otiorhynchus sulcatus*), a pest of many species of hardy stock, the larvae feeding voraciously on the roots. Mealy bugs, *Rhododendron* stem borer not specific to rhododendrons and *Rhododendron* lace bug (*Stephanitis rhododendri*) have all been recorded as colonising rhododendrons.

There is good evidence to associate the feeding habits of the leaf-hopper *Graphocephala fennahi* with the spread of bud blast caused by *Pycnostysanus azaleae* (Peck) E.W.Mason. This is sometimes called the pin cushion fungus as when the fungus fruits on a moribund unopened flower or leaf bud it produces small erect, black, rigid hairs – the pins. These structures produce the asexual spores (conidia) of the fungus, the means by which the fungus spreads. However, the leaf-hoppers carry the propagules as they feed and so help to place the spores close to or within the feeding wound. This fungus also fruits on flower calyces and pedicels, on seed capsules and even on the leaf lamina. Although the last habitat is not uncommon, such sites are rarely recorded probably because people are concentrating on examining the buds. It can persist on a plant, therefore, even when not seen on the buds.

Pycnostysanus is a fungus of poorly maintained collections of rhododendrons where there is a build-up of leaf-hopper populations. In the UK, with a decline in the maintenance of old estates, the fungus flourishes. Control of the fungus is therefore as much to do with controlling leaf-hoppers. Indeed, regular cleaning up under bushes is paramount: the most widely planted species are evergreen and therefore shed leaves on and off throughout the year, providing a good hiding place for hoppers and other harmful insects.

There is no evidence that extensive development of such lichens as *Hypogymnia physodes* (L.) Nyl., which is found in unpolluted areas of the UK, is detrimental to the plant's growth; the only comment to be made is that lichen growth can also harbour unwanted insects.

Fungal problems

Except in certain cases – for example mildew on *Vireya* rhododendrons, *Chrysomyxa rhododendri* de Bary, first recorded on *R. hirsutum* L. in Lanarkshire, Scotland, and more recently *Phytophthora ramorum* Werrs et al. – we are unaware as to whether the fungi involved in *Rhododendron* diseases accompanied the stocks in the early periods of introduction or are local fungal species which have found a suitable host and perhaps even evolved specialised physiological races which can tolerate *Rhododendron* tissues. General quarantine procedures should eliminate any new invasions!

Leaf-spots

A whole array of microfungi have been associated with leaf-spotting and anthracnose. They are generally not life-threatening but can produce very unsightly specimen plants and if left will become very persistent. The causal agents can be divided into two groups: those which possess a sexual stage and those species which until now have been based solely on their asexual structures. The former are all members of the ascomycetes, a group where the spores are borne within a sac or ascus. This group includes *Chaetoapiospora rhododendri* (Tengwall) von Arx, recorded from Scotland on *Rhododendron sinogrande* Balf.f. & W.W.Sm., and the widespread *Mycosphaerella rhododendri* Feltgen; also recorded are *Guignardia philocoprina* (Berk. & Curtis) van de Aa, noted below under its asexual stage of *Phyllosticta*, *Phomatospora gelatinospora* M.E.Barr, *Protoventuria arxii* (Müller) Barr and *Pseudomassaria thistletonia* (Cooke) von Arx. Although *Glomerella rhododendri* (Bris. & Cav.) Spaulding has been recorded elsewhere in *Rhododendron* plantings, it has not been seen in the UK, the only similar fungus being the ubiquitous *G. cingulata* (Stoneman) Spauld & H.Schrenk which is probably the sexual stage of *Colletotrichum gloeosporioides* (Penz.) Penz. & Sacc.; see below.

The second group whose members are termed 'fungi imperfecti' include the rather ubiquitous *Ceuthospora lauri* (Grev.) Grev. with a broad host range, *Monochaetia karstenii* (Sacc. & Syd.) Sutton and the related *Strasseria geniculata* (Berk. & Br.) Höhn., several *Seimatosporium* spp. and *Truncatella angustata* (Pers.) S.J.Hughes. Interestingly, all these fungi possess conidia with one or two, simple or branched, short or long tails on the spore. There are at least five species of *Seimatosporium* occurring on rhododendrons, of which two are widespread (*S. mariae* (Clinton) Nag Raj on *R. californicum* Hook.f. (= *R. macrophyllum* D.Don ex G.Don) and other rhododendrons and *S. vaccinii* (Fckl.) Eriksson on *R. catawbiense* Michx. and *Vaccinium myrtillus* L.); all are also found on other shrubby plants. *Seimatosporium* (formerly *Coryneum*) *rhododendri* (Schwein.) Piroz. (= *Sarcostoma sinicum* Nag Raj), although originally described from rhododendrons, is better known on rosaceous plants; both these last species are now placed in the genus *Sarcostoma*. The other taxa are *S. arbuti* (Bonar) Nag Raj and *S. rosae* Corda, neither of which is host specific. *Monochaetia karstenii* is found on *Camellia* spp. as well as ericaceous plants. *Strasseria* and *Truncatella* are widespread, the former on conifers and rosaceous plants and the latter on a whole range of shrubby ornamentals. *Coleophoma empetri* (Rostr.) Petrak, described as its epithet suggests from *Empetrum*, is also found on *Rhododendron ponticum* L.

The related *Seimatosporium lichenicola* (Corda) Shoemaker & Müller is considered below, as is *Cylindrocladium* which as well as causing leaf and flower blight also produces a twig blight.

Unfortunately, as the sexual spored species of leaf-spot fungus produce their spores in the spring and the asexual forms in the autumn a 12-month vigil is required to collect complete material. It is easily seen in visits to *Rhododendron* collections: those in overcrowded conditions have extensive leaf-spotting. To reduce the risk of infection burn all previously and currently fallen leaves for a year, coupling this with judicious, careful, widespread pruning to increase air movement between the plants. The plants themselves are also often overstressed and can be assisted with a slow-release fertiliser.

Anthracnose and leaf scorch

Probably the commonest microfungus attacking particularly members of the *Ponticum* series is *Gloeosporium rhododendri* Bris. & Cav., now referred to the genus *Colletotrichum*. This produces tell-tale purple spots which collapse in the middle to form a brown papery centre; in extensive attacks whole lamina margins are destroyed. The centres of the spots often fall out and form shot holes, the fragment plus its fungal cargo falling to the ground below. It is impossible to be certain of the identity of the causal organism until the disfigurements are incubated in a damp chamber to encourage the development of fruiting structures, which is the only simple way to identify the culprit. The species which attacks *Rhododendron* may simply be a physiological race of a serious pest known on bananas and citrus fruits. This fungus is the same as *C. azaleae* Ellis & Everhart, described from azaleas, a synonym of *C. gloeosporoides*, a fungus thought to have over 125 synonyms (von Arx, 1970). However, at last molecular studies are coming to our aid in sorting out this confusion and separating out the potential pathogens. Thus Vinnere *et al.* (2002) have demonstrated both *C. acutatum* Simmonds and *C. dematium* (Pers.: Fr.) Grove. *Myxosporium* recorded on *Rhododendron* has much larger conidia but is in this 'consortium'. What is being shown is that there are several closely related species of fungus involved, all with a wide host range – but more important to the grower is the fact that the treatment is the same for all of them.

Several species of *Cryptosporopsis* are hidden amongst the synonyms of *Colletotrichum* (and *Gloeosporium*), with good reasons or not. This genus of 'fungus imperfectus' covers the asexual stages of *Pezicula*, a group of small discomycetes found on twiggy debris. One is known on *Vaccinium* (*P. myrtillina* (Karst.) Karst.) in Britain, although it does occur on *R. ferrugineum* L. elsewhere in Europe, two on rhododendrons from continental Europe (*P. rhododendricola* Rehm on *R. ferrugineum*, *R. hirsutum* and *R. maximum* L.; *P. rhododendri* Remler also on *R. ferrugineum*), and in the USA *P. grovesii* Wehm. occurs on *R. canadense* (L.) Torr. They have been considered saprotrophs but it is interesting that recently a suspected species of *Cryptosporopsis* has been causing concern in *R. ponticum* plantings in south-east England.

Septoria azaleae Voglingo, another 'fungus imperfectus' with long, thin, septate, hyaline conidia (scolecospores), has been intercepted on imported *Rhododendron* material from Europe, but although known from Britain has not become established here. It produces extensive lesions on the edges and lamina of leaves, which finally dry and give the appearance that the plant has been too close to a fire. It has many characteristics in common with *S. rhododendri* Cooke described from Maine, USA, and may even be the same species. It is known from Cornwall on *R. ponticum* but is not widespread. *Pestalotiopsis* (formerly *Pestalotia*) *guepini* (Desm.) Steyaert and *P. sydowiana* (Bres.) Sutton differ in conidial dimensions and both are found on rhododendrons, forming a marginal scorch usually associated with the bad positioning of susceptible plants so that they are receiving high levels of sun and/or drying winds. There are other names found in the literature in *Pestalotiopsis* (as *Pestalotia macrotricha* Kleb. and *P. rhododendri* Guba) but there is some doubt as to the limits of the species circumscription. The genus is related to *Monochaetia* noted above, which is more frequently found on moribund leaves, but in all these 'fungi imperfecti' found fruiting on cast leaves we are ignorant as to where they started their growth! Control is the same as that for leaf-spots but if the spotting persists then a soil drench once the fallen leaves have been removed is recommended.

Mildews

The identity of the mildews on *Rhododendron* has been much confused and the whole scenario is dealt with by Helfer (2003).

Rust fungi

Chrysomyxa ledi var. *rhododendri* (de Bary) Savile has become a widespread species in the British Isles since it was first recorded in 1913. Henderson (2000) gives the most up-to-date list of *Rhododendron* spp. hosts for the British Isles and also offers a much longer list of alternate hosts in the genus *Picea*, based on experimental work. Unlike many other fungi this species, in common with other rust fungi, has two hosts in order to complete its life-cycle. Unfortunately this fungus is a problem to two sectors of the community – foresters as well as *Rhododendron* growers – as it can cause widespread loss of 'needles', especially in such plantation trees as *P. abies* (L.) H.Karst. and *P. sitchensis* (Bong.) Carrière. Over 50 *Rhododendron* spp. or *Rhododendron* hybrids, including intergeneric hybrids with *Ledum* and especially those in *Azaleaodendron*, are listed, in addition to as many cultivars of variable parentage. It is more often a disease of young growth; although the rust fungus can persist in older stands it becomes very prominent on new growth, especially where larger plants have been cut right back to their basal stem and sprouting ensues or where there are volunteer seedlings. Fungicides including benomyl are important in attempting a control but new infections may colonise from either spruce or other *Rhododendron* plantings.

There is no doubt that this obligate parasite was introduced with one or more hosts at the height of the plant expedition period, but is now an established alien and especially widespread in the west and north of the British Isles. Insects may have some role in transferring spores. On one occasion, it was noted during routine checking of cuttings received in quarantine at the Royal Botanic Garden Edinburgh (RBGE), newly arrived from an expedition, that dipterous insects were collecting on *R. arboreum* Sm. leaves. It was not immediately obvious why they were there, but several days later rust lesions appeared at the very spots that the insects were attracted to, presumably by some volatile secondary metabolite of the active mycelium.

In North America *Naohideomyces* (formerly *Pucciniastrum*) *vaccinii* (G.Winter) Sato, Katsuya & Hirats. has been found on *Rhododendron* (*Azalea*) spp. It has not been seen on this genus in the British Isles, although the fungus is not uncommon on *Vaccinium myrtillus* L., *V. oxycoccus* L., *V. uliginosum* L. and *V. vitis-idaea* L.: the alternate hosts are *Tsuga* spp.

Superficial fungi and algal coverings

Two ascomycetous fungi, *Dennisiella* (*Chaetothyrium*) *babingtonii* (Berk.) Bat. & Cif. and *Seuratia millardetii* (Racib.) Meeker, are known to affect rhododendrons. The former is characterised by the mycelium spreading over the surface of the leaf and bristling with dark brown, pointed hairs. The latter, in contrast, has olivaceous black, lobed myc-

elial aggregates which to the untrained eye have more than a passing resemblance to the droppings of insects. The ends of the lobes produce asexual spores in similarly coloured cup-like structures, previously considered an independent fungus called *Atichia glomerulosa* (Ach. & H.Mann) Stein. The sexual stage is found embedded in the thallus. With its superficial similarity to *Collema*, a group of blue-green lichens, it was first considered to be a lichen. Reference to the possible damage caused by thallose lichens has been made earlier. The conidial stage of *Dennisiella* is called *Microxiphium fagi* (Jaap) Matheis. *Euantennaria alaskensis* (Woron) S.J.Hughes (= *rhododendri*), although forming prominent, black, velvety coatings on the branches of *Rhododendron*, apparently does no harm.

These fungi are often associated with honeydew produced by plant-sucking bugs and aphids, and the sugary solutions encourage the growth of adventitious moulds such as the ubiquitous species of *Cladosporium* and *Aureobasidium*. Possibly also because of the presence of such high-energy substances, the growth of bacteria and myxomycetes (slime moulds, which feed on bacteria) have been recorded on azaleas. They do not appear to do damage but are unsightly, and can produce dark, unpleasant smudges on clothing.

The alga *Cephaleuros virescens* Kunze takes up a similar position on the leaf lamina and can cause leaf-spotting and even stem infections. Except for possible examples in quarantined material on both *Rhododendron* and leaves of other woody stock collected on expeditions to tropical and subtropical areas, it has not become established in the UK. It is recorded as a pest in North America.

223

Fungal, lichen and algal coatings commonly occur on the thick cuticles of woody plants in the rainforests of the world but are soon dislodged in the drier atmospheres during cultivation and, although unsightly, do not pose a threat.

Bud blast

The prime fungus causing damage to *Rhododendron* buds before they burst is undoubtedly *Pycnostysanus azaleae*, often called *Briosia azaleae* (Peck) Dearn. in some North American texts; this has been dealt with earlier as the biology of the fungus and its control is much tied up with the fact there is a close insect/fungus condition. There are no sexual spores and spread is dependent on asexual propagules which are transported by leaf-hoppers. The asexual spores (conidia) adhere to the insect and when it alights to feed, the conidia find themselves in a good position to infect the host. Controlling the unwanted insects will contain this disease.

There is another 'fungus imperfectus', *Diplodina eurhododendri* G.Voss, of which we know little except that the conidia are hyaline and two-celled and borne in small, black, flask-shaped structures called pycnidia. It is thought to be the asexual stage of *Cenangella rhododendri* Rehm, also a poorly understood fungus, recorded from continental Europe. If anything can be learnt from what might be considered the related *Erikssonopsis* (formerly *Cenangella*) *ericae* found on dead twigs of *Calluna*, it is that the *Diplodina* could well be only a secondary infection of already damaged buds, perhaps those damaged by nutrient deficiency. *Cylindrocladium* also causes bud death, and will be dealt with below.

Leaf galls

Exobasidium azaleae Peck, *E. japonicum* Shirai, *E. rhododendri* (Fuckel) C.E.Cramer and *E. vaccinii* (Fuckel) Woron. are all examples of members of a genus which induce hypertrophication in the tissues of ericaceous and *Camellia* hosts that they infect. The genus used to be larger, but those species found as parasites of other host genera, e.g. *Laurus*, are now considered to belong to separate and sometimes quite unrelated genera. The naming of the fungi on ericaceous hosts has been fraught with misunderstanding and at one stage all species were known under a single or at most two names. Now experimental work and careful observation shows they are very host-specific fungi.

Both species of *Exobasidium* on rhododendrons are common and widespread and are frequently found attacking potted house plants, the galls often creating either unbelievable interest or horror in the owner. The sexual spores are borne at the apex of a cell called a basidium and therefore these fungi are very different from most of the fungi which have been considered so far. They are basidiomycetes and are related to the familiar mushrooms and toadstools. However, the fruiting body is reduced and covers and invades the tissue of the gall, or leaf-spot if a *Vaccinium*, which the plant forms as a result of chemical interference by the fungus. Unless there is a continual infection these fungi cause little harm as they are so very obvious that hand-picking of the galls, followed by burning, is generally all that is required. The application of a fungicide just as the buds begin to burst is also effective.

Flower damage

Late spring frosts are a prime factor in the damage to *Rhododendron* flowers: individual flowers in the cluster collapse and coalesce together in a wet, drooping head. This is very attractive to dipterous insects and grey mould, the former often carrying spores of potentially important pathogens. An attack by *Ovulinia* will produce the same kind of effect and often with the same results. Indeed, under optimum conditions for the fungus flower collapse takes place rather quickly after opening. It is on these flowers that one can see the black shiny sclerotia of *Ovulinia azaleae* F.A.Weiss, although it is also true that *Botrytis cinerea* Pers.: Fr. will produce similar structures. The sclerotia fall off into the soil so collecting dead flower heads if at all possible is recommended. Once in the soil they are almost impossible to see and it is likely that no amount of soil drench will penetrate these structures. Under favourable conditions these sclerotia will produce asexual spores or in rarer cases the cup-shaped sexual stage – this has not been seen in the British Isles, although I am sure diligent searching will be rewarded with a find.

A closely related pair of fungi found in Europe and North America are *Monilinia alpina* L.R.Batra on the *R. ferrugineum* and *R. hirsutum* consortia and *M. azaleae* Honey on for instance *R. nudiflorum* (L.) Torr. (= *R. periclymenoides* (Michx.) Shinners), two fungi which not only infect the flower as it is developing but also colonise the fruit and mummify the tissue. The fruits fall to the ground and in favourable conditions a small disc-shaped fruiting body is formed. Associated with the fruit infection is an associated leaf blight, and similar control measures are necessary as those traditionally used for controlling *Monilinia* on apple. These two taxa are not recorded from Britain. There are several species of *Monilinia* on members of the Ericaceae; species are host specific.

Shoot die-back

Shoot die-back can be caused either by an aggressive pathogen, or as a result of debilitating conditions. The slow process by which nutrients are transferred to the extremities in cases such as severe root containment may cause die-back simulating pathogen attack.

The widespread opportunistic coral spot fungus, *Nectria cinnabarina* (Tode: Fr.) Fr., and grey mould, *Botrytis cinerea*, are common, with the former often growing on quite old established stems. Judicious pruning helps to control this fungus, as well as keeping in check the resupinate basidiomycete *Stereum rugosum* Pers.: Fr., which forms pale pinkish grey plates on the bark that when growing actively will exude a red liquid on bruising. There is little doubt that both enter through previous wounds and snags and cause gradual, slowly spreading damage to the heart wood. *Botrytis cinerea*, as in other plants, will attack new buds, shoots, leaves, etc. given any opportunity.

Cytospora subclypeata Sacc., a 'fungus imperfectus', has small allantoid conidia which issue as a long, cream-coloured tendril from small, flask-shaped fruiting bodies embedded in darkened tissue called a stroma. It has been recorded in Britain from Devon, areas around Cheshire/Liverpool, Birmingham (Botanic Gardens) and various sites in Scotland. It would appear to be connected to lesions produced by frost cracking and cold damage to the phloem tissue. Although widespread it is not highly aggressive and can be controlled by pruning.

Cylindrocladium scoparium Morgan is considered the asexual stage of the ascomycete *Calonectria scoparia* Terash. This fungus is rarely recorded from Britain, but appears to be an important pathogen in North America, causing flower- and leaf-spot, root rot and wilt, and shoot blight. Control of this fungus has been investigated in studies of root rot of peach and cocoa. *Calonectria kyotensis* Terash is another name found in the literature for this fungus with its asexual stage, *Cylindrocladium floridianum* Sobers. *Cylindrocladium theae* (Petch) Sabram. is a similar pathogen on rhododendrons and as its epithet suggests was described from *Camellia* (Theaceae).

Four other fungi, all ascomycetes, should be mentioned here, especially as one was described from Scotland, indeed by staff at Edinburgh University from one of the RBGE's Specialist Gardens. The first, *Lophodermium vagulum* M.Wilson & N.F.Roberts., was described from material on leaves and twigs of a whole range of Chinese rhododendrons, including *R. croceum* Balf.f. & W.W.Sm. (= *R. wardii* W.W.Sm.) and *R. thomsonii* Hook.f., planted at Dawyck, and has recently been found in Wester Ross and Peeblesshire (both in Scotland); the record from the Isle of Mull is incorrect. The second fungus, *Lophomerum ponticum* Minter, was formerly placed in *Lophodermium* as *L. rhododendri* (Ces. ex Rehm) Sacc. but the true 'rhododendri' is known from Austria and Italy and not the British Isles; all previous records for the UK refer to *L. ponticum*. *Lophomerum* differs from *Lophodermium* in that the lips of the ascomata are not carbonised. In North America there is a third, similar species, *L. schweinitzii* M.Wilson & N.F.Roberts. All are weak parasites and produce a stem canker and leaf blight. *Lophomerum ponticum* is particularly common on *R. ponticum*, where it attacks small shoots and leaves. It is probable that the past records of *Lophodermium melaleucum* on *Rhododendron* refer to *Lophomerum ponti-*

cum; the true *L. melaleucum* grows on *Vaccinium* spp.

The third species to be mentioned is *Lembosina* (formerly *Echidnodes*) *aulographiodes* (Bommer, Rousseau & Sacc.) Theiss., which is not uncommon on living twigs of *Rhododendron* and can also cause a twig blight. The closest relatives of this fungus are found in the tropics. Finally, another leaf and twig inhabitant, especially of depauperate plants, is *Morenoina rhododendri* Ellis, which forms minute, dark, sinuously star-shaped ascomes. The essence of control in these small-scale twig and shoot blights is to encourage healthy growth with applications of slow-release fertiliser and judicious pruning.

Seimatosporium lichenicola Berk. & Br. (= *Coryneopsis* and *Coryneum microsticta*), with its large, three-septate conidia, the lowest cell hyaline in contrast to the other two which are yellowish brown, is known to cause a die-back of cultivated roses. This condition is not uncommon: what is more uncertain is that the same fungus has been involved in a similar die-back in rhododendrons. Unlike the species of this genus described earlier, this one does not possess the tell-tale tails to the spores.

Butt rots

Armillaria is a genus of mushrooms which strikes fear into the gardener. Of the potential suite of species *A. mellea* (Vahl.: Fr.) Kumm. is probably the most well known, although there are other more widespread taxa which might be involved. However, *A. mellea* is a primary parasite. The genus encapsulates a series of basidiomycete fungi with a mushroom morphology, typified by a white spore-deposit, ring on the stem and gills which run down the top of the stem. Sadly, by the time these characteristics are revealed to give a positive identification, namely production of clusters of thin-stemmed, tall yellow fruiting bodies, the host plant is probably dead. However, in Scotland at least, except in remnants of ancient oak forest *A. mellea* is rather uncommon, and in gardens and parkland one finds the much more common *A. gallica* Korh. & Marxmüller (*A. bulbosa* sensu Watling & Kile). This species is an opportunist and is less invasive but takes advantage by entering weakened plants. As it is so common and widespread, growing on a whole range of woody debris, sources of spores and therefore new infections are understandably high. One such entry point is at soil level, from what might be termed 'strimmer' or 'lawn mower disease', where some gardening implement has breached the bark. The bulbous-footed fruiting bodies of this species are produced in ones and twos and less commonly as small but loose clusters, which contrasts markedly with *A. mellea*. They are produced on old wood or buried wood and may travel through soil, as does *A. mellea*, by a vegetative stage colloquially called bootlaces. This colonising stage is round in cross-section in *A. mellea* but flattened in *A. gallica*. It is very probable that in other parts of the world other parallel species of *Armillaria* play similar roles. Unfortunately, generally this kind of butt rot has always been attributed to *A. mellea* without reference to any other species. *Armillaria* will often enter the damaged areas where *Phytophthora* has killed part of the root system. A clear indicator of the presence of *Armillaria* is that when a small portion of bark is peeled back it is found to be loose and reveals beneath a chalk white compact sheet of filaments of the mycelium, which on a warm summer evening when the

eyes get accustomed to the light level will be seen to phosphoresce.

All that can be done to control *Armillaria*, if it is thought to be involved, is to stimulate the growth of the plant to encourage natural compartmentalisation of the fungus. In some cases Armillotox has been found to be successful, although it is highly toxic to plant tissue and could be damaging to the host as well as to the fungus. Better to protect a cherished plant by dropping a thick polythene membrane to 75–100 cm deep encircling the plant, and bending the polythene back facing away from the plant just at the soil surface. There has been much published on this genus of fungi and suggested controls (e.g. Tattar, 1989).

Although examples of root rot caused by *Heterobasidion annosum* (Fr.: Fr.) Bref. have been seen on other ericaceous shrubs in Edinburgh, examples have not been seen on *Rhododendron*, despite it having been recorded on the genus. As in *Armillaria*, control attention should be paid to the growing conditions and position of the plants in the garden, and removal of dead plants close by and any dead roots belonging to adjacent plants. Try never to damage the roots whilst carrying out garden operations.

Root rots, die-backs and collar rots: dreaded phytophthoras!

A trio of *Phytophthora* species are considered major pests in the USA: *P. cinnamomi* Rands, *P. heveae* A.Thomps. and *P. syringae* (Kleb.) Kleb. (leaf-spot), and as their epithets suggest, they were not originally described from *Rhododendron*; they are generalists. The fungi considered earlier attack specific plant parts, both debilitating the plant and making it unsightly; these three species all produce brown staining in the wood but have slightly different attendant symptoms. Thus the first in addition produces wilting coupled with downward-rolled leaves and browning of the apical parts, often in quite noticeable, bright orange-browns. The second produces water-soaked areas on leaves, which spread down the midrib and subsequently into twigs resulting in stem death, whilst the leaves – although still attached – become brown and dry. The third species at first produces irregular necrotic leaf-spots associated in time with blackening branch cankers and finally die-back. *Phytophthora cinnamomi* affects the roots, damaging the nutrient flow by killing sections or whole parts of the root system, although new roots may be stimulated to grow at the junction of the soil and air in the drier environment. Occasionally there is apparently complete recovery, although the dead tissue may play host to a series of pathogenic root fungi. However, when seedlings are attacked several of the same species can cause a seedling blight.

In Europe two other species causing root rot, and in the same genus, can be added: *P. cactorum* (Lebert & Cohn) J.Schroet. and *P. cryptogea* Pethybr. & Laff. The former produces a major die-back with little hope of recovery, whilst the second produces a crown rot in some of the large-leaved rhododendrons such as *R. catawbiense* and *R. maximum*. Without special techniques it is generally impossible to be sure of the causal organism involved: expert attention and artificial culturing is required to be certain. Indeed, in total nine species of *Phytophthora* are known to attack rhododendrons. For a good list of references see Werres *et al.* (2001).

A more recent and potentially more serious disease is that caused by *P. ramorum*, which although described originally from *Rhododendron* and *Viburnum* (Werres *et al.*, 2001) is, through the activities of Dr D. Rizzo, UC – Davis, California, now known to be the same organism which causes 'Sudden Oak Death' (Frankel, 2001). Although it proves to be fatal to some members of the Fagaceae it only produces discoloured leaves and red-purple stem lesions in rhododendrons. The infected plants generally recover but we do not fully understand the details of the epidemiology of this fungus on *Rhododendron*. It was originally isolated from containerised rhododendrons in Germany and the Netherlands but has since been isolated in the field in California associated with diseased *Quercus* spp. Recently the Scottish Executive issued a Plant Health (Scotland) Order 2002 restricting movement of rhododendrons (Anon., 2002).

However, something can be done to reduce possible attack: this is summed up in a single word – hygiene – coupled with extra vigilance and insuring the receipt of healthy stock. As the asexual spores in all the species noted above are motile, they are able to move quite easily through the soil in the soil water, and it is possible for them to be splashed from plant to plant by rain drops. None of the species are restricted to *Rhododendron*, so that infection may come from adjacent plantings of species from other genera or families, meaning care is required throughout the garden. Spread is possible by not washing footwear when an attack is known or suspected, and contaminating the source of water by poor maintenance and procedures. Indeed, water-logged soils are favourable sites for these fungi and every effort should be made to drain the areas well. *Phytophthora* are probably the most important agents in the total die-back of rhododendrons.

Non-British problems

The UK records of *Corticium solani* Prill. & Delacr. (as stat. mycel. *Rhizoctonia solani* J.G.Kühn) on *Rhododendron* have not been reassessed, but I have not personally come across this fungus on this host. It is said to produce a foot and root rot, especially at soil level in young overcrowded plants under nursery conditions, but this may be only because this fungus does just that in other host plant species. It is common and wide-spread on other plant species and should be referred to as *Thanatephorus cucumeris* (A.F.Frank) Donk, but there is a whole host of very closely related fungi and each record must be examined in isolation. However, in North America a disease supposedly caused by this species is what is termed *Rhizoctonia* web-blight. This is when azaleas are exposed to long periods of high humidity and become covered in a hyphal growth associated with leaf-spots and adhering abscised necrotic leaves remaining together. ('Web-blight' is a term used in wider circles for a disease found in coffee nurseries and caused by *Ceratoba-sidium noxia* (Cooke) Talbot (= *Koleroga noxia* (Cooke) Donk).) This is not recorded from the UK.

Two fungi which feature prominently in North American manuals are *Phymato-trichopsis* (formerly *Phymatotrichum*) *omnivorum* (Duggar) Hennebert and *Bot-ryosphaeria quercuum* (Schwein.) Sacc. Both are found on a very wide range of woody plants and are not specific to rhododen-drons. The former may act as a blanket which asphyxiates the host's roots. The latter is known from this country particularly on cultivated roses, causing scab of briers, but it is found more commonly on wild roses.

Although von Arx & Müller (1954)

reviewed the host range of *Botryosphaeria* they adopted a rather narrow concept, and Sivansen (1984) demonstrated that this did not reflect what was found in nature. Thus this species is now one of a whole range of species in the genus which may be found on rhododendrons, namely *B. obtusa* (Schwein.) Shoemaker, known to produce a canker on members of the *Ponticum* group, *B. dothidea* (Moug.: Fr.) Ces. & de Not., *B. quercuum*, and the plurivorous *B. rhodorae* (Cooke) M.E.Barr, which is better known in its asexual stage, *Botryodiplodia theobromae* Pat. (and probably the same as *Gloeospo-rium rhododendricola* Hollós). Farr *et al.* (1996) list an additional three species in this genus but this should not produce a feeling of despair for as far as control measures are concerned they can be treated as one, with control methods adopted from rose cultiva-tors. All species form blackish erumpent, circular or elliptic crusts on the branches of the host, sometimes reaching sizes of 15 mm in diameter. I have not seen this in British samples on *Rhododendron*.

The *Phymatotrichopsis* has been thought to be a complex *Botrytis* sp. but unlike the characteristic species in that genus it forms white mycelial strands in the soil which also entwine around the roots of the host. Amongst the strands are formed spherical white balls of tissue which darken and are called sclerotia. They can act as resting bod-ies and stay in the soil for long periods, as can the mycelial strands waiting for favour-able conditions in which to germinate and infect new stock. When reaching the soil around the plant the mycelial strands form white sheets, some of the constituent hyphae breaching the root tissues, and on the soil become powdered with conidia; this spore-mat slowly becomes tinged ochraceous with age. Drastic action is required to control this pathogen, including soil fumigation. It has been suggested that this is a stage in the development of *Sistotrema brinkmanii* (Bres.) J.Erikss., a resupinate basidiomycete, but this is highly unlikely.

Casuals

A few fungi occur on rhododendrons but not with great regularity; nevertheless their effects may elicit concern. *Stereum rugosum*, which is known to produce a pocket rot of fagaceous trees, forms buff-pink to whitish sheets on branches and trunks, and has been mentioned before. *Psathyrella sarcocephala* (Fr.) Singer, a basidiomycete which forms clusters at the base of rhododendrons, differs from *Armillaria* in producing a chocolate brown spore print. It would appear to have no detrimental effect on the plants with which it grows.

In the literature *Phoma rhododendri* Cooke and *P. rhodora* Cooke (which is really a *Phyllosticta*), *Phoma rhododendri* West. and *Phoma saccardoi* Thüm. are all listed but no recent revision has been attempted to determine whether they are independent species or simply growth forms of a single fungus (see Sutton, 1980). The second and last two grow on *Rhododendron* leaves and *Phoma rhododendri* also grows on twigs. They may be for instance stages in the development of *Phyllosticta concentrica* Sacc., which is itself the asexual stage of *Guignardia*, noted above. Apparently they are very rarely seen or more correctly have never confidently been recorded since their description. They all fruit on fading leaves and this very probably indicates that they are secondary colonists and cause no threat to plant health. Within the same complex is an *Ascochyta*

which has been identified on *Rhododendron* in Scotland, differing from *Phyllosticta* in the two-celled conidia. Significantly, all these fungi have been found sometimes exclusively in the *Rhododendron* collections of either Kew or RBGE or both!

Equally *Diplodia rhododendri* Bellynck found on dead areas of *Rhododendron* leaves in Aberdeen must also be a secondary colonist, although this may be the asexual stage of *Massarina papulosa* (Sacc.) Kunze, which also grows on leaves and twigs and has been shown to have a *Diplodia* stage in its life-cycle. Also, in the large genus *Phoma*, the constituent species of which range in habitat preferences from kitchen paint-work to herbaceous stems, leaves and rotting root-crops, two taxa have been identified: *P. leveillei* Boerema & Bollen and *P. pomorum* Thüm., neither a threat to *Rhododendron* growers. Many of the plants on which these fungi have been found have attracted years of neglect. However, *Phomopsis theae* Curz., with two kinds of asexual propagules, one tiny and the other long and whip-like, has been shown to be a problem in nurseries raising both camellias and rhododendrons; I am unaware of it in the UK. *Phomopsis* spp. are generally asexual stages in members of the genus *Diaporthe*. Farr *et al.* (1996) list three species of *Diaporthe* on *Rhododendron*, one being the ubiquitous *D. eres* Nits., but none are linked with *Phomopsis theae*.

Discussion

Farr *et al.* (1996) list, excluding proven synonyms, just over 500 species of fungi as being recorded growing on rhododendrons; there are almost as many synonyms! This seems to be in discord with my initial statements, but some names admitted by Farr and his colleagues have questionable taxonomic status and many of the fungi listed are known only from South East Asia, especially Japan, and are not, as yet, found in the West. Thus 10 other *Chrysomyxa* species and 22 other *Exobasidium* species are known from these areas.

Amongst the names of fungi described from rhododendrons, 72 species have either the epithet *rhododendri* or *rhododendricola* and 16 the epithet *azaleae*; this gives a false sense of specificity to these organisms, something which is far from the case. Many of the fungi causing problems in *Rhododendron* cultivation have a wide host range, and therefore attention must be paid to the plants which are associated with the rhododendrons.

The list was initially prepared by Farr *et al.* to assist in US plant quarantine activities but I have not experienced the majority of potential fungi during my general plant health studies in Edinburgh, although I have found others in my general field work. Many, especially the agaricoid, corticioid and poroid basidiomycetes which have been listed, are saprotrophs and many of these, and many microfungi, are found only on moribund leaves. In the last category, for instance, there are nearly 50 distinct genera of coleomycetes, some with several species, but it is very unlikely that these cause a threat to healthy *Rhododendron* plants. There are 10 generalists and some which are folicolous and members of the phylloplane, whilst others are widespread root rot fungi, e.g. *Thielaviopsis basicola*, which have an enormous host range. There are 16 mastigomycetous fungi causing wilts, etc., and approximately 10 powdery mildews, but the number of potential major pathogens fades

into the background amongst the 500 plus. It is estimated that in the British Isles probably less than a tenth of this figure might cause potential damage to rhododendrons, with the handful discussed above of concern to growers.

Conclusion

In the discussion above, emphasis has been placed on good hygiene in the garden or nursery: removal from beneath plants of dead flower heads, leaves and other trash which may act as refugia for insects and the resting sites for certain stages of potential fungal pathogens. Encouragement of growth by mulching and application of slow-release fertiliser is essential to maintain plant defences and also those of endomycorrhizal associates. A plant in its own way is better at controlling many of the problems discussed earlier than us throwing everything available, however alien, at the trouble.

Rhododendrons, like most other ornamentals which are planted in our garden, are generally not native and therefore have not evolved in the habitat in which they now find themselves. They should be given all the help possible, as opportunist fungi might also exploit the situation. For instance, in Scotland rhododendrons have a slightly different problem in the west because of the maritime climate: e.g. *Chrysomyxa* and *Dennisiella* are more rarely found in the drier east. What is necessary is the design of a regime of attention which will off-set the fungus, so allowing the continuation of enjoyment of these beautiful plants.

It must always be remembered that ericaceous shrubs have a very special mycorrhizal association, generally formed with the ascomycete *Pezizella ericae* (Read) Korf, although a whole series of unrelated fungi are concerned, e.g. members of the Myxotrichaceae and possibly a club fungus (*Clavaria*).

Chemical means of control should be kept to an absolute minimum as that way their application is more effective when actually used. Care and attention should be given to what one does in pruning and garden tidying, and other ordinary garden activities – where do you plan to have your barbeque?! When mulching, use a fine-grained application to reduce hiding places for insects and always incorporate applications of slow-release fertiliser to ensure strong growth. A commercial leaf-vacuum is ideal to clear away dead leaves, as it will also allow access to the soil if a drench is found to be necessary.

There are good general texts available on control of some of the more important diseases if they finally strike, as the majority of troubles in rhododendrons are not caused by host-specific fungi. Control measures can be modified from those already proven as successful with other plants (e.g. Pirone, 1970). However, hygiene is of paramount importance.

References

ANONYMOUS (2002). The Plant Health (*Phytophthora ramorum*) (Scotland) Order 2002. Scottish Executive, Environment and Rural Affairs Department, 2pp.

VON ARX, J. A. (1970). A revision of the fungi classified as *Gloeosporium*. *Bibl. Mycol.* 24: 1–203.

VON ARX, J. A. & MULLER, E. (1954). Die Gattungen der amerosporan Pyrenomyceten. *Beitr. Krypt. Fl. Schweiz* 11: 1–434.

FARR, D. F., ESTEBAN, H. B. & PALM, M. E. (1996). *Fungi on Rhododendron: A World Reference.* Boone, NC: Parkway Publishers, 192pp.

FRANKEL, S. (2001). *Sudden Oak Death Caused by a New Species, Phytophthora ramorum.* Pest Alert, USDA FS, Pacific Southwest Region, USA, 2pp.

HELFER, S. (2003). Rhododendron powdery mildew – a continuing challenge to growers and pathologists. In: ARGENT, G. & MCFARLANE, M. (eds) *Rhododendrons in Horticulture and Science*, pp. 66–72. Royal Botanic Garden Edinburgh.

HENDERSON, D. M. (2000). *A Checklist of the Rust Fungi of the British Isles.* London: British Mycological Society, 36pp.

PIRONE, P. C. (1970). *Diseases and Pests of Ornamental Plants*, 4th edition. New York: Ronald Press.

SIVANSEN, A. (1984). *The Bitunicate Ascomycetes and their Anamorphs.* Vaduz, Germany: J. Cramer, 701pp.

SUTTON, B. C. (1980). *The Coleomycetes.* Kew: Commonwealth Mycological Institute, 696pp.

TATTAR, T. A. (1989). *Diseases of Shade Trees*, revised edition. London: Academic Press, 341pp.

VINNERE, O., FATEHI, J., WRIGHT, S. A. I. & GERHARDSON, B. (2002). The causal agent of anthracnose of *Rhododendron* in Sweden. *Mycol. Res.* 106: 60–69.

WERRES, S., MARWITZ, R., MAN IN'T VELD, W. A., DE COCK, A. W. A. M., BONANTS, P. J. M., DE WEERDT, M., THIEMAN, K., ILIEVA, E. & BAAYEN, R. P. (2001). *Phytophthora ramorum* sp. nov., a new pathogen on *Rhododendron* and *Viburnum*. *Mycol. Res.* 105: 1155–1165.

The ecology and history of *Rhododendron ponticum* as an invasive alien and neglected native, with impacts on fauna and flora in Britain

Ian D. Rotherham

Sheffield Hallam University, City Campus, Pond St., Sheffield S1 1WB, UK

In the British Isles *Rhododendron ponticum* is a widespread and often pernicious weed – an alien, invasive plant of the most problematic sort. With a native distribution at the extremes of the Mediterranean, the species exhibits a diversity of behaviours. It is very restricted in the Iberian Peninsula, generally to higher ground and/or sheltered valleys, highly invasive and problematic in managed forest areas of the Turkish Black Sea Mountains, but rather limited in the high ground to the west of Turkey close to the Bulgarian border. In northern parts of the Black Sea Mountains there are small pockets of ancient *Rhododendron* forest.

The plant was brought to Britain as seed from Gibraltar in 1763. Spreading rapidly in suitable environments with extensive vegetative layering and huge quantities of very viable seed that disperses widely on the wind, this is now a common and familiar plant over much of the country.

The sheer scale of impact and the potential dominance of the plant make its effect on native plants and animals very significant. The over-riding opinion of nature conservationists and foresters is that this impact is detrimental. Furthermore, most people (much to the disappointment of conservationists) actually like the plant. When it escapes fully to the wild it is capable of wreaking havoc on native communities, but may add value too. In Britain *R. ponticum* is a much-maligned alien, and in Turkey and Iberia, a neglected native.

There is a further twist in the story of *R. ponticum* and this relates to its perception by conservationists and often by the public. For many there is only one 'rhododendron' and *R. ponticum* is it. The negative view of this plant – portrayed as it is as a terrible problem species – has implications not only for the conservation of this species in its own native haunts, but also for the conservation of the wider rhododendrons too. The very real conservation problems posed by alien invasives such as *R. ponticum* present a challenge for all those whose interests cover the wider genus and the enjoyment, awareness and conservation of this complex and beautiful group of plants.

Introduction

Few plants give more joy to the British public and cause more concern for conservationists than *Rhododendron ponticum* L. Some of the species that come close, such as Himalayan balsam (*Impatiens glandulifera* Royle), are also exotic to Britain (Rotherham, 2001a). Native to western and eastern Mediterranean areas, but except as a naturalised alien in Britain and Ireland, absent from north-western Europe, *R. ponticum* was present in the British Isles during interglacial periods (Cross, 1975). For reasons unknown it failed to recolonise after the most recent ice age.

It returned to Britain as seed, brought by plant collectors to Kew Gardens from Gibraltar in 1763 and later perhaps from the Black Sea region (Shaw, 1984). From Kew Gardens the species was dispersed widely across the British Isles. The first site of introduction was probably Lyndhurst in Hampshire. The first plants were established by 1765, with commercial introduction to estates from the 1780s up to the early 1900s, for game cover, wildlife habitat, shelterbelts, ornament and stock for cultivated, hybrid rhododendrons (Cross, 1975; Rotherham, 1983). It successfully colonised diverse vegetation types, usually on sandy or peaty, acid soil (Table 1). The origins of what is considered the 'Rhododendron problem' in Britain, i.e. the massive invasion of woodlands and commercial forests, are discussed by Shaw (1984) and Magor (1989). Colak *et al.* (1998) consider the present situation, highlighting important contrasts in status in native and exotic situations. Recent research (R. Abbott & R. Milne, personal communication) confirms the Iberian origins of most British material, and hybridisation in many colonies.

Rhododendron ponticum occurs as a native species (ssp. *ponticum*) principally in northern Turkey (Plate 1), but it occurs also in the Taurus Mountains in southern Turkey (Uber, 1951; Mattfeld, 1971; Tuncdilek, 1971), the Caucasian States, Lebanon (Clapham *et al.*, 1962) and southern Bulgaria, and as ssp.

Table 1 The major habitat types in which *Rhododendron ponticum* is naturalised.

Semi-natural habitat types (relatively undisturbed)	Artificial habitat types (relatively disturbed)
• Heather moorland	• Coniferous plantations
• Lowland heath	• Deciduous plantations
• Semi-natural deciduous woodland	• Managed heather moorland
• Acidic bogs	• Permanent pasture
• Acidic sand dunes	• Railway/road/motorway embankments
• Unmanaged riverine sites	• Roadsides, cuttings and verges
• Sea cliffs	• Quarries/spoil heaps/post-industrial sites
	• Riverine and reservoir sites
	• Relics of former landscaped gardens, for example in suburban areas
	• Landscaped parks, woods, gardens, etc.

Adapted from Rotherham & Read (1988).

Plate 1 *Rhododendron ponticum* habitat in Turkey's Black Sea Mountains.

baeticum (Boiss. & Reuter) Hand.-Mazz. in south Spain and southern and central Portugal (Walker & Straka, 1970). Its present native distribution is largely a consequence of the Pleistocene glaciations. During the Gortian interglacial it was more widespread as a native in parts of central and western Europe, as demonstrated by pollen records at Hotting-Insburg, Austria (Walter & Straker, 1968) and in Ireland (Watts, 1959).

In northern Turkey, it is found over a wide area from Thrace to the Georgian border, where mean annual precipitation varies from 700 mm to 2000 mm. It occurs in forests, in moist, mountainous areas from sea level up to 2100 m, with an optimum distribution between 600 m and 1600 m, where it forms an integral part of the shrub layer. Following forest management, however, it can become invasive and as a consequence is regarded as a forest weed (Colak, 1997).

Recent work (R. Abbott & R. Milne, personal communication; Abbott & Milne, 1995) now confirms previously held suspicions (Rotherham, 2001b) that the British and Irish plants are predominantly, if not totally, from Spain, Gibraltar and Portugal.

Spreading extensively from its original sites, it now occurs throughout Britain and Ireland from the hyper-oceanic west, where rainfall is in excess of 2000 mm, to the more continental east where rainfall is under 550 mm. It favours acidic soils, moist but not waterlogged conditions, and protection from frost and from drought. It is now perceived as a major weed in forest plantations and sites of conservation interest, especially woodlands on acidic soil in western Britain

235

and Ireland (Cross, 1973, 1975, 1982; Robinson, 1980; Rotherham, 1983, 1986a,b, 1990, 2001b; Judd & Rotherham, 1992).

Until recent years relatively little attention has been paid to native *R. ponticum* in the wild. Abundant and a pernicious weed in disturbed, managed forests in northeast Turkey, it is very restricted, and sometimes rare and vulnerable across much of its range, e.g. in northwest Turkey, Bulgaria, Spain, Gibraltar and Portugal. The last small areas of apparently undisturbed, ancient *R. ponticum* forest in the high mountains of northeast Turkey are one of the most rare and distinctive 'natural' plant communities in Europe and the Balkans. Unfortunately, associated as it is with weed status throughout large areas of Turkey, and in much of its British range, the interest and vulnerability of the undisturbed, native locations have been overlooked. It is now particularly threatened.

In the British Isles, *R. ponticum* behaves much as it does in disturbed native areas, in suitable environments, spreading invasively and aggressively. Now widely established and especially invasive in western England, Wales (especially Snowdonia) (Plate 2), Scotland and Ireland (particularly in the southwest around Killarney), it is a serious threat to native vegetation, and commercial forestry. In Killarney, the native *Arbutus* is considered under threat of displacement, as is holly (*Ilex*) elsewhere in Britain.

In favourable environments *R. ponticum* forms dense, impenetrable thickets, and whilst the massed flowers in open locations are very popular with the public, it causes serious management problems for forestry and wildlife conservation. It is often invasive, spreading by seed and vegetative expansion. Layering freely, it shades out ground flora, preventing natural woodland regeneration

Plate 2 Mass of colonising *Rhododendron ponticum* seedlings in North Wales.

Table 2 Key factors in the invasive behaviour of *Rhododendron ponticum*.

- Prolific production of viable, easily dispersed seeds.

- Capacity for winter photosynthesis under suitable climatic conditions, such as Britain's Atlantic west coast.

- Ability to tolerate severe shading.

- Enhanced growth on soils of low nutrient status due to mycorrhizal infection.

- Apparently diminished herbivore pressure, probably due to high concentrations of simple 'free' phenols in its tissues.

- Possible 'allelopathy' associated with high concentrations of simple 'free' phenols.

- The dense, vigorous growth resulting from the preceding six factors enables *R. ponticum* to overtop, crowd-out and shade its competitors.

- Site disturbance is important in creating regeneration sites to allow invasion.

From Rotherham (1990).

and hindering forestry replanting (Cross, 1975; Shaw, 1984; Rotherham, 1990; Colak *et al.*,1998). This behaviour is typical of many of the larger rhododendrons worldwide.

Establishment in Britain is particularly interesting since this isolated the plant from both its original, native, phytophagous fauna and its mycorrhizal fungus. It has been suggested that invasion of British woodlands, with displacement of native species such as holly (*Ilex aquifolium* L.) is an example of successful penetration of undisturbed communities by an exotic species. However, in most cases disturbance and disruption of the established native vegetation is a significant factor (Rotherham, 2001b).

The consequences of this dramatic spread of *R. ponticum*, especially along the Atlantic coast of the British Isles, are often referred to as the '*Rhododendron* problem'. Research into this topic has been approached in two distinct ways:
1 Work on the underlying ecology of invasion.
2 Investigations into practical control.

Cross (1975) investigated the status of *R. ponticum* in the British Isles, with detailed studies in southwest Ireland. Other local and regional surveys have since been undertaken across Britain (Rotherham, 1986a,b,c). Research has identified eight key factors in the invasive behaviour of *R. ponticum* (Table 2).

Why does *R. ponticum* invade? A strong competitor

Whilst *R. ponticum* performs well under shade, and itself shades out its competitors, how it becomes so dominant has been unclear. The poisoning of competitors ('allelopathy') is a possibly important but controversial aspect of plant interactions, and has been demonstrated in ericaceous plant communities. Also, the critical role of plant biochemistry in plant–herbivore interactions is firmly established, and *R. ponticum* tissue concentrations of 'free' phenols are sufficient for either allelopathic influences or anti-herbivore interactions. Indeed, potentially toxic chemicals, particularly simple 'free' phenols, occur in significant quantities in R. ponticum tissues and have caused poisoning in both vertebrates and invertebrates (see Rotherham & Read, 1988; Judd & Rotherham, 1989).

Laboratory experiments showed interference by R. ponticum on seedlings of competitors, growth being almost totally inhibited. These effects influence competition between R. ponticum and native plant species, and in both field and laboratory are not removed by water or nutrient addition (Rotherham & Read, 1988). Similar observations have been made for heather (Calluna vulgaris (L.) Hull). The agents of these effects are probably 'free' phenols, and highly toxic, short-chain, aliphatic acids (Rotherham, 1983; Rotherham & Read, 1988).

Tissue biochemistry is the key to many aspects of R. ponticum ecology. Phenols are most concentrated in young R. ponticum tissues, which, lacking physical toughness, are most vulnerable to herbivores. Young, emergent leaf buds produce a sticky exudate high in phenols. This may combine physical and chemical deterrents to herbivores at a critical stage in leaf development. Phenol levels fall as tissues age, when protection is through physical toughness of mature leaves. Research has shown 'free' phenol concentrations in R. ponticum tissues considerably higher than in associated woody and herbaceous plants. In vitro concentrations of biologically active flavonoids between 0.1% and 1% can have inhibiting effects, and concentrations in R. ponticum tissues exceed these. High toxin concentrations, not alien status as previously suggested, are responsible for restricted herbivory (Rotherham & Read, 1988; Judd & Rotherham, 1992).

Mycorrhizal roots are important in the growth and establishment of most ericaceous plants and they dramatically increase growth rate of R. ponticum (Rotherham, 1983; Rotherham & Read, 1988). With low-nutrient soils, seedling dry weight was increased by 184% with the addition of the mycorrhizal fungus. Mycorrhizal roots compete more effectively for nutrients, enhancing growth in nutrient-deficient soils invaded by R. ponticum.

It seems that these factors combine to outcompete native plants, the significance of each varying with environmental conditions at individual sites.

How does R. ponticum invade? Control and facilitation

Rhododendron ponticum has two mechanisms of spread: (i) vegetative extension, and (ii) seed. Colonisation by seed and domination by vegetative spread are critical for invasion. However, despite extensive dispersal of viable seed, rapid invasion of a site does not always occur. The facilitation of seedling establishment is vital.

Disturbance creates safe regeneration sites for seedling establishment, and in apparently stable plant communities subtle disturbances may precipitate invasion. Unable to establish in closed herbaceous swards, and vulnerable to desiccation, grazing and competition, the seedlings require damp, bare or mossy ground, and are the 'Achilles heel' for this generally competitive and robust plant. This vulnerability is partly compensated by huge numbers of seeds (> 1 million per bush). Regeneration sites are essential for successful establishment in new areas, disruption of established vegetation giving opportunities for invasion. Like many aliens, R. ponticum invades disturbed environments (roadsides, riversides, railway embankments, abandoned and contaminated post-industrial sites, etc.). Research across the South Pennines region of England has demonstrated that the suitability of areas for colonisation varies immensely. Moorland fringe and woodland, for example, are very susceptible, whilst land under intensive farming and urban influences can be relatively inhospitable. However, the ability of *R. ponticum* to reach and establish in even isolated pockets of suitable habitat is quite amazing. The general mode of dispersal is clearly by wind carrying thousands of tiny seeds over considerable distances.

Seedling *R. ponticum* has established on fragments of suitable habitat in the urban catchment of Sheffield of less than one hectare, even when the nearest significant parent population is several miles away. It may be that the seed is being carried in the typical strong westerly winds off the high ground and raining down over the lowland landscape to the east. The consequence is that if suitable habitat exists, then eventually *Rhododendron ponticum* will find it. This does have implications for attempted control.

Rhododendron ponticum invading established plant communities and successfully establishing in one location but not another raises critical questions.

Invasion: the importance of disturbance and in particular of grazing animals

Grazing animals impact directly on the plant itself and indirectly via their effects on the environment. Browse damage is limited by physical toughness of mature leaves, and by chemical toxins in young tissues. Since the palatability of plant tissues is learnt and not instinctive, some damage to *R. ponticum*

occurs with young, inexperienced animals. Grazing animals also affect *R. ponticum* seedlings where the amount of toxin is insignificant and the leaves relatively soft.

Rabbit grazing restricts invasion, but following release of myxomatosis and the rabbit population crash, spread into disturbed, bare ground is rapid. Grazing may prevent establishment in new areas, but be ineffective in established stands of *R. ponticum*.

The situation is complex, however. In the Irish Killarney oak woods sika deer disturb soil and vegetation, enabling *R. ponticum* invasion. In England's Peak District, sheep and deer grazing produces similar effects, high grazing levels preventing colonisation, but low levels triggering invasion.

Other disturbance by land management (forestry, fires and intensive recreation) may facilitate invasion, with disruption generating safe sites for seedlings. Moorland fires (accidental and deliberate) have been followed by rapid establishment from nearby *R. ponticum* colonies, with complete replacement of heather moor in under 10 years. In urban and urban fringe woods and heaths, recreational disturbance is the critical factor, although site management programmes can unwittingly have a similar impact.

The impact on fauna

When introduced to Britain *R. ponticum* was a vacant niche for herbivorous insects. As a food source it is predictable, apparent, and available all year. It was suggested that taxonomic isolation in Britain limited its pre-adapted, herbivorous fauna, but that in time it would acquire associates from existing native communities. The invasive nature of alien *R. ponticum* was attributed to restricted invertebrate herbivory. However, despite over 200 years and a very extensive occurrence in Britain, this fauna is still very restricted.

Judd & Rotherham (1992) assessed the phytophages associated with *R. ponticum* in Britain. Fourteen phytophagous insects were positively associated with *R. ponticum* in surveys and 17 from the literature, a total of 31. These were in two categories: (i) 26 highly polyphagous indigenous phytophages, and (ii) five host-specific introduced phytophages. Most were common, polyphytophagous insects (those associated with a broad range of host plants) and were found on *R. ponticum* only occasionally. Only two were frequent or numerous.

Seventeen other insects were found, but either feeding or identification was unconfirmed. Some of these may be associates, but many insects had fallen as larvae from the tree canopy, and when reared on *R. ponticum* leaves tended to fail, suggesting poor adaptation.

Other associates included psocids feeding on epiphytic fungi. With predators and parasites, these add to the biodiversity and biomass of associates. Predators also take fungivores, and the fungivores, predators and parasites feed higher predators. 'Tourist' species and casually associated lepidopteran larvae are also taken. Many insects (especially Diptera) and arachnids are trapped by the sticky exudate of young *R. ponticum* shoots. Predators will take these immobilised arthropods, but they themselves risk being trapped.

A comparison with other trees and shrubs in Britain

Thirty-one insects on *R. ponticum* in Britain place it 22nd (with juniper, *Juniperus communis*) in a table of trees and shrubs and their associate phytophages. This list includes planted and naturalised aliens, and natives. For comparison, hawthorn (*Crataegus monogyna* Jacq.) has 204; hazel (*Corylus avellana* L.) 102; and holly (*Ilex aquifolium*) has 10.

Buddleja davidii Franch., another alien shrub (introduced less than 100 years ago and more isolated taxonomically), already has 37. Other species with fewer recorded associates than *R. ponticum* include hornbeam (28), sweet chestnut (11), horse chestnut (7), false acacia (2), holm oak (5), walnut (5), and yew (5).

Comparison with other plants can be misleading. Most insects associated with *R. ponticum* are catholic species found at sites with or without it. Just six (five phytophages and one predator) are host-specific to *Rhododendron*, only three being frequent, numerous and widespread.

Biomass and diversity of associated fauna

Distinction is often made between the number of species associated with a particular host plant and total associated biomass. Sycamore (*Acer pseudoplatanus* L.), for example, produces relatively few associated species but substantial biomass. *Rhododendron ponticum* is impoverished in both. Alien status and taxonomic isolation in Britain could explain this, and *R. ponticum* may develop an adopted fauna. However, after 200 years, this has so far happened only slowly. Further casual associates of common polyphages will arise, but a herbivore having a major impact is unlikely. The small number of invertebrate herbivores associated with *R. ponticum* in Britain is probably not due to its alien status, but is more likely a result of physical and biochemical tissue characteristics. These apparently restrict herbivore activity both in Britain and in its native haunts. Interestingly, little is known of the endemic fauna, except that where native, *R. ponticum* is often invasive and generally unaffected by herbivore activity (Colak *et al.*, 1998). Donald Pigott (personal communication) suggested that seedling predation could be significant in Iberia.

Impact on mammals and birds

Rhododendron ponticum is generally felt to have a negative impact on British wildlife, replacing native vegetation with dense, monospecific *Rhododendron* stands. Substantial declines in breeding birds have been found in infested Welsh and Irish woodlands (e.g. Batten, 1976). However, the impact is not simple and depends on *type* and *quality* of environment, and degree of infestation. Comparisons between unaffected, semi-natural woodlands and those with single-species *R. ponticum* thicket indicate reduced diversity of structure and of associated fauna in the latter.

Some species still utilise dense *R. ponticum*. Warblers such as blackcap and chiffchaff, and the thrush family (robin, blackbird and nightingale) all feed and nest in *Rhododendron* thickets, and northern expansion of chiffchaff has been linked to *R. ponticum*. Insects attracted to the masses of flowers during May and June (Plate 3) provide important food for predatory invertebrates and insectivorous birds. Coinciding with the critical period for birds feeding young, this can be a locally important food source.

Plate 3 *Rhododendron ponticum* flowers.

Rhododendron ponticum provides cover for mammals, nesting and roosting birds, and a useful screen in areas of intense public use. As noted by many ornithologists, dense stands are favoured winter roost sites for huge numbers of finches and thrushes. Furthermore, despite many conservationists wishing otherwise, the quality of cover and the shelter provided is far superior to native shrubs and trees. Shelter from wind chill and good roosting sites secure from predators are major bonuses to small birds in the depths of winter (Rotherham, in prep.).

Mammals are affected by invasion, with the obvious negative effects of displaced native vegetation and some unexpected benefits. Badgers thrive under *Rhododendron*, and sometimes have above-ground setts (E. Neal, personal communication).

Otters in Wales and Scotland use dense, riverside *Rhododendron* cover for holts, and lying-up. Deer (red, fallow and roe) utilise *Rhododendron* cover, as do sika deer in southwest Ireland, and muntjac in southeastern England. (These alien species (sika and muntjac) are also widely regarded as a menace to native wildlife!)

The need for strategic control

Rhododendron ponticum control has two central issues and one supplementary one:
1 Eradication of existing populations.
2 Containment and minimisation of encroachment.
3 Land management in susceptible areas must be modified to minimise future invasion.
Eradication at some sites may be difficult if not impossible, due to massive seed production, a propensity to regenerate from cut stumps and roots, and physical and chemical resistance to herbicides. The terrain of many infested sites makes physical removal by machine very difficult. In addition, on some woodland sites, for example, the damage to woodland archaeology or other features may be a significant deterrent to gross physical removal. For long-term control, land management in susceptible areas must be modified to minimise future invasion. This requires an understanding of the factors responsible for triggering invasion in the first instance, and then a strategic approach: removal and containment. At some sites, huge effort results in removal and control in limited, heavily infested areas, but at the same time, massive invasion occurs elsewhere on the same estates. On the Chatsworth Estate in the English Peak District, 20–30 years on from a major control programme, the target areas are controlled, but the overall infestation is probably five to ten times worse. Embryonic invasion has now become full colonisation on a broad front (Plate 4). Resources are better applied to wider control of young, invasive plants and containment of mature stands.

There are two further aspects to the approach to control: (i) long-term cost, and (ii) public opinion. The first has been demonstrated by the detailed assessments for Snowdonia in Wales, where the costs of removal and control have been estimated at over £50 million. For many reasons, eradication as such is not possible (even if desirable) so we are looking at an ongoing cost for control. So far there has been little indication of a corporate willingness to pay.

The same issues surround the debates about Japanese knotweed and giant hogweed, for example. Some local authorities are taking action but many are not. Ways to help finance this control by economic use of say fuel-wood or charcoal have been consid-

Plate 4 *Rhododendron ponticum* in massed banks at Cordwell in the English Peak District.

ered in Killarney in southwest Ireland, and are now being investigated in Snowdonia (Jenny Wong, personal communication).

There is a hugely important and often overlooked aspect to many exotic invasive plants and animals: people like them! As demonstrated for Himalayan balsam (Rotherham, 2001a) a major factor in their spread has been that people deliberately carry seed to new locations sometimes hundreds of kilometres apart, and occasionally across several countries, and liberally introduce them to hundreds of new locations. This is happening faster than the conservationists can catch up with them. In the case of balsam, which arrived in Britain in 1836, this deliberate spread can be documented from the 1840s onwards to the present day.

In the English Peak District and in the Welsh Snowdonia National Park, people travel many miles as tourists to see the stunning displays of *R. ponticum* in flower. Local people took several years of persuasion to agree to the Snowdonia control programme because they felt it would threaten their economic interests in tourism (Jenny Wong, personal communication). These are important issues that are frequently overlooked in the conservation debate.

Conclusions

In both native situations and as an alien, *Rhododendron ponticum* can be a problematic weed. However, the need both to conserve native *R. ponticum* and to address key aspects of its history and ecology, in native and exotic situations, is highlighted by recent research. If *R. ponticum* had recolonised the British Isles after the last ice age, our anthropic view of its native/exotic status and conservation merit would be very different. Like many exotic species, 'wild' *R. ponticum* is firmly established in Britain. Like it or not, it is here to stay, and managers must learn to co-exist with it, with containment not eradication. Acceptance of aliens strikes at the core of much conservation training, but is at least pragmatic. Long-term and targeted management but not necessarily eradication is a realistic approach.

The problems of managing *R. ponticum* in commercial forests are similar in both native and exotic locations. Control is seen as necessary and desirable, but this is often without any thorough survey, or assessment of whether control is feasible. This approach has been exported from Britain to the native environments of *Rhododendron*, where it is applied without question and may be a threat to a unique and endangered species. This imminent threat to ancient *Rhododendron* forest in Turkey and other native locations gives serious cause for concern.

Management of exotic species (such as *R. ponticum* in Britain) requires a strategic approach. Control, whilst effective, may be difficult and expensive, and may damage woodland ecology and archaeology. For long-term effectiveness (particularly avoiding recolonisation by *R. ponticum* when control stops), changed site management is necessary. Long-term, ongoing, control measures are needed. The initial invasion into woodland by young *R. ponticum* is the most effective stage for control, but most management attacks mature, established thickets. These are difficult and expensive to treat, damage to the original site has already occurred, and wildlife interest (badger setts, bird roosts, etc.) may be affected. Containing established colonies and controlling invasive

areas, together with careful changes to wider landscape management to discourage spread, is more effective.

The public are often neglected in the debate on exotic species, yet they have a major and ongoing impact in the active dispersal of more and more aliens. They also have an impact as tourists in taking their leisure in amongst the aliens in the landscape. Many British locations for this are landscaped parks and gardens of the 1700s and 1800s. Rhododendrons are often fundamental in the landscape and structure planting, the cultivars and species (including *R. ponticum*) as integral to the historic landscape as the planted trees and built structures. Here, eradication for nature conservation is often a misconception. These are imposed landscapes that often removed or supplanted the original history and ecology in their development. Here *R. ponticum* is now an alien in a landscape of aliens.

Management and control of *R. ponticum* should be undertaken where appropriate, the misapplication of eradication under a guise of conservation should be moderated, and resources concentrated where most vitally needed. This needs to be set alongside proactive educational work, in both Britain and countries such as Turkey and those of Iberia, to more effectively set down the case and the solution. This approach to *R. ponticum* should also be fully embedded in an increased awareness of the importance of the wider *Rhododendron* genus and the broader conservation issues that this raises.

References

ABBOTT, R. J. & MILNE, R. I. (1995). Origins and evolutionary effects of invasive weeds. In: *Proceedings of the 1995 BCOC Symposium, No. 64, 'Weeds in a Changing World'*, pp. 53–64.

BATTEN, L. A. (1976). Bird communities of some Killarney woodlands. *Proc. Royal Irish Acad.* B76(19): 285–313.

CLAPHAM, A. R., TUTIN, T. G. & WARBURG, E. F. (1962). *Flora of the British Isles*. Cambridge: Cambridge University Press.

COLAK, A. H. (1997). *Investigations on the silvicultural characteristics of Rhododendron ponticum L.* Unpublished PhD thesis, University of Istanbul, Istanbul, 181pp.

COLAK, A. H., CROSS, J. R. & ROTHERHAM, I. D. (1998). *Rhododendron ponticum* in native and exotic environments, with particular reference to Turkey and the British Isles. *Practical Ecology and Conservation* 2(2): 34–41.

CROSS, J. R. (1973). *Ecology and control of Rhododendron ponticum L.* Unpublished PhD thesis, University of Dublin.

CROSS, J. R. (1975). Biological flora of the British Isles: *Rhododendron ponticum* L. *J. Ecol.* 63: 345–364.

CROSS, J. R. (1982). The invasion impact of *Rhododendron ponticum* in native Irish vegetation. *J. Life Sci. R. Dubl. Soc.* 3: 209–220.

JUDD, S. & ROTHERHAM, I. D. (1989). *Invertebrates associated with Rhododendron ponticum L. in Britain: ecological reasons for an impoverished fauna*. Royal Entomological Society of London, Northern Region Meeting with Sorby Natural History Society and Sheffield City Museum, November 1989 (unpublished).

JUDD, S. & ROTHERHAM, I. D. (1992). The phytophagous insect fauna of *Rhododendron ponticum* L. in Britain. *The Entomologist* 111(3): 134–150.

MAGOR, W. (1989). History of *Rhododendron*. *The Cornish Garden (Journal of the Cornwall Garden Society)* 32: 15.

MATTFELD, J. (1971). *Situation of East Thrace under the light of Plant Geography* (translation from German to Turkish by M. Selek). University of Istanbul Publication No.

1544, Faculty of Forestry Publication No. 159, Istanbul, 37pp.

ROBINSON, J. D. (1980). *Rhododendron ponticum*: a weed of woodlands and forest plantations seriously affecting management. In: *Proceedings of the Weed Control in Forestry Conference*, pp. 89–95.

ROTHERHAM, I. D. (1983). *The ecology of Rhododendron ponticum L. with special reference to its invasive and competitive capabilities*. Unpublished PhD thesis, University of Sheffield.

ROTHERHAM, I. D. (1986a). The introduction, spread and current distribution of *Rhododendron ponticum* in the Peak District and Sheffield area. *Naturalist* 111: 61–67.

ROTHERHAM, I. D. (1986b). *Rhododendron ponticum* L. in the Sheffield area. *Sorby Record* 24: 19–24.

ROTHERHAM, I. D. (1986c). The spread of *Rhododendron ponticum* at three sites in the Sheffield area. *Sorby Record* 24: 25–28.

ROTHERHAM, I. D. (1990). *Factors facilitating invasion by Rhododendron ponticum*. British Ecological Society Conference on the Biology and Control of Invasive Plants, Cardiff, September 1990.

ROTHERHAM, I. D. (2001a). Himalayan Balsam – the human touch. In: BRADLEY, P. (ed.) *Exotic Invasive Species – Should We Be Concerned?*, pp. 41–50. Proceedings of the 11th Conference of the Institute of Ecology and Environmental Management, Birmingham, April 2000. Winchester: IEEM.

ROTHERHAM, I. D. (2001b). *Rhododendron* gone wild. *Biologist* 48(1): 7–11.

ROTHERHAM, I. D. & READ, D. J. (1988). Aspects of the ecology of *Rhododendron ponticum* with reference to its competitive and invasive properties. In: 'The practice of weed control and vegetation management in forestry, amenity and conservation areas'. *Aspects Appl. Biol.* 16: 327–335.

SHAW, M. W. (1984). *Rhododendron ponticum* – ecological reasons for the success of an alien species in Britain and features that may assist in its control. In: 'Weed control and vegetation management in forests and amenity areas'. *Aspects Appl. Biol.* 5: 231–242.

TUNCDILEK, N. (1971). *Southern West Asia (Physical Environment)*. University of Istanbul Publication No. 1675, Institute of Geography Publication No. 65, Istanbul, 221pp.

UBER, M. A. (1951). *Floristic-Systematic Journeys in Anatolia for Plant Collection Purposes and their Value in Plant Geographical Aspect* (translation from German to Turkish by A. H. Demiriz). Biology 1(3) Istanbul, pp. 97–109.

WALTER, H. & STRAKA, H. (1968). *Die Vegetation der Erde in öko-physiologischer Betrachtung. Band II. Die gemassigten und arktischen Zonen*. Stuttgart, 1001pp.

WALTER, H. & STRAKA, H. (1970). *Arealkunde*. Einfuhrung in die Phytologie, 111/2, Floristisch-historische Geobotanik, Stuttgart, 478pp.

WATTS, W. A. (1959). Interglacial deposits at Kilbeg and Newtown, Co. Waterford. *Proc. Royal Irish Acad.* B60: 79–134.

Further information

- www.shu.ac.uk/sybionet
- www.uk-econet.com
- www.shu.ac.uk/wildtrack

Plant-hunting along the Nu Jiang (Salween River) in Yunnan Province, China

Steve Hootman

Rhododendron Species Foundation, PO Box 3798, Federal Way, WA 98063, USA

A summary is presented of the observations made on three recent expeditions (1997, 2000 and 2001) into the mountainous regions bordering the Salween River (Nu Jiang) along the Myanmar (Burma) frontier in western Yunnan Province, People's Republic of China. These expeditions focused on the collection of data and scientific (herbarium) specimens of the diverse flora found in this region. Primary genera of interest included *Rhododendron* and other Ericaceae, as well as those woody and herbaceous plants closely associated with these groups in their natural habitats. Several species were collected representing substantial extensions of known geographic ranges. In addition, several taxa were collected and introduced into horticulture for the first time.

Introduction

Over a period of five years I was a participant in a series of three expeditions to the remote and mountainous regions along both sides of the Salween River (Nu Jiang) in western Yunnan Province, China. This region is bordered to the west by Myanmar (Burma) and to the north by Xizang (Tibet). On the border with Myanmar lies the mountain range known as the Gaoligong Shan, which is bordered on the east by the Salween River. This region was our primary focus for exploration as it is known to be very rich in diversity of *Rhododendron* species as well as other interesting plant groups. Until now, in the modern era of plant exploration, this region had been relatively untouched by western plant-hunters. This was due in a large degree to the very remoteness that allows its extraordinary flora to exist.

In co-operation with botanists from the Kunming Institute of Botany, I joined Peter Cox, Dr David Chamberlain and other knowledgeable *Rhododendron* experts and plantsmen in pursuit of the incredible flora of the Salween region. In our quest for herbarium specimens, photographic documentation, and, above all, knowledge, we hoped to find many species that had not been collected or even observed by western botanists or horticulturists since the early part of the 20th century. Indeed, the opportunity for observing a large number of *Rhododendron* species in this region is such that, during one day's trek, on 3 June 2000, we observed 40 distinct taxa of *Rhododendron*.

Our primary area of focus was to be the extreme northwestern corner of Yunnan where a tributary of the Irrawaddy runs parallel to the Salween. This tributary, lying to the west of the Salween and forming a deep gorge between the west slope of the Gaoligong Shan and the range of peaks along the Myanmar border, is known as the Dulong or Taron River. Two secondary areas of focus were the mountains along both sides of the Salween River in NW Yunnan and the Hpimaw Pass into Myanmar far to the south in central Yunnan. This region had been visited by many of the early plant-hunters but most of it had not been explored since that time. As expected, travel in this area was difficult if not impossible in some instances. Poor roads on steep slopes with large quantities of precipitation combine to make driving anywhere in the region problematic. As it turned out, access to the higher slopes was also very difficult and plans often had to be altered accordingly.

Expedition to the Salween (Chamberlain, Cox, Hootman, Hutchison), 1997

In the autumn of 1997, during our first visit to the Salween River region, we did not actually manage to get to the Hpimaw Pass or the Dulong River but still managed to find many exciting plants. *Rhododendron* taxa from numerous subsections are represented in this region, with subsections *Neriiflora*, *Maddenia* and the big-leaf species especially abundant and diverse.

Among the lepidote or scaly-leaved *Rhododendron* species we noted many members of subsection *Maddenia*, including *R. taggianum* Hutch., *R. dendricola* Hutch., *R. nuttallii* Booth, *R. maddenii* ssp. *crassum* (Franch.) Cullen and *R. megacalyx* Balf.f. & Kingdon-Ward. These species typically grew on cliffs or as epiphytes in the forest trees. A very exciting find was a fine form of *R. edgeworthii* Hook.f. with incredibly thick orange-brown indumentum and very large

flowers (CCHH#8016). We also found plants of the rare *R. chrysodoron* Tagg ex Hutch., a deep yellow-flowered member of subsection *Boothia* with striking, smooth and peeling, dark red-brown bark. This is the first collection (CCHH#8109 and #8116) of this species since Frank Kingdon-Ward found it in 1953 (#20878 from 'the Triangle' of Upper Myanmar). It occurred primarily on vertical cliffs or as an epiphyte on forest trees at an altitude of around 2100–2300 m.

Another yellow-flowered species that had us all very excited was *R. monanthum* Balf.f. & W.W.Sm. (Plate 1). Although we first found it lying alongside the road in a jumble of rocks and forest debris deposited in a landslide, it grew primarily as an epiphyte on the trunks and large branches of forest trees. As far as any of us were aware, this was the first introduction of this species into cultivation (CCHH#8133 and #8208). What was particularly interesting was that, although it was early October, the plants were covered with yellow flowers. Subsequent observations in the wild and in cultivation have confirmed this unusual and (so far) completely unique phenomenon – *Rhododendron monanthum* is a true autumn-blooming taxon. This member of subsection *Monantha* is a dwarf species with small, elliptic to ovate, grey-green leaves and smooth, coppery brown bark. The deep yellow to greenish-yellow, tubular-campanulate flowers appear in October and November. These are solitary or more commonly in a terminal inflorescence of up to four flowers.

An unusual epiphytic lepidote (CCHH#8106 and #8126) that really caught our attention was a species with a remarkable resemblance to the well-known *R. edgeworthii*. The plants were generally smaller

Plate 1 *Rhododendron monanthum* (Gaoligong Shan, east side, 3100 m).

Plate 2 *Rhododendron seinghkuense.*

in stature with a correspondingly smaller, but more rounded, leaf and a much thinner indumentum on the lower leaf surface. We guessed that we were looking at the rare *R. seinghkuense* Kingdon-Ward (Plate 2), a species represented in cultivation by only one clone (from KW#9254 – collected on the Myanmar–Tibet frontier in 1931) grown and propagated by Peter and Kenneth Cox at Glendoick Gardens. This species is remarkable in that it has yellow flowers of a most distinctive appearance. They are rather small in comparison with the flowers of *R. edgeworthii* but are nonetheless quite attractive, with a rotate-campanulate corolla, a short and sharply deflexed style and an indumentum of long brown hairs on the ovary. My personal view is that this taxon arose from an ancient merging of *R. edgeworthii* with a member of subsection *Boothia* (a small group having some species with flowers that are both rotate and yellow with short, sharply deflexed styles). The most likely parent would be *R. chrysodoron*, which is found at similar elevations, lower than either *R. megeratum* Balf.f. & Forrest 1920 or *R. sulfureum* Franch. 1887, the other more or less obvious possible choices as a parent.

The appearance of the expected deep yellow flowers on several seedlings grown from this new collection verified our guess as to their correct identity when they first appeared in the spring of 2001. These plants differ from KW#9254 in that the flowers of the former are larger and deeper yellow in colour with a much more prominently hairy ovary. We found *R. seinghkuense* to be quite common in NW Yunnan where it was always observed either on vertical cliffs and boulders or literally hanging from the moss-covered trunks of large trees. It was typically found at lower elevations than *R. edgewor-*

thii and will probably be correspondingly more tender. The stunning new growth and heavily indumented stems and leaves of this species earn it its place in mild climate gardens and the cool conservatory.

Rhododendron genestierianum Forrest is another lepidote species we found (CCHH#8080 and #8119) to be quite common in the temperate broad-leaved forests of the Salween River region. It typically grew as scattered specimens in the forest but I observed one amazing population forming the entire mid-level canopy under a hemlock forest in the Gaoligong Shan of NW Yunnan. This large grove consisted of small trees up to 7 m in height, many with single trunks 20 cm in diameter. Needless to say, the display of glossy purple-black, smooth and exfoliating bark was stunning. I would have to rate this small forest of the unique *R. genestierianum* as one of the greatest displays in the genus that I have ever observed, in the wild or in cultivation. This species is rarely seen in gardens as it is rather tender and flushes its new growth very early in the season, leading to damage from spring frosts. It is quite unique, however, in several ways. The inflorescence comprises up to 15 small, reddish-purple bells covered with a whitish glaucous bloom. The undersides of the deep glossy green leaves are also covered with a bright white waxy coating and are most attractive in colour when they first emerge in shades of reddish-purple. This is another species that had not been collected from the wild since Frank Kingdon-Ward's 1953 expedition to 'the Triangle' of Upper Myanmar (KW#20682).

A completely unexpected and indeed remarkable find (CCHH#8162) was that of *R. cinnabarinum* Hook.f., far to the east of its known range. The plants had stunning glaucous blue leaves similar in appearance to the foliage of *R. cinnabarinum* ssp. *xanthocodon* (Hutch.) Cullen (Concatenans Group) but on average were much more narrow in shape. With such narrow leaves, the plants in this area technically fall within the parameters of the Himalayan subspecies *cinnabarinum* (at least until flowers are observed), which is not recorded again until central Bhutan! In my opinion, it makes much more sense to consider these populations as part of subspecies *xanthocodon* (Concatenans Group) as this is the closest taxon geographically (excluding the quite distinct ssp. *tamaense* (Davidian) Cullen). Subsequent observations of other populations in nearby areas in 2000 and 2001 show that this undetermined taxon is locally widespread and can be quite common where it occurs. It is interesting that this species is not recorded from this area although Kingdon-Ward, Forrest, Rock and others went through in the early part of the 20th century.

Among the more interesting elepidotes observed and collected were tremendous forests of *R. sinogrande* Balf.f. & W.W.Sm., and, at lower elevations, large specimens of the much rarer big-leaf *R. protistum* Balf.f. & Forrest. The latter species occurred in small groves or as scattered individuals within the evergreen broad-leaved forest. Specimens up to 20 m in height were not uncommon at elevations around 2000 m. Although this species is considered to be quite rare by the Chinese, it is common in the temperate rainforests along the Salween River. As one gains elevation in these mountains, *R. protistum* is replaced by the magnificent *R. sinogrande*, which in turn is replaced by *R. arizelum* Balf.f. & Forrest in the coniferous forests below the timberline.

Plate 3 *Rhododendron gongshanense.*

Within the *R. sinograndе* zone, from around 2100 to 2500 m, a strange plant appeared that had us all baffled. It seemed to be a member of subsection *Glischra* as it had similar foliage to the species in that group. The first large plant we observed was around 8 or 9 m in height with an almost equal spread. It was growing directly off the path amongst the forest trees. It had long, narrowly lanceolate leaves around 20 cm in length. These were rugose with a recurved margin and a brownish indumentum of tufted hairs beneath. Several names were debated, including *R. vesiculiferum* Tagg, a rare species known from this region. None quite fit, however, and we labelled the collection as *species nova* (new species) and gave it the number CCHH#8110. Several other large specimens were observed along the trail, as were numerous small seedlings in the rocks and cliff faces. The identity of this taxon remained a mystery until 1999 when Volume III of *Rhododendrons of China* was published. There on page 46 was our 'new species' – *R. gongshanense* T.L.Ming (Plate 3), described and named as a member of subsection *Irrorata* from the exact area we had explored in 1997 – the Gaoligong Shan of Gongshan County, Yunnan. Until our initial expedition into this region, this newly described species had been unavailable for study in the West and little if anything was known of it. Photographs published with the text show an inflorescence of around 20 tubular-campanulate flowers opening deep red and fading to rich rose-pink with nectar pouches. In addition, the new growth is a stunning reddish-bronze in colour.

Further observations of this species in the wild and in cultivation lead us to think that it is part of a cline including the closely related species *R. tanastylum* Balf.f. & King-don-Ward, *R. ramsdenianum* Cowan and *R. kendrickii* Nutt. It should make a fine new species for mild gardens but its early flowers and new growth will prevent it from becoming a good garden plant in many areas.

A short list of some of the other elepidote species of interest would include *R. kyawi* Lace & W.W.Sm., a fine but tender species closely related to *R. facetum* Balf.f. & Kingdon-Ward but larger in all its parts, with red flowers in mid- to late summer; the yellow-flowered *R. campylocarpum* ssp. *caloxanthum* (Balf.f. & Forrest) D.F.Chamb.; and an extremely narrow-leaved form of *R. arboreum* ssp. *delavayi* var. *peramoenum* (Balf.f. & Forrest) D.F.Chamb. with leaves similar in shape to the foliage of *R. roxieanum* Forrest var. *oreonastes* (Balf.f. & Forrest) Davidian.

British-American Salween Expedition, 2000

For our second expedition, in the spring of 2000, we did manage to cross a pass over the Salween/Irrawaddy Divide, thus getting much closer to the gorge of the Dulong than in 1997. Our one afternoon in the Irrawaddy drainage and several days in nearby areas yielded many interesting *Rhododendron* species. In addition to those collected on the first expedition we found the following lepidotes. A species observed growing almost exclusively on cliffs was the distinct and fragrant maddenia *R. megacalyx*. The forms observed in flower along the new Dulong road, from the town of Gongshan west to the Dulong Gorge, were a fine rich pink in colour (BASE#9532). This should make a fine addition to the normal range of white to white-flushed pink forms typically seen in cultivation. Several other species representing subsection *Maddenia* were found,

Plate 4 *Rhododendron zaleucum* aff. (Gaoligong Shan, above the Dulong Valley, 2800 m).

including the unparalleled *R. nuttallii* which was seen in virtually every region we entered that spring. This highly ornamental species was in glorious full bloom with trumpets up to 15 cm in length, and was almost always growing on inaccessible steep cliff faces high overhead.

Another area explored on this expedition was the mountain range known as the Biluoxue Shan forming the Mekong/Salween Divide on the east side of the Salween River. Here, while following a rocky streambed up to a pass, we found yet another member of subsection *Maddenia* (BASE#9577). This was *R. fletcherianum* Davidian which was a complete surprise as this species was known only from further north in southeastern Tibet (last collected in 1932 by Joseph Rock). The few specimens we could find had a low, mounded habit and were growing on boul-ders and steep mossy cliffs at around 2900 m. The scene of pale yellow flowers set against deep green moss glowing in the misty light is one we will not soon forget.

An interesting and rather mysterious species that was quite common in forest openings was one we called *R. zaleucum* Balf.f. & W.W. Sm. affinity (BASE#9651) (Plate 4). This had the general appearance of the distinct *R. zaleucum*, but lacked the intense glaucous white on the undersides of the leaves so characteristic of that species (this may be *R. lateriflorum* Fang & Zhang 1983). The large, white, funnel-shaped flowers were flushed with rose on the tube and were similar in appearance to the flowers of *R. zaleucum*. Indeed, they were so large that they appeared to belong to a member of subsection *Maddenia* when viewed from a distance.

253

During our brief visit to the drainage of the Irrawaddy River (the Dulong side of the mountains), we found two other exciting lepidote species. These were two that we had hoped to find as they are species known primarily from the Irrawaddy drainage. The first find as we descended from the pass was an outstanding form of R. *tephropeplum* Balf.f. & Farrer, another species last collected by Frank Kingdon-Ward from 'the Triangle' of Upper Myanmar in 1953. The only plants observed (BASE#9649) were all growing in a small colony amongst boulders in a fairly open situation. The large, rose purple flowers were finer than any seen in cultivation and the bold, olive green leaves shone brilliantly in the sunshine. The newly emerging shoots and leaves were a dark blackish-purple – an altogether stunning display. As we continued to descend we began to enter the coniferous zone once again. Growing beneath the large conifers was a tremendous assemblage of plants including many different rhododendrons. Possibly the most exciting among these was R. *cinnabarinum* ssp. *tamaense* (BASE#9652), another taxon we had not planned on seeing. This was the first documentation of this rare species outside of Myanmar where it was last collected by Kingdon-Ward in 1953 on the Triangle Expedition. The plants were in full flower and, although the flowers were a bit paler in colour than the form originally introduced from Upper Myanmar, their lavender-purple coloration was superb in contrast with the sea-green colour of the new foliage. In addition to its disjunct range and purple flowers, this subspecies of the well-known Himalayan R. *cinnabarinum* is distinct in that it is deciduous to semi-deciduous. It is rarely seen in cultivation but well worth growing for its lovely flowers and bright blue-green leaves.

An as yet unnamed taxon (BASE#9655) was found at the lowest point we reached in the Irrawaddy drainage that year. I observed a single mature specimen growing epiphytically on a large maple with scattered smaller specimens nearby. This did not match anything else that we had seen or that I was familiar with other than it appeared to be a member of subsection *Glauca* or, possibly, *Boothia*. The scales on the undersides of the leaves were quite different from anything else, appearing to be quite distant from each other with a somewhat white glaucous coating. I thought it might possibly be R. *luteiflorum* (Davidian) Cullen, a species known from nearby Upper Myanmar, but at 2460 m it was much too low in elevation. The appearance of flowers should indicate what this is or at least where to place this possible new species.

Among the tremendous variety of different elepidotes observed on this expedition, we found the following to be the most interesting or exciting in addition to those we had seen in 1997. A species common throughout the region in forests and on cliffs was R. *araiophyllum* Balf.f. & W.W.Sm., a member of subsection *Irrorata*. This species is rarely seen in cultivation as it is tender and, like all *Irrorata* species, eventually forms quite a large plant, rendering it unsuitable for container or indoor culture. This was the first introduction (BASE#9529 and #9698) of this species into cultivation since 1925 and, with any luck, plants from this new collection will make it into the taxonomic collections and public gardens in those mild regions that can accommodate such a treasure. One of the most exciting plants of the entire expedition was a species observed as a single specimen on the east side of the Salween in the Biluoxue Shan. This small tree was identified immediately as *R. hylaeum* Balf.f. & Farrer by Peter Cox, who has seen many of its close relatives in the wild. *Rhododendron hylaeum* (BASE#9551 – first collection since 1931), a very rare species in cultivation, is a member of subsection *Thomsonia* with striking smooth, reddish-grey bark. Later in the expedition, while exploring the Irrawaddy side of the mountains to the west of the Salween, we observed dozens of specimens of this tree-like species. They were incredibly beautiful with their twisting curving trunks rising smoothly into the mist of the virtually non-stop rain.

Two species rarely seen in collections that we managed to locate but failed to introduce were *R. pocophorum* Balf.f. ex Tagg and the westernmost member of subsection *Argyrophylla* – *R. coryanum* Tagg & Forrest.

The former species is a magnificent large-growing member of subsection *Neriiflora* with waxy red flowers and lovely reddish-brown indumentum. The latter is extremely rare in gardens but makes a lovely specimen when it covers itself with masses of small white flowers.

As the expedition came to an end we headed south where we were finally able to reach the Hpimaw Pass on a newly paved road just north of the city of Liuku. We had not succeeded in reaching this pass into Myanmar in 1997 due to the road being impassable. Conditions were much improved in the spring of 2000 and we spent two full days exploring both sides of the 3200 m pass. At the very top we found an amazing assemblage of *Rhododendron* species including *R. mallotum* Balf.f. & Kingdon-Ward (first collection

Plate 5 *Rhododendron basilicum* with *R. calostrotum* ssp. *calostrotum* (Ziben Shan, Yangtze–Mekong Divide, W Yunnan, 3130 m).

255

since 1924 – BASE#9672). This species had been collected by Peter's father Euan Cox with Reginald Farrer (Farrer#815) in 1919 from this exact location, and Peter was very happy to have the opportunity to retrace his father's footsteps. Other interesting species observed on the Hpimaw were *R. sidereum* Balf.f., *R. sinogrande*, *R. arizelum*, a fine *R. campylogynum* Franch. (Myrtilloides Group) growing on vertical cliffs, many large plants of the rich red *R. facetum* in full bloom, a very good form of *R. zaleucum*, the rare big-leaf *R. basilicum* Balf.f. & W.W.Sm. (Plate 5), and a stunning deep orange-red form of *R. dichroanthum* Diels ssp. *scyphocalyx* (Balf.f. & Forrest) Cowan.

We spent the final two days in the field on a peak known as Ziben Shan near the town of Caojian on the eastern side of the Salween. This mountain is famous for its population of a big-leaf species that Chinese botanists have named *Rhododendron gratum* T.L.Ming 1981. Our observations of this 'new' taxon indicate that it is nothing more than a form of the variable *R. basilicum*. Growing with and even underneath these tree-like rhododendrons were carpets of *R. calostrotum* Balf.f. & Kingdon-Ward ssp. *calostrotum* in full bloom. It was an unusual and memorable site. Another exciting find on this mountain was a member of subsection *Maddenia* – *R. pseudociliipes* Cullen, an epiphytic or cliff-dwelling species with small, narrowly elliptic leaves and only one or two flowers per inflorescence. As far as we know, this is the first introduction (BASE#9697) of this little-known species to cultivation.

Dulong Jiang/Gaoligong Shan Expedition to Yunnan, 2001

In co-operation with the Kunming Institute of Botany, a third expedition was organised to the Salween River region in the autumn of 2001 to collect herbarium specimens and, hopefully, some small quantities of seed. This expedition was ultimately successful in finally entering the Dulong Gorge, and we were able to more fully explore several of the areas covered on the previous two expeditions. In addition, we were able to investigate many unexplored but botanically rich areas along the newly opened road from the town of Gongshan into the Dulong Gorge. Unfortunately, once in the gorge of the Dulong, we encountered extreme difficulties in accessing the higher elevations along the river. Various factors contributed to this but the primary cause was a lack of a good track up to any of the passes. Also, the incredible steepness of the slopes along both sides of the Dulong, in conjunction with the dense growth of vegetation, gave little hope of finding suitable clearings (away from the river itself) to accommodate the camping requirements of our 10-person group.

The third expedition found most of the same material noted previously with the addition of the following *Rhododendron* species: *R. martinianum* Balf.f. & Forrest, *R. sanguineum* ssp. *didymum* (Balf.f. & Forrest) Cowan, and a probable new species in subsection *Boothia*. The only population of *R. martinianum* located was along the Dulong road at around 2850 m, well down from the pass into the Irrawaddy drainage. This was the first time (DGEY#042) that this species had been seen in the wild or collected since 1953 when Frank Kingdon-Ward found it in 'the Triangle' of Upper Myanmar (KW#21557). The small shiny leaves were almost sessile on the stems and the plants were no more than 60 cm in height, with a few specimens even providing an off-season display of white-flushed pink flowers. The small colony was growing on an exposed

Plate 6 Upper reaches of the Dulong River.

rocky slope near a stream with *R. calostrotum* ssp. *keleticum* (Balf.f. & Forrest) Cullen and *R. sanguineum* ssp. *didymum*. The *R. sanguineum* ssp. *didymum* (DGEY#043, last collected in 1949) displayed the persistent perulae, small shiny leaves and glandular capsules indicative of this distinct subspecies.

Once in the actual gorge of the Dulong, many interesting species were observed although we were limited to relatively low elevations as described previously. Some species of note would include: *R. nuttallii*, *R. dendricola* (*taronense*), *R. kyawi* and *R. protistum*. Another species (DGEY#071), which was growing on boulders in a thicket with *R. chrysodoron*, had us baffled. Like *R. chrysodoron*, this was obviously a member of subsection *Boothia*. However, it did not match anything we had seen or any descriptions of species that it might possibly be and,

with Jens Nielsen in agreement, I labelled the specimens 'probable *species nova*'. Although distinct, it is perhaps closest to *R. sulfureum* but, at 1850 m, far too low in elevation to be this species.

As on the initial two expeditions, we made a short visit to the Hpimaw Pass in the southern Gaoligong Shan. Along with the incredible assortment of species discussed previously, we added *R. euchroum* Balf.f. & Kingdon-Ward and the rare *R. caesium* Hutch. to the list. *Rhododendron euchroum* is a member of subsection *Neriiflora* that had not previously been introduced into cultivation. It is known only from the Myanmar border regions in central Yunnan and was collected very infrequently in the early days of plant exploration. It shows some affinities with *R. albertsenianum* Forrest but has a distinct appearance with a low, mounded habit and a relatively thick, brownish indumentum on the undersides of the leaves. Comparisons with the specimens in the Herbarium of the Royal Botanic Garden Edinburgh verified our guess as to the identity of this strange new plant. *Rhododendron caesium* is another species only rarely seen in collections. It is probably represented in cultivation only by the original collection (Forrest#26798). This species is a member of subsection *Trichoclada* and somewhat similar in appearance to *R. viridescens* Hutch. 1933 but with a glaucous lower leaf surface and less bristly stems. Finally, we were able to collect herbarium specimens and a few seeds of the fantastic deep orange-red form of *R. dichroanthum* ssp. *scyphocalyx* observed in flower the year before (first collection since 1925 – DGEY#407).

Concluding remarks

Overall, I think all parties involved would have to rate these three expeditions as highly successful. In terms of specimens collected and knowledge gained, we greatly advanced our understanding of the relationships between the species in some of the more taxonomically confusing subsections. Just as importantly, we expanded the known geographic range of many species, even adding new records to the flora of China in some instances. Several species known only from herbarium specimens or botanical descriptions were photographed in their native habitats and, when proper, introduced into horticulture.

Although we covered a great deal of territory in our three expeditions, there is much yet that needs to be done. The Chinese government is shutting down much of the logging in the remaining forests of Yunnan but large areas of the adjacent forests in Upper Myanmar are being logged in their place. New roads are being developed into the most remote regions, bringing all of the resulting destruction accompanying even those projects undertaken with the best of intentions.

Key to expedition acronyms

CCHH = Chamberlain, Cox, Hootman, Hutchison – Expedition to the Salween, 1997.

BASE = British-American Salween Expedition, 2000.

DGEY = Dulong Jiang/Gaoligong Shan Expedition to Yunnan, 2001.

Rhododendron maximum in the USA: Similarities to Rhododendron ponticum in Britain and ecological mechanisms for community effects

Erik T. Nilsen & Jonathan Horton

Biology Department, Virginia Tech, Blacksburg, VA 24061, USA

Rhododendron maximum is a dominant subcanopy shrub in the southern Appalachian forest. The area of forest occupied by thickets of *R. maximum* has increased significantly over the past 15 years. Moreover, it is suggested that canopy tree recruitment is inhibited under thickets of *R. maximum* in a similar, but less dramatic, manner to the inhibition caused by *R. ponticum* in the UK. In this project we determined that canopy tree seedlings are inhibited by *R. maximum* and explored the most likely mechanisms of inhibition. Seed rain and seedling germination from the seed bank were unaffected by the presence of a thicket of *R. maximum*. However, we found that seedlings of red oak, eastern hemlock, and black cherry had lower survivorship in forest with a thicket of *R. maximum* compared with forest without a thicket. Leachates from leaves, litter, humus, or throughfall had no ecologically significant toxicity. Light availability, both canopy openness and sun fleck frequency, was significantly lower in forest with an *R. maximum* thicket. Seedlings growing under a thicket of *R. maximum* did not have any additional acclimation to low radiation compared with seedlings growing in forest without a thicket of *R. maximum*. Water availability, some nutrient resources (particularly cations), and mycorrhizal association were all significantly lower in forest with a thicket of *R. maximum* compared with forest without a thicket. All these studies point to a resource competition model for canopy tree seedling inhibition by *R. maximum*. As a result canopy tree seedlings cannot persist under *R. maximum* thicket for as long as they can in forest without a thicket. These results may be useful for understanding the mechanisms by which *R. ponticum* inhibits canopy tree seedling recruitment in the UK.

Introduction

The southern Appalachian Mountains are dominated by the oak–hickory forest type (formerly oak–chestnut) and contain a number of commercially important timber species (oak, hickory, yellow poplar, yellow and white pine, birch, hemlock, etc.). This forest has experienced major disturbances such as human habitation, agriculture, chestnut blight, Dutch elm disease, gypsy moth, and dogwood anthracnose. The dynamics of the forest have been amazingly resilient to these intense disturbances. Another less obvious factor that is influencing the natural dynamics of the southern Appalachian forest in the long term is the presence and spread of ericaceous subcanopy evergreen species such as *Rhododendron maximum* L.

It is well documented that *R. maximum* inhibits the recruitment of canopy tree seedlings (Wahlenberg, 1950; Barden, 1979; Phillips & Murdy, 1985; Clinton *et al.*, 1994; Lei *et al.*, 2002), and it is likely that *R. maximum* reduces the growth of mature canopy trees (Monk *et al.*, 1985). A number of studies on other species also indicate that subcanopy evergreen plants (including *R. ponticum* L.) can inhibit the recruitment of canopy tree seedlings (Niering & Egler, 1955; Fuller & Boorman, 1977; Cross, 1981; Veblen, 1982; Taylor & Qin, 1992). Although the effect of *R. maximum* on Appalachian forest is less dramatic than the effect of *R. ponticum* on forests of the UK, it is clear that subcanopy evergreen shrubs in the southern Appalachian Mountains are increasing in area and occupying high quality forestry sites (Dobbs, 1995). For example, the forests in the Coweeta basin (LTER site) have been undergoing a transition since the disappearance of American chestnut in the overstory followed by an increase in evergreen shrubs (*R. maximum*, *Kalmia latifolia* L.) in the understory resulting in a significant reduction in recruitment of canopy tree seedlings (Day *et al.*, 1988). Regeneration of temperate forest trees is highly dependent upon the establishment and persistence of their seedlings (Canham, 1985, 1988). Therefore, in order to understand the long-term dynamics of this ecosystem and to appropriately manage the southern Appalachian forest it is critical to understand the basic mechanisms by which the evergreen shrubs inhibit recruitment of seedlings and growth of canopy trees.

Over the past five years we have been studying the mechanisms by which *R. maximum* inhibits seedling recruitment at the pre- and post-germination stages. The central objective of this project was to determine the relative importance of plausible mechanisms of seedling inhibition by *R. maximum*. The major hypotheses we tested were that the presence of a thicket of *R. maximum* in the understory (compared with forest without a thicket) is associated with (i) a reduction in canopy tree seed rain, (ii) a reduction in canopy tree seedling emergence from the seed bank, (iii) an allelopathic effect of leachates from *R. maximum* on seed germination and growth, (iv) a reduction in mycorrhizal development on seedlings, (v) a reduction in light availability, (vi) a reduction in water availability, and (vii) a reduction in nutrient resources. We believe that the results of this study will pertain to, or be useful for understanding, the mechanisms by which *R. ponticum* inhibits canopy tree seedling recruitment in the UK.

Species description and methods

Rhododendron maximum L. ranges from central Georgia to southern Canada along the Appalachian–Allegheny mountain ranges. This evergreen shrub reaches 3 m in height and can form dense thickets in certain areas. In the southern part of its range *R. maximum* is most abundant along stream courses and on north-facing slopes.

Seed rain and seed bank

Seed rain was determined using circular fine-mesh (1 mm aperture) seed traps of 1 m^2 (c.1 m above the forest floor) placed in the vicinity of the experimental plots at five locations 200–500 m apart. At each location, three traps (3–5 m apart) were placed under a thicket of *R. maximum* and another three in adjacent forest outside the thicket. The traps were placed without regard to location of overstory species. From 11 May 1996 to 1 December 1997, contents of the seed traps (seed and litter) were collected monthly. Total seed count was made for all tree species except *Betula lenta* L. where seed number per trap was estimated by applying the number of seeds per dry weight in three subsamples of the fine sieved fraction to the total weight of the fine fraction.

Seed bank samples were collected in the same vicinity as the seed trap locations in April 1997. Care was taken to maintain the substrate structure (litter, organic, and top mineral layer), and samples were placed in trays and incubated in a greenhouse at 50% full sun, 25°C, and kept moist. The samples were maintained for six months to allow for full germination of viable seeds before a total seedling count by species was made.

Allelopathy

Leachate sampling design

Throughfall was collected with standard wet–dry collectors throughout the year. Following storm events during each month of the year, the throughfall was collected and immediately frozen (–25°C) until used in bioassays. Forest substrate was collected at five randomly selected locations in both the sites with *R. maximum* (+*Rm*) and those without (–*Rm*). Collections were made in the late spring (the time of most seed germination and seedling establishment). Forest floor substrate was separated into litter and partially decomposed organic layer and pooled by forest type (+*Rm* or –*Rm*). Leachates used in bioassays were made by mixing specified masses of fresh litter or organic layer with distilled water in sealed plastic bags. The standard ratio of litter or organic layer to water was 1:5, and controls were distilled water alone in the plastic bags. The mixtures were incubated at 25°C for 24 hours in a growth chamber and periodically shaken, before filtering through Whatman no. 3 filter paper. Leachates were kept in the refrigerator until used within five days. The pH and osmolality of leachate solutions were measured before all experiments with a pH meter (Accumet Model 610, Fisher, Pittsburgh, Pennsylvania) and a vapour pressure osmometer (Model 5520, Wescor, Logan, Utah), respectively.

Bioassay of solution toxicity

Two bioassay species, lettuce (*Lactuca sativa* L. var. 'black seeded Simpson' lot no. 371, Wyatt-Quarles Seed Co., Garner, North Carolina) and cress (*Lepidum sativum* L. var. 'Upland' lot no. 1278, Wyatt-Quarles Seed Co.)) were used to assay potential inhibition of seed germination and radicle elongation by leachates and throughfall. These species were selected because they germinate rapidly and have been used in many bioassay experiments in the past (Rice, 1979). In each bioassay experiment, two layers of Whatman no. 1 filter paper were inserted in a sterile Petri dish. The filter paper was moistened with 3 ml of leachate or throughfall solution at room temperature. One lot number of lettuce and cress seeds was used for all experiments. In seed germination tests, 30 lettuce or cress seed were randomly placed in the Petri dishes; the dishes were sealed with Parafilm, and placed in a growth chamber at 25°C, 60% relative humidity (RH), and a 12/12 hours day/night (photosynthetic photon flux density = 90 µmol photons m^{-2} s^{-1}).

Germination percentage was counted in all Petri dishes daily at noon for seven days (no more germination occurred after this time). The experiment was replicated five times.

Root elongation was studied by placing 20 seeds in two rows down the middle of the Petri dish at five replicates per treatment. Each seed was oriented so that the roots grew out towards the edge of the plate. To permit repeated census at the same time each day an image was made of each bioassay plate with the use of an image analysis program on a Mac platform (NIH IMAGE). Length of root for each seedling was measured using a calibrated scale placed in each Petri dish. Mean root length was determined for each plate on each sampling date.

Native seed germination experiments

Individual lots of seed for seven species native to the research site were obtained from seed suppliers as follows: eastern hemlock – *Tsuga canadensis* (L.) Carr, northern red oak – *Quercus rubra* L., and red maple – *Acer rubrum* L. seeds from Sheffield's Seed Co. Inc., Locke, New York; yellow poplar – *Liriodendron tulipifera* L. from F.W. Schumacher Co. Inc., Sandwich, Massachusetts; white pine – *Pinus strobus* L., black tupelo – *Nyssa sylvatica* Marshall, and black birch – *Betula lenta* L. from Herbst Tree Seed, Inc., Fairview, North Carolina. Initial germination trials indicated that germination rates of three species (*Liriodendron tulipifera*, *Acer rubrum*, and *Betula lenta*) were too low (5–10%) to be useful in these experiments. Therefore, only four species (*Quercus rubra*, *Pinus strobus*, *Nyssa sylvatica*, and *Tsuga canadensis*) were analysed at the end of the experiment. In each zip-loc bag, 50 seeds of each species were placed in 250 g of substrate and

brought to field capacity with distilled water. Three substrate types were used: forest floor substrate (litter and organic layer) from sites with a thicket of *R. maximum* (+*Rm*), forest floor substrate from sites without a thicket of *R. maximum* (–*Rm*), and vermiculite. The experiment was replicated three times for each species. Seeds were imbibed for three months at 3°C and then transferred into a growth chamber at 25°C, 60% RH, and a 12/12 hours day/night. Germinating seeds were counted each week for the next five months. Moisture lost from bags was replaced as needed on a weekly basis.

Field seedling experiment

A total of six rectangular plots (10 m × 20 m) were randomly located in a forest site that contained a thicket of *R. maximum*. Three plots were randomly located outside the thicket and three plots were located inside the thicket. The thicket of *R. maximum* had a patchy distribution in the main forest area. Therefore, all plots with *R. maximum* (+*Rm*) were not located in one section of the greater forest site. The number and species of overstory trees associated with each main plot (within and 10 m beyond the main plot) was similar (17–24 individuals). In each main plot, fifteen 2 m × 2 m subplots were established, resulting in a total of 90 plots, 45 within the *R. maximum* thickets (+*Rm*) and 45 in forest free of *R. maximum* (–*Rm*). The 15 subplots were randomly assigned into five substrate manipulation treatments with the four combinations of forest and *R. maximum* litter, and forest and *R. maximum* organic layer, plus a control with no disturbance to the substrate. In the autumn of 1995, litter and the organic layer were collected from the subplots, bulked, and redistributed back to randomly assigned subplots. Each of the five treatment types is represented by three replicate subplots per block. The aim of the substrate manipulation was to evaluate the possible inhibitory effects of the *R. maximum* litter and organic layers on the growth and survival of seedlings of three common tree species: *Quercus rubra*, *Prunus serotina*, and *Tsuga canadensis*. The experimental plots were allowed to settle over winter before seeds or seedlings were planted in the following spring.

Northern red oak (*Quercus rubra*) acorns were collected in the vicinity of the experimental plots in March and April 1996 and planted in the subplots in mid-April 1996. The naturally stratified acorns – some acorns had already begun radicle extension – were planted into the plots at 16 acorns per subplot. Black cherry (*Prunus serotina*) seeds of local provenance (Cherokee Seed Company, Murphy, North Carolina) were first soaked in concentrated H_2SO_4 for 30 minutes and then stratified in two separate lots of forest and *R. maximum* organic substrate at 4°C for four months before planting at 16 seeds per subplot. Because of their unlikely probability of germinating *in situ*, hemlock (*Tsuga canadensis*) seedlings (same source as the cherry) were raised by first stratifying in separate forest and *R. maximum* organic substrates for two months then sowing in greenhouse flats. Seedlings were maintained under shade cloth (10% full sunlight) for one month before they were transplanted at nine seedlings per subplot in early June 1996.

Light environment measurements

Microclimatic parameters were measured continuously in two forest sites (with and without *R. maximum*) using permanently installed sensors. For each site, we measured PPFD (photosynthetic photon flux density)

using a quantum sensor (Li-Cor model 190s, Lincoln, Nebraska), air temperature at 20 cm height (copper constantan thermocouple), and RH with a shielded RH sensor (Campbell Scientific, Inc., model 217, Logan, Utah). Data were taken for all sensors every minute and the minimum, maximum, and average for every 10 minute interval were recorded in a datalogger (Campbell Scientific, Inc., model 21X microdatalogger). We also assessed the spatial heterogeneity of light and temperature by setting up arrays of photo-diodes (individually calibrated against an Li-190s quantum sensor) and thermocouples and detected no significant discrepancies between the permanent sensors and the range of values produced by the arrays (data not shown).

To evaluate seedling response to the light environment at the subplot level, canopy hemispherical photographs at the 90 subplots were taken in July 1996. The camera was positioned over the centre marker of the subplot, and the top of the lens was 80 cm from the forest floor. The images were analysed using a software program (FEW 4.0) developed by M. Ishizuka (personal communication). We derived direct and diffuse site factors for each subplot.

Leaf and seedling characteristics

To assess the effect of *R. maximum* on seedlings, a range of leaf and whole plant level properties were measured on seedlings harvested from the subplots on 15 September 1996 and 6 September 1997. While *Quercus rubra* and *Tsuga canadensis* were sampled in both years, *Prunus serotina* seedlings were sampled only in the second year. One seedling per species was randomly taken from each subplot.

Seedlings were lifted with a largely intact root system, placed on ice, and transported to the laboratory where they were processed. Leaf area, plant height, and stem basal diameter were recorded for each seedling. The root was separated from the stem and kept frozen, and later examined for mycorrhizae and dried. Stem and leaves were dried at 72°C until constant in weight. Specific leaf area was calculated as the fraction of total leaf dry weight divided by leaf area. Leaves and the stem of individual seedlings were ground in a Wiley mill and analysed for carbon and nitrogen content on a dry weight basis using a CHN analyser.

Statistical analysis

The three replicate main plots under $+Rm$ and $-Rm$ canopy types were randomly selected and not contiguous. Edaphic and environmental characteristics indicate no systematic bias of the replicate plots assigned to each canopy type (data not shown). However, because the main block types of the two canopy types are not strictly experimental treatments but naturally occurring heterogeneity in the forest environment, to determine the effect of canopy type on a selected response character we used the following sums of squares (SS of Type III ANOVA) ratio to calculate the F-statistic with d.f. $=1,4$:

$$F = \frac{SS_{canopytype}/1}{SS_{mainplot}/4}. \qquad (1)$$

Analysis of variance and correlation between survivorship and environmental parameters were made using SAS (SAS Institute, Cary, North Carolina 1988). The effects of substrate manipulation treatments examining possible allelopathic effects were analysed using orthogonal contrasts. We

did not find significant effects of substrate manipulation on seedling survivorship (Nilsen et al., 2001) and this topic will not be discussed further.

Results and discussion

Pre-emergent factors that could lead to the suppression of canopy tree seedlings under R. maximum include reduced seed input into areas with shrubs or reduced seed viability in the seed bank in soils under shrubs, or both. Competition for below-ground resources between mature trees and shrubs could reduce seed production in trees co-occurring with shrubs. However, we found no differences in viable seed input between forest sites with and without R. maximum for several species collected over two years at the Coweeta Hydrologic Laboratory in the southern Appalachian Mountains, USA (Lei et al., 2002). Seed input of one species, Nyssa sylvatica, was significantly higher in sites with R. maximum. Competition between shrubs and trees may reduce the viability of seeds produced by trees, and other factors such as the accumulation of toxins may reduce seed viability in the soil under shrubs. However, we found no difference in numbers of seedlings of four overstory species germinating from seed bank samples collected at forest sites within and without R. maximum thickets (Lei et al., 2002).

Reduced germination of tree seeds has been observed under R. maximum. Clinton & Vose (1996) designed an experiment including low light plots under R. maximum (created with shade cloth), and open forest sites with ambient understory light. The authors found reduced germination of Acer rubrum under R. maximum while germination was significantly higher in the

Table 1 Mean germination percentages for four species native to the southern Appalachian Mountains when imbibed in forest floor substrate from sites without (−Rm) or with (+Rm) a thicket of *Rhododendron maximum*. The control treatment was pure vermiculite. Numbers in parentheses are standard errors of the mean.

Treatment	Species			
	Quercus rubra	*Nyssa sylvatica*	*Pinus strobus*	*Tsuga canadensis*
− Rm	42.0 (4.1)	22.7 (1.9)	44.3 (3.7)	15.2 (5.4)
+ Rm	45.7 (1.3)	22.0 (1.6)	46.3 (2.1)	16.6 (3.1)
Control	36.5 (1.5)	16.3 (1.3)	49.3 (1.2)	28.0 (1.6)

other two treatments. Reduced germination under *R. maximum* in this study was likely due to reduced soil water availability under *R. maximum* compared with the other treatments. Additionally, the porous humus layer beneath *R. maximum* creates conditions of low litter moisture and prevents seeds from receiving adequate moisture from the mineral soil (Clinton & Vose, 1996).

Allelopathy from chemicals in *R. maximum* litter and humus could be another explanation for reduced germination of tree seeds under shrubs. Leachates produced from *R. maximum* litter reduced germination and root elongation in one test species (cress) but not the other (Nilsen *et al.*, 1999). However, we found no difference in germination for three tree species imbibed in leachates made from litter and organic matter and throughfall collected in forest sites with and without *R. maximum* (Table 1; Nilsen *et al.*, 1999). Litter and humus type (forest floor or *R. maximum*) had no effect on canopy tree seed germination, photosynthesis, or growth in field trials (Nilsen *et al.*, 1999; Semones, 1999).

Allelochemicals from *R. maximum* may have a negative effect on important mycorrhizal species, which could affect successful mycorrhization in tree seedlings.

We found that leaf material from *R. maximum* suppressed the growth of one important mycobiont, but had no effect on two other species (Nilsen *et al.*, 1999). However, *R. maximum* leaves were equally or less inhibiting to mycorrhizal growth than leaves from four canopy tree species. It is possible that *R. maximum* allelochemicals may build up in the soil over time, although Beckage *et al.* (2000) found that upon removal of *R. maximum*, seedling densities in cleared areas quickly increased to that of areas where *R. maximum* had not been present. These results suggest no residual toxicity remaining in the substrate following removal of living *R. maximum* plants. Since the results of our study showed no significant toxicity of the substrate, the results of Beckage *et al.* (2000) are in agreement.

There is little evidence in this study that *R. maximum* thickets present a pre-emergence 'filter' to the establishment of canopy tree seedlings. However, in one other study there was evidence that reduced water availability under *R. maximum* may reduce the germination rate of tree seeds relative to forested areas without *R. maximum*. However, the majority of information available indicates that most of the seedling inhibition by *R. maximum* occurs post-emergence. What factors could inhibit tree seedling growth and survival under *R. maximum*? Clinton & Vose (1996) suggest that low light is most likely the primary factor, but others include greater litter depth and reduced litter quality, competition for below-ground resources between seedlings and shrubs, and allelopathy.

The predominant limiting resource to seedling growth under *R. maximum* is undoubtedly light. In Clinton & Vose's (1996) study, seed germination was similar in open forest plots and in plots where light availability was reduced with shade cloth to simulate the light environment under *R. maximum*. Within three weeks, seedling survival in shade cloth plots began to decline at a much faster rate than in forest control plots, and it was equal to survival under *R. maximum* within eight weeks. Light in hardwood forest understories is typically less than 10% of full sunlight (Clinton & Vose, 1996), and light is often limiting to plant growth in forest understories (Chazdon,

1988; Kobe *et al.*, 1995; Walters & Reich, 1996). *Rhododendron maximum* further attenuates light in the forest understory, reducing light levels 50–80% relative to forest understory without shrubs (Clinton & Vose, 1996; Beckage *et al.*, 2000; Nilsen *et al.*, 2001).

Overall canopy tree seedling survivorship after two years was lower in forest sites with (+*Rm*) than in sites without (−*Rm*) a thicket of *R. maximum* (Figure 1; Nilsen *et al.*, 2001; Lei *et al.*, 2002), but what are the mechanisms that lead to lower survivorship under shrubs? Seedlings growing under *R. maximum* had significantly lower leaf area and leaf mass than those growing in forest sites without *R. maximum* (Lei *et al.*, 2002). Additionally, stems of seedlings under shrubs were significantly taller but had less biomass than open forest seedlings, suggesting a shift in carbon allocation in these seedlings associated with lower light (Lei *et al.*, 2002). Root biomass was also reduced in seedlings growing under *R. maximum* (Walker *et al.*,

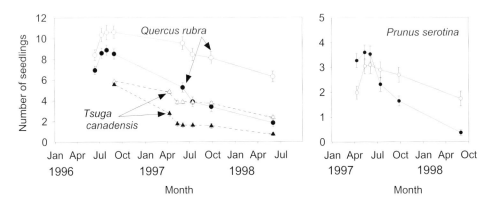

Figure 1 Survivorship of *Quercus rubra*, *Tsuga canadensis*, and *Prunus serotina* seedlings in the experimental plots over a three-year period in forest with or without a subcanopy of *Rhododendron maximum*. Open symbols refer to seedlings growing in forest without a thicket of *R. maximum* and closed symbols refer to seedlings growing under a thicket of *R. maximum*. *Prunus serotina* seedlings only began germinating in the spring of 1997. Error bars refer to one standard error of the mean. (From Lei *et al.*, 2002.)

1999). Typically, plants experiencing resource limitation will allocate more growth to those tissues that acquire the limiting resource. However, our findings suggest that shifts in allocation in seedlings growing under dense shrub cover are insufficient to improve seedling survival under dense *R. maximum* canopies. Instead, physiological adaptations to low light may be more important for seedling survival under shrubs.

Semones (1999) found similar low light acclimation in hardwood seedlings growing in forested areas with and without *R. maximum*. Light response curve parameters (quantum yield, light compensation point, light saturation, and dark respiration rates) were characteristic of shade acclimated plants, and there was no significant difference in these parameters between seedlings growing within and without *R. maximum* thickets. It seems that seedlings are not able to further acclimate to the lower light levels experienced under *R. maximum* because understory light levels in these temperate forests are already at or below the light compensation point for most forest tree species. However, seedlings in the open forest had significantly higher maximum photosynthetic rates, suggesting a higher photosynthetic capacity.

Light levels in southern Appalachian forests are very low during the growing season, with mean maximum PPFDs of 2–4 μmol m^{-2} s^{-1} and 10–15 μmol m^{-2} s^{-1} for sites with or without a thicket of *R. maximum*, respectively (Nilsen *et al.*, 2001). Diurnal diffuse light levels are higher in forested sites without than in sites with *R. maximum*. In low light understory habitats, sun flecks, or higher intensity direct beam radiation penetrating through holes in the canopy, can be an important resource for seedlings (Pearcy, 1983; Chazdon & Fetcher, 1984; Chazdon, 1988). In the southern Appalachian Mountains, sun flecks are shorter, of lower intensity, and much less frequent in forests with a thicket of *R. maximum* than in understory sites without a thicket of *R. maximum* (Figure 2). The reduced occurrence of sun flecks could lead to an inability to maintain a positive carbon balance for seedlings growing under *R. maximum*. Reduced carbon gain could lead to reduced growth and allocation to both leaves and roots, which would further limit both above- and below-ground resource acquisition, causing seedlings to enter a resource limitation spiral eventually leading to death.

There is ample evidence that low light is the primary limiting factor to seedling growth, but it is not the only one. Germination and initial seedling survival of *Acer rubrum* seeds was lower in plots with low light under *R. maximum* than in plots with low light generated by shade cloth (Clinton & Vose, 1996). We found that seedling survival was correlated with light availability as assessed by canopy photograph analysis; however, this relationship differed between seedlings growing under *R. maximum* and those growing in forest understories without *R. maximum* (Nilsen *et al.*, 2001). In both studies, soil water content was lower in forest plots with a thicket of *R. maximum* than in forest sites without a thicket of *R. maximum* (Figure 3). Also, many of the sites where *R. maximum* grows are low in nitrogen and phosphorus, so competition for other below-ground resources may be an additional factor affecting seedling survival under shrubs. This competition could reduce seedling nutrient uptake, further reducing the photosynthetic capacity of seedlings. Ectomycorrhizae are

ERIK NILSEN & JONATHAN HORTON

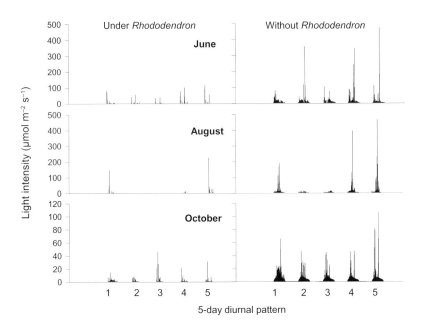

Figure 2 Representative variation in diurnal cycles of photosynthetically active radiation in the subcanopy of a southern Appalachian forest. Each horizontal pair of panels represents a five-day period in June, August, or October when light intensity was measured simultaneously at two sites. Steep vertical spikes in the day cycles illustrate the frequency and intensity of sun flecks. (From Nilsen *et al.*, 2001.)

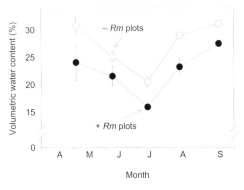

Figure 3 Soil moisture content at a depth of 0–15 cm in southern Appalachian forest sites with (+*Rm*) or without (−*Rm*) a thicket of *Rhododendron maximum*. Each point refers to a mean of 45 samples. Error bars refer to one standard deviation. (From Nilsen *et al.*, 2001.)

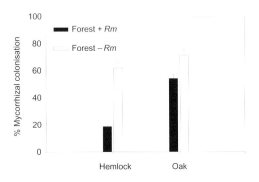

Figure 4 Inhibition of mycorrhizae on first-year seedlings of two tree species planted in forest with and without *Rhododendron maximum* in the subcanopy. Both groups are statistically significantly different (*P* < 0.1). (After Walker *et al.*, 1999.)

important for optimal accumulation of nitrogen and phosphorus when nutrient availability is low. If *R. maximum* interfered with mycorrhization of seedlings, this could further exacerbate resource limitations and place additional limitations on seedling growth and survival.

We found reduced mycorrhizal colonisation on *Quercus rubra* and *Tsuga canadensis* growing under *R. maximum* when compared with seedlings growing in forest sites without *R. maximum* (Figure 4; Walker *et al.*, 1999). In this study, there was also a distinct shift in the morphotypes colonising root tips under *R. maximum*, with a higher colonisation by *Cenococcum geophilum*, a generalist mycobiont that is less effective than many other mycobionts in promoting seedling growth (Marx *et al.*, 1978). Walker & Miller (2002) found no differences in sporophore diversity or community composition between forest sites with and without *R. maximum*, therefore reduced colonisation and a shift in the mycobiont-community assemblage colonising seedling root tips under *R. maximum* was not due to differences in the fungal community between forest types. What caused the reduced mycorrhizal colonisation under *R. maximum*? We found that *R. maximum* leaves could inhibit mycobiont growth *in vitro* (Nilsen *et al.*, 1999), but there was little or no evidence of allelopathy or accumulated biotoxins in the soil affecting mycorrhization in the field (Walker *et al.*, 1999). Competition between ectomycorrhizae associated with tree seedlings and ericoid mycorrhizae associated with *R. maximum* is a possibility, but was not specifically addressed by Walker *et al.* (1999). A likely explanation for reduced mycorrhizal colonisation under *R. maximum* is that seedlings under *R. maximum* are

light limited and as a result carbon starved. Many seedlings under *R. maximum* were unable to maintain positive daily carbon gain (Semones, 1999). These carbon-starved seedlings are likely allocating carbon to increased leaf area and stem growth in order to capture more light. They lack the carbon necessary to promote root growth and support beneficial mycobionts.

Competition for below-ground resources may exacerbate seedling stress imposed by low light under shrubs. There is evidence that *R. maximum* influences some below-ground resource (Monk *et al.*, 1985; Clinton & Vose, 1996), reducing some nutrients in the soil by sequestering them in the evergreen canopy of the shrubs (Monk *et al.*, 1985). We found that pH was slightly lower under *R. maximum* than in forest plots without it (Nilsen *et al.*, 2001), but all soils in the study were acidic (pH ranged from 4.5 to 5.2 for all plots). Soil carbon and nitrogen content did not differ between forest types (data not shown). Extractable nitrate and phosphorus were often undetectable, while extractable ammonium ranged from 0.5 to 2.0 ppm, and did not differ between forest types. Nitrogen mineralisation rates ranged between 3 and 10 mg N kg^{-1} soil, but the average N mineralisation rate did not differ between forest types. Extractable soil cation concentrations (K, Ca, Mg) were significantly lower in areas with *R. maximum* than in areas without it (Nilsen *et al.*, 2001), likely due to sequestration in *R. maximum*'s leaves (Monk *et al.*, 1985).

The research reviewed in this paper has been compiled into the following model of seedling inhibition under evergreen shrubs (Figure 5). Low light is the primary limitation to seedling performance, resulting in reduced carbon gain. Reduced carbon gain leads to

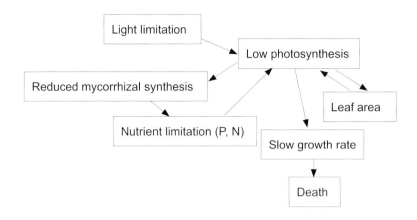

Figure 5 Theoretical model of seedling inhibition under *Rhododendron maximum*.

low leaf area, which limits seedlings' ability to capture sufficient light to maintain positive carbon gain. When seedlings are starved for carbon, they are unable to allocate sufficient carbon below ground for root growth and to attract and support necessary mycorrhizal symbionts. This results in an inability to acquire sufficient below-ground resources (water and nutrients), further lowering the photosynthetic capacity and ability to utilise sun flecks. The combined effect of both the above- and below-ground resource limitation is a slow or negative growth rate and eventually death of seedlings growing under *R. maximum*.

Summary and comparison with *R. ponticum*

We have outlined the evidence supporting our model of inhibition of tree seedlings under *R. maximum* in the southern Appalachian Mountains, USA. It is likely that the mechanisms of inhibition are similar in the case of *R. ponticum* in the British Isles. *Rhododendron ponticum* typically grows on acidic soils and forms dense thickets

that strongly attenuate sunlight, allowing less than 2–10% to penetrate to the forest floor (Cross, 1975). It is likely that seedlings growing under *R. ponticum* are under similar light limitation as those growing under *R. maximum*. Like *R. maximum*, *R. ponticum* forms a dense humus layer (Cross, 1975), which could limit seedlings' ability to access adequate soil moisture. Soils under *R. ponticum* tend to have lower cation exchange capacity and lower cation concentration than uninvaded soils (Cross, 1975; Mitchell *et al.*, 1997). *Rhododendron ponticum* litter produces many polyphenols that mobilise cations and cause them to be leached from the soil (Cross, 1975).

While allelopathy is not a likely factor in the inhibition of seedlings under *R. maximum*, there is evidence that *R. ponticum* produces more allelochemicals. Seed bank studies have shown that there are fewer viable seeds in soils under *R. ponticum* than in areas without *R. ponticum* (Mitchell *et al.*, 1998). It has been suggested that root exudates and allelochemicals from *R. ponticum* litter may accumulate in the soil and inhibit seedling germination, growth, and survival (Cross, 1975). Mitchell *et al.* (1999) found greater re-vegetation success in restoring areas invaded by *R. ponticum* when both the shrub and the litter layer were removed. While little research has been conducted addressing the mechanisms of seedling inhibition associated with *R. ponticum*, it seems likely that both allelopathy and low light limitation interact to reduce tree seedling viability under this shrub.

Acknowledgements

This research was funded by the US Department of Agriculture National Research Initiative. Many thanks are given to Dr Thomas Lei, John Walker, Shawn Semones, and Orson Miller for countless hours collecting and analysing data.

References

BARDEN, L. S. (1979). Tree replacement in small canopy gaps of *Tsuga canadensis* forest in the southern Appalachians. *Oecologia* 44: 141–142.

BECKAGE, B., CLARK, J. S., CLINTON, B. D. & HAINES, B. L. (2000). A long-term study of tree seedling recruitment in southern Appalachian forests: the effects of canopy gaps and shrub understories. *Can. J. For. Res.* 30: 1617–1631.

CANHAM, C. D. (1985). Suppression and release during canopy recruitment of *Acer saccharum*. *Bull. Torrey Bot. Club* 112: 134–145.

CANHAM, C. D. (1988). Growth and canopy architecture of shade-tolerant trees: the response of *Acer saccharum* and *Fagus grandifolia* to canopy gaps. *Ecology* 69: 786–795.

CHAZDON, R. L. (1988). Sunflecks and their importance to forest understory plants. *Adv. Ecol. Res.* 18: 1–63.

CHAZDON, R. L. & FETCHER, N. (1984). Photosynthetic light environments in a lowland tropical rainforest of Costa Rica. *J. Ecol.* 72: 553–564.

CLINTON, B. D. & VOSE, J. M. (1996). Effects of *Rhododendron maximum* L. on *Acer rubrum* L. seedling establishment. *Castanea* 612: 38–45.

CLINTON, B. D., BORING, L. R. & SWANK, W. T. (1994). Regeneration patterns in canopy gaps of mixed-oak forests of the southern Appalachians: influences of topographic position and evergreen understory. *Amer. Midl. Nat.* 132: 308–319.

CROSS, J. R. (1975). Biological flora of the British Isles: *Rhododendron ponticum* L. *J. Ecol.* 63: 345–364.

CROSS, J. R. (1981). The establishment of *Rhododendron ponticum* L. in the Killarney oakwoods, S.W. Ireland. *J. Ecol.* 69: 807–824.

DAY, F. P., PHILLIPS, D. L. & MONK, C. D. (1988). Forest communities and patterns. In: SWANK, W. T. & CROSSLEY, D. A., JR. (eds) *Forest Hydrology and Ecology at Coweeta*, pp. 141–149. New York: Springer-Verlag.

DOBBS, M. M. (1995). *Spatial and temporal distribution of the evergreen understory in the southern Appalachians*. MS thesis, University of Georgia, Athens, GA, 100pp.

FULLER, R. M. & BOORMAN, L. A. (1977). The spread and development of *Rhododendron ponticum* L. on the dunes at Winterton, Norfolk, in comparison with invasion by *Hippophae rhamnoides* at Salt Fleetby, Lincolnshire. *Biol. Conserv.* 12: 82–94.

KOBE, R. K., PACALA, S. W., SILANDER, J. A. & CANHAM, C. D. (1995). Juvenile tree survivorship as a component of shade tolerance. *Ecol. Appl.* 5: 517–532.

LEI, T. T., SEMONES, S. W., WALKER, J. F., CLINTON, B. D. & NILSEN, E. T. (2002). Effects of *Rhododendron maximum* thickets on tree seed dispersal, seedling morphology, and survivorship. *Int. J. Plant Sci.* 163: 991–1000.

MARX, D. H., MORRIS, W. G. & MEXAL, J. G. (1978). Growth and ectomycorrhizal development of loblolly pine seedlings in fumigated and nonfumigated nursery soil infested with different fungal symbionts. *For. Sci.* 24: 193–203.

MITCHELL, R. J., MARRS, R. H., LE DUC, M. G. & AULD, M. H. D. (1997). A study of succession on lowland heaths in Dorset, southern England: changes in vegetation and soil chemical properties. *J. Appl. Ecol.* 34: 1426–1444.

MITCHELL, R. J., MARRS, R. H. & AULD, M. H. D. (1998). A comparative study of the seedbanks of heathland and successional habitats in Dorset, Southern England. *J. Ecol.* 86: 588–596.

MITCHELL, R. J., MARRS, R. H., LE DUC, M. G. & AULD, M. H. D. (1999). A study of restoration of heathland on successional sites: changes in vegetation and soil chemical properties. *J. Appl. Ecol.* 36: 770–783.

MONK, C. D., MCGINTY, D. T. & DAY, F. P. (1985). The ecological role of *Kalmia latifolia* and *Rhododendron maximum* in the deciduous forests of the southern Appalachians. *Bull. Torrey Bot. Club* 112: 187–193.

NIERING, W. A. & EGLER, F. E. (1955). A shrub community of *Viburnum lentago*, stable for twenty-five years. *Ecology* 36: 356–360.

NILSEN, E. T., WALKER, J. F., MILLER, O. K., SEMONES, S. W., LEI, T. T. & CLINTON, B. D. (1999). Inhibition of seedling survival under *Rhododendron maximum* (Ericaceae): could allelopathy be a cause? *Amer. J. Bot.* 86: 1597–1605.

NILSEN, E. T., CLINTON, B. D., LEI, T. T., MILLER, O. K., SEMONES, S. W. & WALKER, J. F. (2001). Does *Rhododendron maximum* L. (Ericaceae) reduce the availability of resources above and belowground for canopy tree seedlings? *Amer. Midl. Nat.* 145: 325–343.

PEARCY, R. W. (1983). The light environment and growth of C_3 and C_4 tree species in the understory of a Hawaiian forest. *Oecologia* 58: 19–25.

PHILLIPS, D. L. & MURDY, W. H. (1985). Effects of rhododendron (*Rhododendron maximum* L.) on regeneration of southern Appalachian hardwoods. *For. Sci.* 31: 226–233.

RICE, E. L. (1979). Allelopathy, an update. *Bot. Rev.* 45: 15–109.

SEMONES, S. W. (1999). *Inhibition of canopy tree seedlings by Rhododendron maximum L. (Ericaceae) thickets in an eastern deciduous forest.* PhD dissertation, Virginia Polytechnic Institute and State University, Blacksburg, VA, 173pp.

TAYLOR, A. H. & QIN, Z. (1992). Tree regeneration after bamboo die-back in Chinese *Abies–Betula* forests. *J. Veg. Sci.* 3: 253–260.

VEBLEN, T. T. (1982). Growth patterns of *Chusquea* bamboos in the understory of Chilean *Nothofagus* forest and their influences in forest dynamics. *Bull. Torrey Bot. Club.* 109: 474–487.

WAHLENBERG, W. G. (1950). From bush to pine. *South. Lumberman* 180: 40–41.

WALKER, J. F. & MILLER, O. K. (2002). Ectomycorrhizal sporophore distributions in a southeastern Appalachian mixed hardwood/conifer forest with thickets of *Rhododendron maximum*. *Mycologia* 94: 221–229.

WALKER, J. F., MILLER, O. K., LEI, T., SEMONES, S., NILSEN, E. & CLINTON, B. D. (1999). Suppression of ectomycorrhizae on canopy tree seedlings in *Rhododendron maximum* L. (Ericaceae) thickets in the southern Appalachians. *Mycorrhiza* 9: 49–56.

WALTERS, M. B. & REICH, P. B. (1996). Are shade tolerance, survival and growth linked? Low light and nitrogen effects on hardwood seedlings. *Ecology* 77: 841–853.

POSTERS

An investigation into the performance of rhododendrons on lime-tolerant rootstocks

Paul D. Alexander

RHS Garden Wisley, Woking, Surrey GU23 6QB, UK

Introduction

Rhododendron species and cultivars are usually thought to favour acidic soil conditions but several taxa have been reported to be lime-tolerant (e.g. Rankin, 1997; Kinsman, 1999). Preil & Ebbinghaus (1994) reported how attempts were being made to breed lime-tolerant *Rhododendron* rootstocks at the Federal Centre for Breeding Research on Cultivated Plants at Ahrensburg (Germany). These 'lime-tolerant' rootstocks were available in limited numbers to amateur gardeners from 1998 and more widely from 1999. They are retailed under the trade name 'Inkarho' (**In**teressengemeinschaft **Ka**lktoleranter **Rho**dodendronunterlagen) through a growers' consortium. The consortium generously donated 200 rhododendrons, 100 on 'Inkarho' rootstock and 100 on 'Cunningham's White' rootstock, to the Royal Horticultural Society (RHS) for this experiment.

Method

A field trial was established at the Deers Farm field site (RHS Garden Wisley) consisting of a series of 24 isolated beds. These were constructed by initially digging two trenches which then had a series of boards erected within them delimiting each bed (3 m wide × 4 m long × 0.5 m deep). These were then lined with plastic sheeting to prevent any contamination by wood preservatives, a 15 cm layer of gravel was placed at the base of each bed, and then the soil was refilled in two layers.

The 24 beds are grouped in blocks of four with the pH of the soil randomly corresponding to one of four pH levels (pH 5, 6, 7 and > 7). There are thus six beds at any one pH level. The pH was altered by applications of $CaCO_3$ (lime) or sulphur powder when the beds were constructed (no further applications to date) and the pH is monitored monthly. Within each bed all 12 plant types are represented once (in total six plants of any one type experienced the same soil pH). The pH 5 beds are being used as the 'control'.

Plants

Following the bed construction the plants listed in the table on p. 275 were randomly planted in each bed in their rootball on 19 March 2001. In addition, shelter plants were planted around the whole site and in the gaps between beds so that all experimental plants were surrounded by rhododendron on all sides.

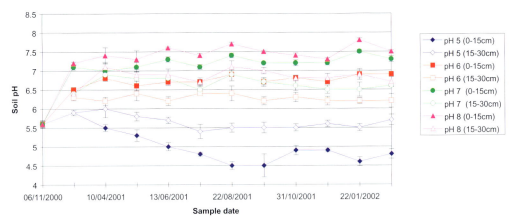

Figure 1 Soil pH (± standard error) of the Inkarho beds following lime and sulphur applications ($n = 6$).

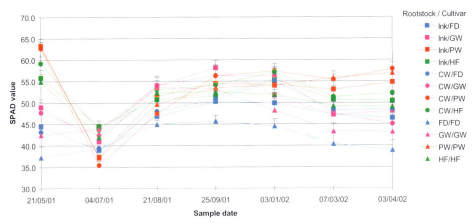

Figure 2 Mean SPAD value (± standard error) for each plant type at soil pH 5 ($n = 6$).

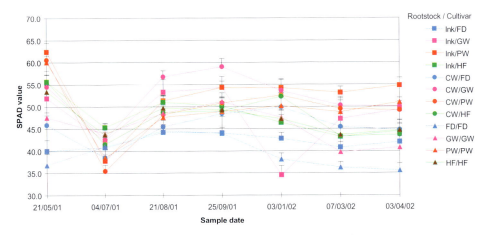

Figure 3 Mean SPAD value (± standard error) for each plant type at soil pH 6 ($n = 6$).

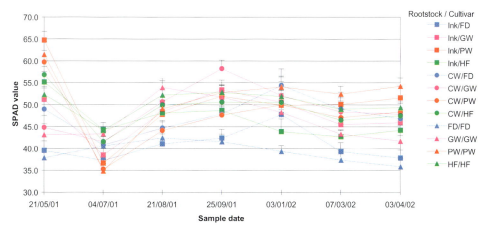

Figure 4 Mean SPAD value (± standard error) for each plant type at soil pH 7 ($n = 6$).

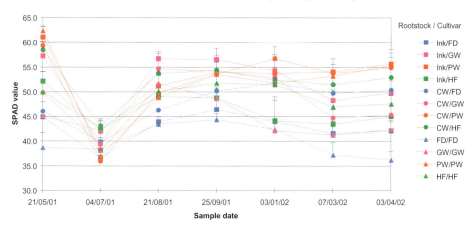

Figure 5 Mean SPAD value (± standard error) for each plant type at soil pH > 7 ($n = 6$).

Rootstock	Cultivar
'Inkarho'	'Furnivall's Daughter'
'Inkarho'	'Gomer Waterer'
'Inkarho'	'Percy Wiseman'
'Inkarho'	'Hachmann's Feuerschein'
'Cunningham's White'	'Furnivall's Daughter'
'Cunningham's White'	'Gomer Waterer'
'Cunningham's White'	'Percy Wiseman'
'Cunningham's White	'Hachmann's Feuerschein'
'Furnivall's Daughter'	'Furnivall's Daughter'
'Gomer Waterer'	'Gomer Waterer'
'Percy Wiseman'	'Percy Wiseman'
'Hachmann's Feuerschein'	'Hachmann's Feuerschein'

Plant assessments

The symptom most associated with ericaceous plants in alkaline soil is chlorosis. This is simply the yellow discoloration of the leaves brought about by an interruption in the synthesis of chlorophyll. The most common cause of chlorosis is nutrient deficiency, either through a lack of nutrients in the soil or their unavailability (often associated with alkaline soils). In order to avoid evaluator bias in determining the degree of chlorosis a SPAD meter was used.

SPAD (**S**oil **P**lant **A**nalysis **D**evelopment) meters have been successfully used in determining chlorophyll differences in a range of plants including tropical and subtropical fruit trees (Schaper & Chacko, 1991) and sugar beet (Uchino & Kanzawa, 1995). The SPAD meter emits light in two wavelength regions and calculates the SPAD value from the light transmitted through the leaf. These values are proportional to the amount of chlorophyll in the plant leaf. The technique is non-destructive and the results are reproducible. Thus the SPAD meter will enable us to scientifically monitor changes in chlorophyll content of the leaves in an attempt to identify any deterioration in plant health.

Work to date

SPAD values, plant height, plant width and soil pH have been regularly recorded but to date no statistical analysis has been undertaken.

The results of soil pH analysis illustrate that the pH of the beds has been considerably altered (see Figure 1). To date only one application of either lime or sulphur has been made; a further application is now being considered in order to maintain the pH 5 and pH > 7 beds.

SPAD recordings to date would appear to suggest that soil pH has had little effect on plant health. Figures 2–5 illustrate the mean recordings for each plant on each treatment. Every plant mean is the mean of 24 leaf readings.

The results would also appear to suggest that rootstock has had little effect on the plants' health to date (no trends being evident), but there may be a slight cultivar effect. These results would support the theory that as the plants were planted with a rootball of soil they would take up to two years to grow out of this into the experimental soil.

Conclusion

The experiment is designed to continue for a further four years in order to assess the performance of the plants in what are considered to be 'less than favourable' soil conditions. To date there would appear to be little effect of soil pH on plants health.

Acknowledgements

I would like to thank Miss Emily Reid (Research Assistant) for help in sampling and analysis, the Curatorial Department at RHS Garden Wisley for their help in constructing the beds (particularly Rob Fitzmaurice, Andy Rudnicki and Jim England) and Georgina Godwin-Keene (CABI Bioscience) for assistance in production of the poster.

References

Kinsman, D. (1999). Rhododendrons and associated plants growing in soils overlying limestone. *The New Plantsman* 6(1): 21–22.

PREIL, W. & EBBINGHAUS, R. (1994). Breeding of lime-tolerant *Rhododendron* rootstocks. *Acta Horticulturae* 364: 61–70.

RANKIN, D. (1997). Non-conformist rhododendrons. *Newsl. Scottish Rhododendron Soc.* 40: 5–6.

SCHAPER, H. & CHACKO, E. K. (1991). Relation between extractable chlorophyll and portable chlorophyll meter readings in leaves of eight tropical and sub-tropical fruit-tree species. *J. Plant Physiol.* 138: 674–677.

UCHINO, H. & KANZAWA, K. (1995). Evaluation of yellowing intensity of sugar beet leaves infected with rhizomania by using a handheld chlorophyll meter. *Ann. Phytopathol. Soc. Japan* 61: 123–126.

Dormancy, flowering and chlorophyll fluorescence of a rhododendron and azaleas in the greenhouse

David J. Ballantyne

Department of Biology, University of Victoria, Victoria, B.C., V8W 3N5, Canada

Two azalea cultivars ('Vuyk's Scarlet' and 'Noordthiana') and a *Rhododendron* cultivar ('Molly Ann') were grown outside in Victoria, British Columbia, and each month from October 2000 to March 2001 plants were brought into a 10°C night greenhouse to be forced into flower. After 5 hours in the greenhouse, chlorophyll *a* fluorescence values (F_v/F_m, F_o, F_m and F_v) were determined. Dates of and times to anthesis were also measured. When plants were moved into the greenhouse later than October, they flowered earlier than if they had entered the greenhouse in October, indicating that flower bud dormancy was being broken. This was especially pronounced between October 2000 and December 2000. At the same time, between October 2000 and November 2000 there were decreases in F_v/F_m, F_m and F_v in all three cultivars, and between October 2000 and December 2000 in 'Molly Ann' rhododendrons. Chlorophyll levels remained relatively constant from October 2000 to March 2001. Decreases in chlorophyll fluorescence may indicate the completion of part of the dormancy breaking process.

Introduction

The objectives of this study included:

1 The determination of changes in chlorophyll *a* fluorescence in three cultivars of *Rhododendron* (two azaleas and a *Rhododendron*) during winter dormancy and fluorescence.

2 The determination of a possible relationship between changes in chlorophyll *a* fluorescence, breaking of dormancy and earlier flowering after plants are moved into a 10°C night greenhouse.

3 The determination of the effectiveness of a 10°C night temperature on the greenhouse forcing of two hardy azaleas and a *Rhododendron*.

Materials and methods

Two azalea cultivars ('Vuyk's Scarlet' and 'Noordthiana') and a *Rhododendron* cultivar ('Molly Ann') were grown in a field near Victoria, British Columbia, Canada. The plants were propagated as cuttings during September 1999. Five plants each of 'Vuyk's Scarlet' and 'Noordthiana', and four plants of 'Molly Ann', were brought into a double polycarbonate greenhouse held at 10°C nights every month from October 2000 to March 2001. Temperatures in the greenhouse were about 5°C warmer than outside from October through March. Current greenhouse growing techniques have been discussed by Leach (1961) and Larson (1993).

Chlorophyll *a* fluorescence values (F_v/F_m, F_o, F_m and F_v) were determined with an ADC Fim 1500 Fluorescence Induction Monitor on the youngest fully expanded leaves. F_v/F_m indicates the photochemical yield of photosystem 2, and has frequently been used to determine photosynthetic responses (Hall & Rao, 1994; Fernandez-Blaco *et al.*, 1998; Kao *et al.*, 1998). F_o indicates the level of open reaction centres of photosystem 2, F_m indicates the level of closed reaction centres of photosystem 2, and F_v indicates a drop in fluorescence of photosystem 2 due to photosynthesis (Hall & Rao, 1994). Fluorescence values were measured after 2 hours of darkness in the 10°C greenhouse, and 5 hours after plants had been moved into the greenhouse. Chlorophyll levels were determined with a Minolta SPAD Chlorophyll Meter (Tenga *et al.*, 1989; Hoel & Solbaugh, 1998).

Results

When plants were moved into the greenhouse later than October, they flowered earlier than if they had entered the greenhouse in October (Table 1). This was especially pronounced in 'Molly Ann', and to a lesser extent in 'Vuyk's Scarlet'. This indicates that the flower buds on these plants were dormant until they had been exposed to low temperature outside. After the plants had been exposed to sufficient low temperature to break dormancy, the flower buds were presumably quiescent until the higher temperatures in spring induced them to flower. In all three cultivars there was a decrease in F_v/F_m of the leaves after October (Table 1). This decrease in F_v/F_m was accompanied by decreases in both F_m and F_v. There was no decrease in F_o. Chlorophyll levels 5 hours after the plants were brought into the greenhouse remained about the same in the leaves of all three cultivars from October until March.

Discussion

'Molly Ann' seemed to have a higher degree of dormancy than 'Vuyk's Scarlet' or 'Noordthiana', but both 'Molly Ann' and 'Vuyk's Scarlet' flowered earlier if they were outside longer in the winter. Even 'Noordthiana' flowered in a shorter period of time in the greenhouse if the plants remained outside longer (Table 1). Thus, to decrease forcing time it is desirable to allow the plants to remain outside for a period of time during the winter, or to use refrigeration (Larson, 1993).

Chlorophyll *a* fluorescence (F_v/F_m) decreased as plants were exposed to more low temperatures outside. This decrease in F_v/F_m could be due to a change in the amount of dormancy, a change in the environment outside, or simply to aging of the leaves. Changes in F_v/F_m, which can be rapidly and reliably determined, might allow determination of the level of dormancy in the plants,

Table 1 Azalea and rhododendron flowering and chlorophyll fluorescence (F_v/F_m). SD, standard deviation; LSD, least significant difference.

Cultivar	Date to green-house	Date of anthesis (± SD in days)	Days to anthesis in greenhouse (± SD in days)	F_v/F_m (± SD)
'Vuyk's Scarlet' (azalea)	20/10/00	March 8 ± 2.6	139.6 ± 2.6	0.80 ± 0.02
	17/11/00	March 3 ± 3.5	88.2 ± 3.5	0.63 ± 0.08
	18/12/00	Feb 19 ± 1.9	63.6 ± 1.9	0.72 ± 0.04
	12/01/01	March 6 ± 2.2	54.2 ± 2.2	0.72 ± 0.07
	08/02/01	March 20 ± 3.6	45.4 ± 3.6	0.74 ± 0.03
	08/03/01	April 7 ± 0.8	29.8 ± 0.8	0.66 ± 0.05
	Remained outside	May 11 ± 1.1		
LSD$_{0.05}$		4 days	3.9	0.04
'Noordthiana' (azalea)	20/10/00	March 10 ± 3.4	141 ± 3.4	0.82 ± 0.04
	17/11/00	March 3 ± 6.1	106.6 ± 6.1	0.61 ± 0.07
	18/12/00	March 11 ± 3.0	83.6 ± 3.0	0.59 ± 0.05
	12/01/01	March 16 ± 3.2	64.4 ± 3.2	0.70 ± 0.05
	08/02/01	March 30 ± 1.8	51.2 ± 1.8	0.66 ± 0.07
	08/03/01	April 14 ± 1.4	37 ± 1.4	0.67 ± 0.03
	Remained outside	May 25 ± 1.4		
LSD$_{0.05}$		8 days	7.8	0.07
'Molly Ann' (rhododendron)	20/10/00	April 23 ± 19.6	185.5 ± 19.6	0.85 ± 0
	17/11/00	April 14 ± 8.3	138.5 ± 8.3	0.75 ± 0.03
	18/12/00	March 13 ± 8.3	86 ± 8.3	0.69 ± 0/06
	12/01/01	March 7 ± 2.6	54.7 ± 2.6	0.77 ± 0.04
	08/02/01	March 20 ± 1.0	41.2 ± 1.0	0.67 ± 0.03
	08/03/01	April 5 ± 3.5	28 ± 3.5	0.65 ± 0.06
	Remained outside	May 8 ± 3.4		
LSD$_{0.05}$		13 days	12.6	0.06

and could be used as a means of estimating when the plants may be moved to the greenhouse for forcing. Forcing in a 10°C night greenhouse takes somewhat longer than in a 15–16°C greenhouse, but still speeds up flowering time compared with the plants remaining outside (Table 1; Larson, 1993).

References

FERNANDEZ-BLACO, L., FIGUEROA, M. E. & DAVEY, A. J. (1998). Diurnal and seasonal variations in chlorophyll *a* fluorescence in two Mediterranean grassland species under field conditions. *Photosynthetica* 35: 535–544.

HALL, D. O. & RAO, K. K. (1994).

Photosynthesis, pp. 175–179. Cambridge: Cambridge University Press.

HOEL, B. A. & SOLBAUGH, K. A. (1998). Effect of irradiance on chlorophyll estimation with the Minolta SPAD-502 leaf chlorophyll meter. *Ann. Bot.* 82: 389–392.

KAO, W.-Y., TSAI, T.-T. & CHEN, W.-H. (1998). Response of photosynthetic gas exchange and chlorophyll *a* fluorescence of *Misanthus*

floriculus (Labill) Warb. to temperatures and irradiance. *J. Plant Physiol.* 152: 407–412.

LARSON, R. A. (1993). *Production of Florist Azaleas*. Portland, OR: Timber Press, 136pp.

LEACH, D. G. (1961). *Rhododendrons of the World*, pp. 377–383. New York: Scribners.

TENGA, A. Z., MARIE, B. A. & ORMROD, D. A. (1989). Leaf greenness meter to assess ozone injury to tomato leaves. *HortScience* 24: 514.

In vitro polyploidy induction in *Rhododendron simsii* hybrids

Tom Eeckhaut, Gilbert Samyn & Erik van Bockstaele

DvP-CLO, Department for Plant Genetics and Breeding, Caritasstraat 21, 9090 Melle, Belgium

Three different *in vitro* protocols for the induction of tetraploid *Rhododendron simsii* Planch. hybrids were evaluated. The most successful technique was a daily application of mitotic poison on the apical meristem of seedlings. However, this technique was labour intensive. Direct sowing on media enriched with these compounds requires less work but is less efficient. To induce ploidy chimeras starting from an existing genotype, mitotic inhibitors can be added to multiplication media.

Methods

Method 1: treatment of *in vitro* seedlings

Azalea seeds ('Nina' × 'Dogwood') were sown on Anderson-based medium; after three weeks cotyledons had emerged and seedlings were treated. A colchicine, oryzalin or trifluralin solution was transferred between the cotyledons daily for either three or seven days. Among the oryzalin and trifluralin treated seedlings, tetraploids and ploidy chimeras were found (Figure 1 shows typical fluorescence profiles of diploids, tetraploids and chimeras).

Treatment	% dead seedlings	Tested seedlings	2x/4x chimeras	Tetraploids
COL 0.05% 3d	10	90	0	0
COL 0.25% 3d	40	15	0	0
COL 0.05% 7d	24	19	0	0

POSTERS

ORY 0.01% 3d	62	38	9	2
ORY 0.05% 3d	72	7	1	0
ORY 0.01% 7d	68	8	1	0
TRI 0.01% 3d	64	36	17	4
TRI 0.05% 3d	84	4	1	0
TRI 0.01% 7d	76	6	1	0

Method 2: direct sowing on induction medium

Azalea seeds ('Nina' × 'Dogwood') were sown on Anderson-based medium enriched with 0, 0.3, 1, 3 or 10 μM oryzalin or trifluralin. The ploidy number of the surviving seedlings was determined after 10 weeks. Out of 176 seedlings only one tetraploid and one diploid/tetraploid chimera were retrieved. It can therefore be concluded that this method is obviously less effective than the treatment of seedlings.

Treatment	% non-germi-nating seeds	Tested seed-lings	2x/4x chimeras	Tetraploids
Control	4	–	–	–
ORY 0.3 μM	12	22	0	0
ORY 1 μM	12	22	0	0
ORY 3 μM	8	23	1	0
ORY 10 μM	16	21	0	0
TRI 0.3 μM	8	23	0	0
TRI 1 μM	8	23	0	0
TRI 3 μM	20	20	0	0
TRI 10 μM	12	22	0	1

Method 3: application of mitosis inhibitors in multiplication medium

'Laura Ashley' and 'Madame Troch' were multiplied on basal medium + 5 mg l^{-1} 2iP (isopentenyladenosine) + 0, 0.5, 1, 2, 5 or 10 μM oryzalin. After 12 weeks, the formation of new shoots was evaluated. These shoots were tested flow cytometrically. Oryzalin was obviously very limiting for the multiplication rate. No tetraploids were found, although the number of chimeras was fairly high.

'Laura Ashley'	Induced plantlets/explant	Tested plants	2x/4x chimeras	Tetraploids
ORY 0 μM	2.83	0	0	0
ORY 0.5 μM	2	9	0	0
ORY 1 μM	2.17	12	1	0

281

ORY 2 µM	1.83	9	2	0
ORY 5 µM	0.67	4	0	0
ORY 10 µM	0.17	1	0	0

'Madame Troch'	Induced plantlets/ explant	Tested plants	2x/4x chimeras	Tetraploids
ORY 0 µM	4.11	0	0	0
ORY 0.5 µM	2	12	1	0
ORY 1 µM	1.11	4	3	0
ORY 2 µM	0.83	2	0	0
ORY 5 µM	0.61	3	0	0
ORY 10 µM	0.5	1	0	0

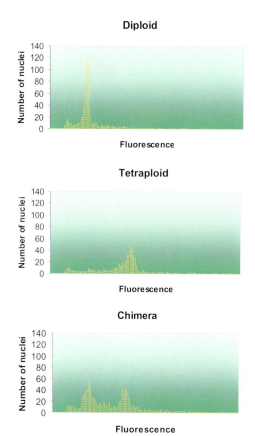

Figure 1 Typical fluorescence profiles of diploids, tetraploids and chimeras.

Discussion

Although polyploidy research has been performed within the *Rhododendron* genus (Kehr, 1986; Van Overschelde, 1989), the possible interest of polyploids for azalea breeding is still unknown since all available cultivars are diploid (Heursel & De Roo, 1981; De Schepper, 2001). Morphologically, chimerical 'Laura Ashley' or 'Madame Troch' shoots are not different from diploid ones, though flowering still needs to be observed. Among the tetraploid seedlings, some have more circular leaves. The most efficient method to obtain tetraploid azaleas, though it is rather labour intensive, is the treatment of seedlings. However, from a commercial point of view the creation of tetraploid forms of existing genotypes looks more promising. Since the *in vitro* regeneration of tetraploid plants from chimerical picotee sports is possible (De Schepper, 2001), tetraploids can probably be developed from the 'Laura Ashley' and 'Madame Troch' chimeras.

References

DE SCHEPPER, S. (2001). *Molecular analysis of the induction of sports in azalea (Rhododendron simsii hybrids)*. Thesis submitted in fulfilment of the requirements for the degree of Doctor (PhD) in Applied Biological Sciences, Cell and Gene Biotechnology, UG.

HEURSEL, J. & DE ROO, R. (1981). Polyploidy in evergreen azaleas. *HortScience* 16: 765–766.

KEHR, A. E. (1996). Polyploids in *Rhododendron* breeding. *J. Amer. Rhododendron Soc.* 46: 215–217.

VAN OVERSCHELDE, M. (1989). *Essais de polyploïdisation dans le cadre de la sterilité dans le genre Rhododendron*. Mémoire en vue de l'obtention du grade d'ingenieur agronome, UCL.

Plate 1 (a) Mother plant flower (diploid); (b) tetraploid flower and (c) chimera flower from the cross described.

A molecular phylogeny of lepidote rhododendrons based on a low copy number nuclear gene

Loretta Goetsch & Ben Hall

Botany and Genome Sciences, University of Washington, 6209 30th Avenue, Seattle, WA 98115, USA

Introduction

Cullen's taxonomic revision of 1980 recognised within the subgenus *Rhododendron* the sections *Rhododendron*, *Pogonanthum* and *Vireya* (Cullen, 1980). While the species comprising these sections occur primarily in SE Asia, section *Rhododendron* is represented also by species in North America (*R. minus* Michx.), Europe (*R. ferrugineum* L.

and close relatives) and circumpolar North America and Eurasia (*R. lapponicum* (L.) Wahlenb.). More recent studies of Kron & Judd (1990) concluded on morphological grounds that the widely distributed species of former genus *Ledum* also are properly included within section *Rhododendron* of subgenus *Rhododendron*.

A broad-scale molecular phylogenetic analysis of the genus *Rhododendron* based on chloroplast gene sequences strongly supported the monophyly of subgenus *Rhodo-dendron*, excluding *Ledum* spp. (Kurashige *et al.*, 2001). These authors concluded that section *Rhododendron* is paraphyletic, with two sister clades, one also containing all species of section *Vireya*, the other containing in addition a species of section *Pogonan-thum*. Our phylogenetic analysis, based on a rapidly evolving nuclear intron sequence, agrees with the main conclusions reached by Kurashige *et al.* (2001). Support for the splitting of section *Rhododendron* into clades with respective affinities to sections *Pogo-nanthum* and *Vireya* is especially robust.

The RPB2 gene in angiosperms

RPB2 is a nuclear gene found in all eukaryotes. Its function is to encode the second-largest subunit of RNA polymerase II, the enzyme that transcribes the vast majority of nuclear genes. In plants generally, RPB2 consists of 25 exons of total length 3500 base pairs (bp) separated by 24 introns, most of which are relatively short (~ 100 bp). While the positions of these introns within the coding sequence are conserved all the way from bryophytes to angiosperms, the intron sequences themselves are quite variable (Denton *et al.*, 1998). The longer (> 300 bp) introns in RPB2 genes provide a very useful source of sequence data for phylogenetic analysis, since they can be amplified by PCR primers located in the more conserved exons on either side. In two major angiosperm groups, the Ericales and Asteridae I, the nuclear genome contains two very distinct RPB2 genes that have evolved as separate lineages. The RPB2-d paralog is the major vegetatively expressed gene, while RPB2-i

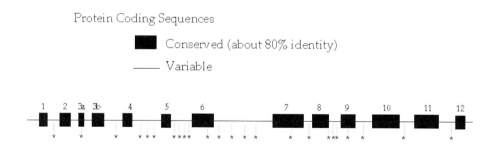

Figure 1 RPB gene structure in plants.

is expressed mainly in anthers (Luo *et al.*, unpublished results).

Both RPB2-i and RPB2-d have been completely sequenced for *Rhododendron macrophyllum* D.Don ex G.Don (McConaughy *et al.*, unpublished results). Across their 25 exons, the two genes differ in sequence by 10%, while the intron sequences are totally different in length and content for the two genes. The exon sequence differences make it possible to design PCR primers that amplify only one of the paralogous genes. We have used specific primers based upon the *R. macrophyllum* RPB2-i sequence to amplify an intron of the gene from a wide variety of lepidote species. In the schematic diagram of RPB2 (Figure 1), the 24 conserved intron positions are indicated by *. We present here phylogenetic analysis of RPB2-i intron 2 sequences obtained by PCR amplification from rhododendron genomic DNA using paralog-specific primers.

Results

The second intron of RPB2-i was PCR amplified and sequenced from 38 taxa, comprising 37 lepidotes and *R. macrophyllum*. These sequences were aligned with Clustal X, then subjected to phylogenetic analysis using PAUP* with the heuristic search option (DeBry & Olmstead, 2000) for maximum parsimony. The resulting tree is shown in Figure 2. Numbers above the branches indicate support from 500 bootstrap replicates. For the analysis of these data, the retention index was 0.904.

Conclusions

Additional outgroup taxa need to be sequenced before our data can be fully analysed. Using either *R. macrophyllum* or *R.*

tomentosum (Stokes) Harmaja + *R. hypoleucum* (Kom.) Harmaja as the outgroup, our preliminary conclusions are the following:

1 The RNA polymerase II intron used in these studies, although only 679 base pairs in length, provides phylogenetic resolution at the subsection level and, to some extent, at the species level. Nucleotide distances in this data set ranged from 0.025 to 0.40.

2 The species in section *Rhododendron* are clearly divided into two sister groups, each of them well supported. We refer to these groups as V, which contains all *Vireya* species studied, and P, which contains both *Pogonanthum* species. These correspond roughly to the VIREYA and RHOD clades resolved in the phylogeny of Kurashige *et al.* (2001). A more detailed comparison of the two respective trees is made difficult by differences in taxon sampling between our study and that of Kurashige *et al.* (2001).

3 In group V, five *Vireya* species from three subsections make up the clade with 97% bootstrap support. The subsection *Pseudovireya* species *R. kawakamii* Hayata and *R. sororium* Sleumer were outside this clade, but included in the broader group V, along with three *Triflora* species and two from *Cinnabarina*, as well as representatives of subsections *Uniflora* and *Heliolepida* and monotypic subsections *Camelliflora*, *Afghanica*, *Genesteriana* and *Campylogyna*.

4 Group P included, besides *R. sargentianum* Rehder & E.H.Wilson and *R. trichostomum* Franch. from section *Pogonanthum*, section *Rhododendron* representatives from subsections *Tephropepla* (two), *Moupinensia* (two), *Maddenia* (four, comprising a well-supported clade) and *Lapponica* (*R. orthocladum* Balf.f. & Forrest).

285

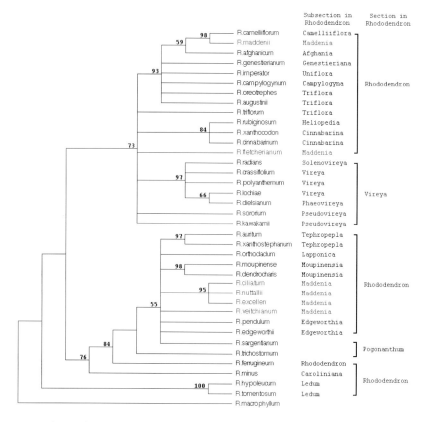

Figure 2 Tree obtained in the analysis of RPB2-i intron 2 in 37 lepidote rhododendrons and *R. macrophyllum*.

Acknowledgements

We would like to thank Steve Hootman for guidance and help in obtaining specimens from the Rhododendron Species Foundation Botanical Garden.

References

CULLEN, J. (1980). A revision of *Rhododendron*. I. Subgenus *Rhododendron* sections *Rhododendron* and *Pogonanthum*. *Notes Roy. Bot. Gard. Edinburgh* 39: 1–207.

DEBRY, R. & OLMSTEAD, R. G. (2000). A simulation study of reduced tree-search effort in bootstrap resampling analysis. *Syst. Biol.* 49: 171–179.

DENTON, A. L., MCCONAUGHY, B. & HALL, B. D. (1998). Usefulness of RNA polymerase II coding sequences for estimation of green plant phylogeny. *Mol. Biol. Evol.* 15(8): 1082–1085.

KRON, K. A. & JUDD, W. S. (1990). Phylogenetic relationships within the Rhodoreae (Ericaceae) with specific comments on the placement of *Ledum*. *Syst. Bot.* 15: 57–68.

KURASHIGE, Y., ETOH, J. I., HANDA, T., TAKAYANAGI, K. & YUKAWA, T. (2001). Sectional relationships in the genus *Rhododendron* (Ericaceae): evidence from *mat*K and *trn*K intron sequences. *Plant Syst. Evol.* 228: 1–14.

Rhododendrons and their Standards in the RHS Herbarium (WSY)

Susan Grayer & Diana Miller

RHS Garden Wisley, Woking, Surrey GU23 6QB, UK

Standards fix the name of *Rhododendron* cultivars and act as a reference point when names become confused. Standard Portfolios include additional material such as related specimens, images, references, history and any other information which may help in identification. The RHS Herbarium at Wisley has identified over 2000 Standards of which 500 are of *Rhododendron* cultivars, and is aiming to create an international database linking other institutions holding Standards.

A vast number of new *Rhododendron* cultivars have been introduced over the years, but until the mid-20th century few were defined by more than the briefest descriptive entry in a nursery catalogue. These entries alone are rarely sufficient to identify a cultivar with certainty, so similar cultivars can easily become confused and be sold under the same name. Alternatively, one cultivar can acquire several different names – perhaps through poor labelling, the 'rediscovery' and renaming of existing plants, or deliberate changes to something more attractive.

The genus *Rhododendron* is fortunate in having a very efficient International Cultivar Registration Authority (ICRA) under the auspices of the Royal Horticultural Society (RHS). Established in 1955, the first Rhododendron Register was published in 1958, followed by supplements of additional cultivars. These ensure that every new name is registered with a description. Registrants are encouraged to send illustrations or material for pressing but this is not a condition for acceptance of a valid cultivar name. This material provides a valuable addition to the description and may be designated a Standard (Miller, 1995).

A Standard is the herbarium specimen or illustration of a cultivar which forms a permanent record of the distinguishing characteristics of that cultivar.

Standards fix a name and act as a reference point when plants become confused. Standards can be used to help resolve any naming problems, to decide if a 'new' plant is really unique, and potentially can contribute DNA samples for molecular studies. Any additional information and material such as related specimens, images, history, references, award details, pollen samples or even frozen DNA and the results of analyses, may help in the identification of a cultivar and constitute the Standard Portfolio (e.g. Plate 1), as encouraged by the International Code of Nomenclature for Cultivated Plants (Trehane *et al.*, 1995).

Plate 1 Standard Portfolio of *Rhododendron* 'Maurice Daffarn', raised by J.H. Mangles before 1884.

Benefits of Standards

There are many benefits in producing a Standard when a new cultivar is named (Miller, 1999; Miller & Grayer, 2001). One of the most obvious is that there is no uncertainty about its authenticity. In the past there has been no equivalent to a botanical type specimen for cultivars (Miller & Grayer, 1999). However, the first edition of the Code, in 1953, recommended that dried specimens of new cultivated plants be prepared for future reference. The suggestion for more formal 'cultivated type' specimens was proposed by Chris Brickell, botanist and former Director General of the RHS, at an International Symposium in the Netherlands in 1986.

The concept of Standards and Standard Portfolios was expanded by Diana Miller, Keeper of the RHS Herbarium, at the Second International Symposium on the Taxonomy of Cultivated Plants in Seattle in 1994, and formalised in the sixth edition of the Code in 1995 (Miller, 1995).

Standards will feature strongly at the Fourth Symposium which is to be held in Toronto in August 2002. This will be followed by a revision of the Code in which there may be greater emphasis on the relevance of Standards in horticulture. The RHS Herbarium, specialising in cultivated ornamental plants, is in a unique position to advance the concept of Standards and also to encourage similar institutions to follow suit.

Sources of current *Rhododendron* Standards

Standard specimens of rhododendrons have been and are being obtained from various sources. Traditionally many new British cultivars were exhibited before RHS Committees. About 70 years ago, it was RHS practice to employ botanical artists to depict award plants, many of which have proved to be Standards. Paintings have since been superseded by the preparation of herbarium specimens and photographic images. Rhododendrons are also given awards after Trials at Wisley, which have been carried out since about 1930. Many of these plants have then entered the trade and remained popular. Other sources include herbarium material and photographs of original plants maintained in national or historic collections. New cultivars from contemporary breeders are generally acquired through the RHS Rhododendron Registrar, through a Trial or Committee entry, or from the breeder directly. This is an ongoing process. Raisers of *Rhododendron* cultivars are being encouraged to send samples of their plants to the RHS Herbarium.

How older Standards were traced retrospectively

All *Rhododendron* specimens and most images in the RHS Herbarium have been assessed since 1998. Initially decisions had to be made about how Standards would be recognised retrospectively. Each specimen and illustration was examined. Information in the RHS *Proceedings*, entry forms of plants exhibited before RHS Committees, international registers, nursery catalogues, monographs and other *Rhododendron* publications was studied. Often exhibitors or raisers were contacted for further information. Any specimens or images sent to the RHS Rhododendron Registrar were included where appropriate.

Once it was determined that the specimen could be designated a Standard, it was filed in a labelled green-edged folder, to distinguish it from a botanical type specimen. If an illustration constituted the Standard, it was marked appropriately. Copies of all relevant information and articles were taken and filed with the specimens together with references to illustrations or even living plants. All this material and information constitutes the Standard Portfolio of a cultivar (Miller & Grayer, 1999), which may be augmented in the future if more information becomes available. Records were made on the RHS Horticultural Database, making all these details readily accessible. Today 501 Standards of *Rhododendron* cultivars have been identified in the collection of the RHS Herbarium.

Disseminating the information

Botanical type specimens of new species may be deposited in any herbarium in the world and can be notoriously difficult to track down. This has led the RHS Herbarium to create a searchable website enabling anyone to find conclusive information about Standard cultivars held at Wisley (see www.rhs.org.uk/research). This website directs the user to the location where Standards and their Portfolios are kept. The search result gives details of whether a Standard is an image or pressed specimen, the dates of its preparation and designation, together with any further information about additional material in the Herbarium, and references. Illustrations are added whenever possible.

A further aim is to create an international database with links to other institutions holding Standards so that details of cultivars deposited in any part of the world may be retrieved. In addition, the list of current designated Standards in the RHS Herbarium will be published in the *Extracts from the Proceedings of the RHS*, Vol. 126 (summer 2002).

Acknowledgements

We would like to thank Dr Alastair Culham, University of Reading for his assistance.

References

MILLER, D. M. (1995). Standard Specimens for Cultivated Plants. In: *Proceedings of the Second International Symposium on the Taxonomy of Cultivated Plants. Acta Horticulturae* 413: 35–39.

MILLER, D. M. (1999). Raising Standards. *The Garden* 124(4): 282–283.

MILLER, D. M. & GRAYER, S. R. (1999). Standard Portfolios in the Herbarium of the Royal Horticultural Society. In: ANDREWS, S., LESLIE,

A. & Alexander, C. (eds) *Proceedings of the Third International Symposium on the Taxonomy of Cultivated Plants*, pp. 397–399. Royal Botanic Gardens, Kew.

Miller, D. M. & Grayer, S. R. (2001). Setting the Standard for cultivated plants. *The New Plantsman* 8(2): 112–126.

Trehane, P., Brickell, C. D., Baum, B. R., Hetterscheid, W. L. A., Leslie, A. C., McNeill, J., Sponberg, S. A. & Urugtman, F. (1995). *The International Code of Nomenclature for Cultivated Plants*. Regnum Vegetabile 133: 1–175.

Rhododendrons on limestone

Anthony J. McAleese & David W. H. Rankin

Department of Chemistry, University of Edinburgh, West Mains Road, Edinburgh EH9 3JJ, UK

Rhododendrons can't easily be cultivated on alkaline soils. But in China many flourish in the limestone mountains. Are they really growing in contact with the alkaline soil? We analysed many samples of root, soil and rock to try to find answers to this question.

It is often presumed that the plants cannot actually be growing with their roots in contact with the limestone, and various explanations have been proposed:

1 Rhododendron plants really grow on organic soil, out of contact with the limestone.

This is not true! The soils range from very acidic (pH < 4) to as alkaline as is possible on limestone (pH 8.4). Available calcium concentrations in soils range from 200 to 210,000 ppm – up to almost pure limestone.

2 The limestone is dolomitic – with a high magnesium content to counter the effects of the calcium.

This again is not true! The magnesium carbonate level is low, 0.2 to 1.6. In any case, magnesium and calcium do not compete.

3 The limestone is hard and insoluble.

This also is not true! It is soft (although nothing like as soft as chalk) and soon dissolves to saturate water.

4 Heavy rainfall ensures no transport of dissolved limestone into the overlying soil.

This is not true either! Analysis of soil samples shows calcium to be present and available in the root zone.

The plants do grow with their roots in contact with limestone. Lime-loving companion plants confirm the presence of limestone in the root soil.

Analysing leaf samples shows how availability of metals, particularly iron and manganese, affects their growth.

Manganese (Mn) availability appears to be critical

Manganese availability is usually low on alkaline soils. The normal concentrations in leaves of woody plants are 20–300 ppm (Figure 1). With too much they are poisoned; too little and photosynthesis fails.

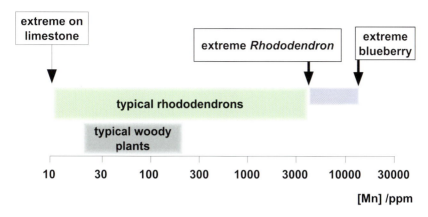

Figure 1 Manganese concentrations in the leaves of *Rhododendron* and other plants.

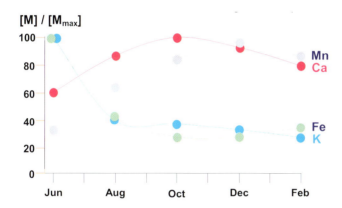

Figure 2 Concentrations of metals in *Rhododendron* leaves through a year of growth (averages for 10 plants).

Rhododendrons on alkaline soils may be down to 10 ppm in leaves so that manganese deficiency limits growth, but availability in Chinese soils is much higher than expected. Rhododendrons often have high manganese concentrations in their leaves. This may frequently be 2500 ppm (4500 maximum) and may be up to 15,000 ppm in blueberries (*Vaccinium* sp.).

Does decay of these leaves supply Mn for plants on limestone?

Rhododendron plants accumulate large amounts of manganese in their leaves. This uptake by leaves was tracked through a year of growth (Figure 2). The results show that iron and potassium are taken only into young leaves but manganese and calcium continue to accumulate in the leaves throughout the growing season. Unlike iron and potassium the excess is not removed and remains in the leaf when it falls. Manganese is therefore available in decaying leaves for other plants to absorb.

The key to cultivating rhododendrons on alkaline soil is thus the recycling of manganese.

Scanning electron microscopy in *Vireya* rhododendron research

Maureen C. Warwick

Royal Botanic Garden Edinburgh, 20A Inverleith Row, Edinburgh EH3 5LR, UK

The subsectional classification of *Rhodo-dendron* section *Vireya* (Blume) Copel.f. is based on the morphology of a number of characters including the type of scales or multicellular epidermal trichomes on the undersurface of the leaf, and features on the margins of the floral bud scales. While it is possible to see these features using a hand lens or light microscope, the use of a scanning electron microscope allows sharp images at much higher magnifications. This greatly facilitates the study of these plants and, indeed, can provide new information.

As part of the research on *Vireya* rhododendrons at the Royal Botanic Garden Edinburgh (RBGE), a survey was carried out in which a scanning electron microscope was used to examine the undersurface of the leaves of most species. A record was made of the scales (multicellular epidermal trichomes), which consist of a stalk, a central region and a marginal flange. The general distribution of the scales was photographed, together with a close-up of a single typical scale. These were recorded at the same magnifications as far as possible, to enable easier comparison. The margins of the floral bud scales were also examined, as was the distribution of stomata and other parts of the plants which were particularly of interest, such as the ovary. Seeds were also examined: the long tails so characteristic of many species increase their buoyancy, enabling them

to float long distances. These tails are longest in epiphytic species; those which occur in montane habitats do not need their seeds to disperse widely and many have no tails.

Both dried material from herbarium specimens and fresh material from RBGE's *Vireya* rhododendron collection was used. One of the advantages of using fresh material is to see the leaves and scales in a more natural state, even after they are coated with gold palladium for the scanning process. Scales may be altered on pressed and dried herbarium specimens, which also may be old, dirty and missing scales from being handled over many years. Some herbarium specimens

Plates 1–4 (1) *Rhododendron phaeochitum* F.Muell.: general view of the underside of the leaf showing many missing scales. (2) *R. phaeochitum*: detail of a single scale. (3) *R. konori* Becc.: a seed typical of section *Vireya*, with a long tail at each end; image taken with a light microscope. (4) *R. retusum* (Blume) Benn.: a scanning electron micrograph of a seed with 'wings'.

Plates 5–8 The four different types of scale on the leaf undersurface. (5) *Rhododendron malayanum* Jack: overlapping scales with a large, central cushion and a broad, marginal flange. (6) *R. meliphragidum* J.J.Sm.: individual scale which is almost entire and sessile, with a narrow marginal flange and a broad, central cushion. (7) *R. rarum* Schltr.: stellate and dendroid scales, on top of an epidermal tubercle. (8) *R. superbum* Sleumer: scales with a broad, marginal flange and a small centre.

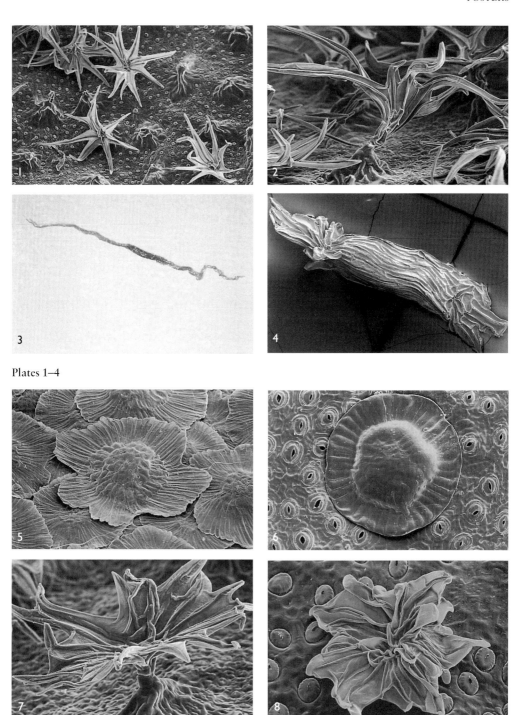

Plates 1–4

Plates 5–8

293

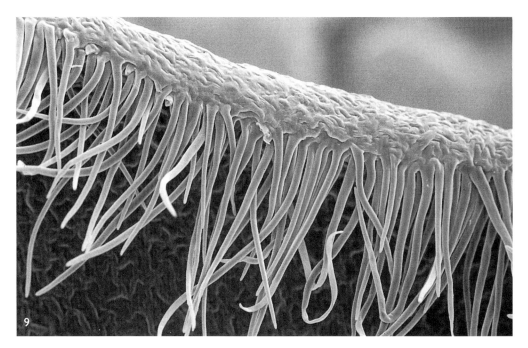

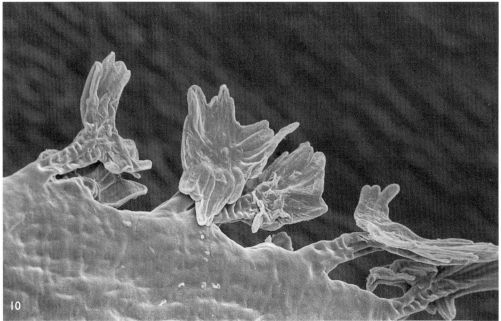

Plates 9 and 10 The bracts (perulae) of the floral bud. (9) *Rhododendron vanderbiltianum*: margin with simple hairs. (10) *R. vidalii* Rolfe: margin edged with scales.

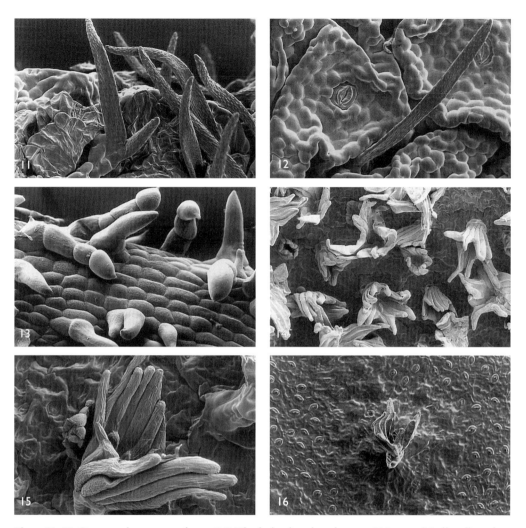

Plates 11–13 Ovary and stamen surfaces. (11) *Rhododendron brookeanum* H.Low ex Lindl. collected on Mt. Kinabalu in Sabah: ovary with scales and short, simple or branched hairs. (12) *R. brookeanum*, collected at Palawan in the Philippines: ovary without scales; with both long and short hairs, some of them branched. (13) *R. verticillatum* H.Low ex Lindl.: glandular hairs on filaments.

Plates 14–16 Comparing the scales and their density on an immature leaf of *R. edanoi* Merr. & Quis. with those on a mature leaf. (14) Immature leaf showing scales. (15) Single immature leaf scale. (16) Mature leaf scale.

require various methods of cleaning (Warwick & Helfer, 1990) before they can be viewed clearly, and as this can occasionally damage the specimen it is always better to use fresh material whenever available.

Reference

WARWICK, M. C. & HELFER, S. (1990). Scanning electron microscopy in *Vireya* rhododendron research. *Notes Roy. Bot. Gard. Edinburgh* 46(3): 365–374.

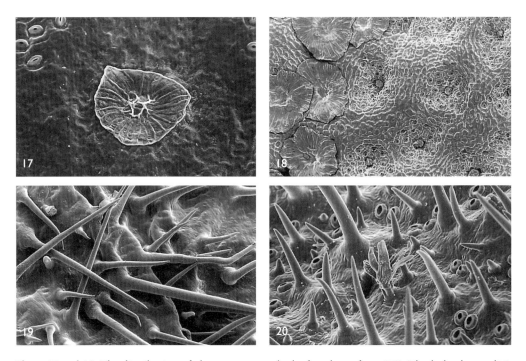

Plates 17 and 18 The distribution of the stomata on the leaf undersurface. (17) *Rhododendron salicifolium* Becc.: showing large areas around the leaf scale with no stomata. (18) *R. apoanum* Stein: underside of leaf with some scales having been removed to show the stomata grouped underneath the scale in small depressions.

Plates 19 and 20 Rhododendrons with simple hairs on the leaf undersurface. (19) *Rhododendron stapfianum* Hemsl. ex Prain: dense hairs of variable lengths. (20) *R. verticillatum*: simple hairs intermixed with a scale.

Breeding and selection for lime-tolerant rhododendrons

F. M. Wilson, T. R. Marks, C. H. A. Bates, C. M. James, P. E. Meyers, J. B. Rose & K. R. Tobutt

Horticulture Research International, East Malling, West Malling, Kent ME19 6BJ, UK

Introduction

Rhododendrons are popular garden plants but most are 'intolerant of lime', with a few notable exceptions such as *R.* 'Cunningham's White'. To introduce lime tolerance into a wider range of desirable cultivars would be valuable, as much of southern and eastern Britain has calcareous soils.

Our aim is to develop markers to assist breeding for lime tolerance. Seedlings have been raised from crosses between *R. hirsutum* L. (lime tolerant or calcicole) and *R. ferrugineum* L. (lime intolerant or calcifuge). These are two alpine species of subsection *Rhododendron* and are known to hybridise naturally, giving rise to *R. × intermedium* Tausch. These seedlings have been screened for lime tolerance using a novel *in vitro* method, and DNA markers for lime tolerance are currently being developed. In addition we have developed an *in vitro* method for regenerating putative somaclones displaying a higher degree of lime tolerance.

Microsatellite markers

A microsatellite is a short repetitive sequence of DNA that varies in length (allelic variation) in different species and varieties. The variants can be visualised by separating the DNA on a gel. In this case, with three different microsatellites different alleles can be observed in various *Rhododendron* species and varieties. Microsatellites that are close to (or coupled with) genes for lime tolerance are potential markers for this trait.

We have generated a genomic DNA 'library' from *R. hirsutum*, enriched for microsatellite-containing regions. Sequencing of this library has allowed identification of a number of microsatellites, which are currently being screened for their potential use as markers.

Selected microsatellites will be used to screen the progeny of *R. hirsutum × R. ferrugineum*. Segregation patterns will be linked to lime tolerance by comparison with *in vitro* response to high pH.

In vitro screening of progeny for lime tolerance

Seeds from controlled crosses (see above) were germinated *in vitro* to allow clonal multiplication, and rooted into a 'Sorborod' (paper plug). They were aseptically grown under photosynthetic conditions in a liquid medium where pH was accurately controlled with a non-toxic buffer. Studies have shown that high pH is the principal factor in lime intolerance (manifested here as suppression of growth, and the onset of chlorosis).

By growing rooted shoots in solutions of increasing pH, selection pressure for survival and growth at higher pH can be created. Necrosis of new leaves and chlorosis of established leaves (green = 0, white = 4) are useful indicators of intolerance.

This pH intolerance is expressed most strongly in new growth, and plants are smaller at higher pH. This aids the selection of tolerant genotypes from intolerant siblings.

Somaclones

In an attempt to produce somaclones from specific *Rhododendron* clones, auxin (10 μM IBA) and cytokinin (3 μM CPPU) have been used to regenerate shoots at different levels of pH. Regenerants from all treatments were grown on *in vitro* at pH 7, and potentially tolerant regenerants have been produced from *R. × intermedium* regenerated at pH 7.

Summary

- Seedlings have been generated from the cross between *R. hirsutum* (calcicole) × *R. ferrugineum* (calcifuge).
- pH can be used to screen for lime intolerance *in vitro*, and screening of the progeny indicated that within the population there is a range of tolerance.
- A microsatellite-enriched DNA 'library' has been generated from *R. hirsutum*, and screening of microsatellites to look

for markers linked to lime tolerance is currently underway.

- Microsatellite markers will be used to link segregation patterns with lime tolerance in the progeny. These markers can then be used to assist in breeding for lime tolerance in rhododendrons.

- Somaclones resistant to high pH will be trialled to assess their tolerance to lime under field conditions.

Acknowledgements

This project was funded by DEFRA (grants HH1012SHN and HH1026SHN).

A partial cytotaxonomic revision of *Rhododendron* subsections *Heliolepida* and *Triflora*

Sally Whyman

National Museum & Gallery Wales, Cathays Park, Cardiff CF10 3NP, UK

A partial cytotaxonomic revision of *Rhododendron* subsections *Heliolepida* and *Triflora* has been attempted using the available, living, wild-source collections. This is the first time chromosome counts have been attempted on these species since the late 1940s, and the first time such data have been correlated with morphology to arrive at taxonomic conclusions. Extensive polyploidy has been confirmed, with ploidy levels varying from diploid ($2n = 26$) to decaploid ($2n = 130$). *Rhododendron brevistylum* Franch. and *R. heliolepis* Franch. have been found to be hexaploid and decaploid, respectively, and have been redefined morphologically. *Rhododendron augustinii* Hemsl. has been found to contain both tetraploids and a hexaploid. Chromosome counts are confirmed or reported for the first time for several other species within the two subsections.

Registrants

Rhodo '02 Conference (Rhododendrons in Horticulture and Science), 17–19 May 2002, Royal Botanic Garden Edinburgh

Anderson, Alan, *Aberdeen, Scotland*
Anderson, Paul, *California, USA*
Anderson, Marty, *Illinois, USA*
Argent, Dr George, *Royal Botanic Garden Edinburgh, Scotland*
Arora, Prof. Rajeev, *Iowa State University, USA*

Ballantyne, David, *Victoria University, Canada*
Bennell, Alan, *Royal Botanic Garden Edinburgh, Scotland*
Bennett, Eve, *Royal Botanic Garden Edinburgh, Scotland*
Berg, Warren, *Washington, USA*
Binney, David, *Taurango, New Zealand*
Blair, John, *Royal Botanic Garden Edinburgh Trustee, Scotland*
Bodenham, John, *Plymouth, England*
Bomford, G. R. A., *Redditch, England*
Brown, Gillian, *Canberra, Australia*
Brown, Jane, *Peterborough, England*
Buffin, Michael, *Hillier Gardens, England*
Buxton, Anna, *Edinburgh, Scotland*
Byatt, Mary, *Elgin, Scotland*
Byatt, David, *Elgin, Scotland*

Callard, Chris, *London, England*
Camelbeke, Koen, *Haacht, Belgium*
Campbell, Barbara, *California, USA*
Campbell, William, *Gargunnock Estate, Scotland*
Cavender, Richard, *Oregon, USA*
Cavender, Karen, *Oregon, USA*
Chamberlain, Dr David, *ex-Royal Botanic Garden Edinburgh, Scotland*

Clark, Clarice, *Washington, USA*
Collier, David, *Oregon, USA*
Collier, Mrs Kathleen, *Oregon, USA*
Collier, Thomas, *Oregon, USA*
Corby, Terry, *London, England*
Cox, Kenneth, *Glendoick, Scotland*
Cox, Peter, *Glendoick, Scotland*
Crosbie, Colin, *The Royal Horticultural Society, England*
Cullen, Dr James, *Cambridge University, England*
Cherry, Robert, *New South Wales, Australia*
Cherry, Derelie, *New South Wales, Australia*

Daneri, Dee, *California, USA*
De Zhu, Dr Li, *Kunming, China*
Dixon, Prof. Geoffrey, *Strathclyde University, Scotland*
Douglas, Ian, *Fife, Scotland*

Edwards, Dr Alun, *Devon, England*
Edwards, Mrs Marianne, *Devon, England*
Eeckhaut, Tom, *DvP-CLO, Melle, Belgium*
Evans, Dr Harry, *CABI, Berkshire, England*
Evans, Philip, *Devon, England*

Fairweather, Chris, *Beaulieu, England*
Farthing, Duncan, *Yorkshire, England*
Flanagan, Mark, *Windsor Great Park, England*
Forster, Brent, *New Zealand Rhododendron Association, New Zealand*
Foster, Clive, *Wakehurst Place, England*
Fox, T. Stephen, *Derbyshire, England*

Gardiner, Jim, *The Royal Horticultural Society, England*

Goetsch, Loretta, *Washington University, USA*

Grant, Mike, *The Royal Horticultural Society, England*

Grasing, Bob, *Oregon, USA*

Grayer, Susan, *The Royal Horticultural Society, England*

Hall, Tony, *Royal Botanic Gardens, Kew, England*

Hammond, John, *Lancashire, England*

Hammond, Mrs Margaret, *Lancashire, England*

Harbison, Mrs Caroline, *Argyll, Scotland*

Heasman, Matthew, *Glasgow, Scotland*

Helfer, Dr Stephan, *Royal Botanic Garden Edinburgh, Scotland*

Hooper, Jonathon, *Argyll, Scotland*

Hootman, Steven, *Rhododendron Species Foundation, Washington, USA*

Hopkins, Steve, *Oregon, USA*

Hudson, Tom, *Cornwall, England*

Hutchison, Sir Peter, *Stirling, Scotland*

Jack, Dr Robert, *Lanark, Scotland*

Jorgensen, Marilyn, *Washington, USA*

Jurgens, Michael, *Berkshire, England*

Jurgens, Mrs Jori, *Berkshire, England*

Justice, Douglas, *Canada*

Kerby, Ross, *ex-Royal Botanic Garden Edinburgh, Scotland*

Knottenbelt, Mrs Morna, *Wirral, England*

Kratz, Heinz, *Würzburg, Germany*

Kron, Dr Kathy, *Wake Forest University, USA*

Latham, Dr Alfred, *Liverpool University, England*

Lear, Michael, *Oxford, England*

Leslie, Alan C., *The Royal Horticultural Society, England*

Lima, Henry (Chip), *California, USA*

McFarlane, Marjory, *Royal Botanic Garden Edinburgh, Scotland*

MacLachlan, Malcolm, *Cornwall, England*

McLean, Dr Brenda, *Wirral, England*

McLean, Dr Robin, *Wirral, England*

McNamara, Steve, *National Trust for Scotland – Branklyn Garden, Scotland*

MacPherson, *Cluny, Cornwall, England*

McPhun, Simon, *National Trust for Scotland – Inverewe, Scotland*

Main, John, *ex-Royal Botanic Garden Edinburgh, Scotland*

Martin, Rachael, *Exbury, England*

Mitchell, David, *Royal Botanic Garden Edinburgh, Scotland*

Monthofer, Martin, *Bremen, Germany*

Moser, Erhard, *Chemnitz, Germany*

Murch, Mrs Penny, *Worksop, England*

Nanney, Dave, *Virginia, USA*

Nanney, Mrs Leslie, *Virginia, USA*

Newman, Hilary, *Kent, England*

Nilsen, Prof. Erik, *Virginia, USA*

O'Halloran, Sally, *Glasnevin, Ireland*

Oziewicz, Beverley, *Toronto, Canada*

Park, Mike, *Surrey, England*

Paterson, David, *Royal Botanic Garden Edinburgh, Scotland*

Philipson, Melva, *Greytown, New Zealand*

Pollard, Tom, *Cumbria, England*

Pollard, Mrs Ilene, *Cumbria, England*

Postan, Lady Cynthia, *Cambridge, England*

Pradhan, Rebecca, *Thimphu, Bhutan*

Putschky, Magrit, *Hannover, Germany*

Rae, David, *Royal Botanic Garden Edinburgh, Scotland*

Rankin, Prof. David, *Edinburgh University, Scotland*

Rankin, Philip, *Edinburgh, Scotland*

Raven, Mrs Faith, *Argyll, Scotland*
Revell, Bill, *Cumbria, England*
Richardson, Garratt, *Washington, USA*
Robinson, Dr Mike, *Sussex, England*
Robinson, Mrs, *Sussex, England*
Rotherham, Dr Ian, *Sheffield University, England*
Rowe, Mrs Carol, *Argyll, Scotland*
Rowe, James, *Argyll, Scotland*
Rutherford, Fran(cis), *Washington, USA*

Sanderson, Graham, *Windsor Great Park, England*
Scariot, Valentina, *Torino, Italy*
Schepker, Dr Hartwig, *Bremen, Germany*
Selcer, Donald, *California, USA*
Selcer, Mrs Edy, *California, USA*
Shuster, Karen, *Vancouver, Canada*
Sinclair, Ian, *Ardkinglas Estate, Scotland*
Sinclair, June, *Washington, USA*
Skagseth, Pål, *Rådal, Norway*
Smith, E. White, *Oregon, USA*
Smith, Mrs Lucie S., *Oregon, USA*
Smith, Graham, *Pukeiti, New Zealand*
Smith, Mrs Dora, *Pukeiti, New Zealand*
Speed, G. R. Mac, *Bedfordshire, England*
Spethmann, Prof. Dr Wolfgang, *Hannover University, Germany*
Spoelberch, V. Philippe de, *Haacht, Belgium*
Steele, Ray, *Devon, England*
Steele, Mrs Anne, *Devon, England*
Steele, Ann, *National Trust for Scotland – Clarkston, Scotland*
Stewart, Mrs Mary, *Edinburgh, Scotland*
Stewart, Ian, *Edinburgh, Scotland*

Talbot, Rob, *Royal Botanic Gardens, Kew, England*
Tan, Bian, *Strybing Arboretum, USA*
Thornley, Michael, *Rhu, Scotland*
Thornley, Mrs Sue, *Rhu, Scotland*
Thornton, Dr Ray E., *Southampton, England*

Thornton, Mrs Norma, *Southampton, England*
Townsend, Ray, *Royal Botanic Gardens, Kew, England*

Vetås, Kåre, *Sabøvågen, Norway*
Voss, Donald, *Virginia, USA*

Walker, Chris, *Yorkshire, England*
Walsh, Timothy, *California, USA*
Walsh, Mrs June, *California, USA*
Walter, Roy, *Wigtownshire, Scotland*
Warren, Mrs Josephine, *Devon, England*
Warwick, Maureen, *Royal Botanic Garden Edinburgh, Scotland*
Watling, Prof. Roy, *ex-Royal Botanic Garden Edinburgh, Scotland*
Weese, Mrs Margt. de, *British Columbia, Canada*
Werbeck, Michael, *Bremen, Germany*
Westhoff, Ms Julia, *Bremen, Germany*
Wheeler, Mrs Eileen, *Pembrokeshire, Wales*
Wheeler, Peter, *Pembrokeshire, Wales*
White, Keith, *Rhododendron Species Foundation, USA*
White, Lady Josephine, *Devon, England*
Whyman, Sally, *Cardiff, Wales*
Wilkins, Maurice, *National Trust for Scotland – Arduaine, Scotland*
Wilson, Fiona, *Horticulture Research International, Kent, England*
Wylie, Gordon, *Oregon, USA*
Wylie, Mrs Linda, *Oregon, USA*

Young, Donald, *New Zealand Rhododendron Association, New Zealand*
Young, Mrs Helen, *New Zealand Rhododendron Association, New Zealand*
Younger, David A. H., *Argyll, Scotland*

Zeigler, Brenda, *Oregon, USA*

Rhodo '02 Conference participants

Index to scientific and cultivar names

Page references to illustrations are underlined